Neutron Scattering in Condensed Matter Physics

SERIES ON NEUTRON TECHNIQUES AND APPLICATIONS

Editors: J. L. Finney and D. L. Worcester

Vol. 1: Modulation Spectrometry of Neutrons with Diffractometry Applications
by Pekka Hiismäki

Vol. 2: Single Crystal Neutron Diffraction from Molecular Materials
by Chick C. Wilson

Series on Neutron Techniques and Applications – Vol. 4

Neutron Scattering in Condensed Matter Physics

Albert Furrer
Joël Mesot
Thierry Strässle

ETH Zurich & PSI Villigen, Switzerland

World Scientific

NEW JERSEY • LONDON • SINGAPORE • BEIJING • SHANGHAI • HONG KONG • TAIPEI • CHENNAI

Published by

World Scientific Publishing Co. Pte. Ltd.
5 Toh Tuck Link, Singapore 596224
USA office: 27 Warren Street, Suite 401-402, Hackensack, NJ 07601
UK office: 57 Shelton Street, Covent Garden, London WC2H 9HE

Library of Congress Cataloging-in-Publication Data
Furrer, Albert.
Neutron scattering in condensed matter physics / by albert Furrer, Joël Mesot and Thierry Strässle.
p. cm. -- (Series on neutron techniques and applications ; v. 4)
Includes bibliographical references and index.
ISBN-13 978-981-02-4830-7
ISBN-10 981-02-4830-X
ISBN-13 978-981-02-4831-4 (pbk)
ISBN-10 981-02-4831-8 (pbk)
1. Neutrons--Scattering. 2. Condensed matter. I. Strässle, Thierry. II. Mesot, Joël. III. Title.
539.7213--dc22

2010275854

British Library Cataloguing-in-Publication Data
A catalogue record for this book is available from the British Library.

Cover Illustration: The scene depicts an incoming and outgoing neutron scattered on a single crystal of ZnS, to visualize the moving neutron the incoming and outgoing neutrons are shown in decreasing degrees of opacity.

Preface

This book grew out of lectures presented by the authors at the ETH Zurich for master students in condensed matter physics in their last year of study. The first author started these lectures in 1979, continued by the second author in 2003, and assisted by the third author in 2007 up to the present. Initially, the lectures were largely based on the excellent book entitled *Introduction to the Theory of Thermal Neutron Scattering* by G.L. Squires (Cambridge University Press, Cambridge, 1978). During all these years, however, the contents of the lectures have been gradually improved as a reaction to the feedback by the students, new developments in the neutron scattering techniques have been included, and the presentation of experimental results has been extended to take account of the rapid progress in the application of neutron scattering to novel aspects of condensed matter physics. As a result of all these developments, the lectures cover today a broad spectrum of relevant topics in condensed matter physics studied by neutron scattering techniques which can hardly be found in existing textbooks. The authors therefore appreciated the request by World Scientific to upgrade the existing lecture notes to a comprehensive book entitled *Neutron Scattering in Condensed Matter Physics.*

We emphasize that the book is written by experimentalists with some interest in theory, thus it is mainly addressed to experimentalists, but also to theoreticians interested in experiments. It should assist the readers in planning and analyzing neutron scattering experiments in the vast field of condensed matter physics. No previous knowledge of the theory of neutron scattering is assumed, but a reasonable familiarity with the basic concepts of both condensed matter physics and quantum mechanics is expected for a proper understanding of the text. The book is confined to the major issues in condensed matter physics (excluding soft condensed matter), but it does not cover all the applications. Nevertheless, we believe that the readers can tackle some specialized topics not treated in the book by applying the ideas and procedures described in the appropriate chapters. In conclusion, the book is a comprehensive guide for both lecturers and students of introductory courses at the graduate and post-graduate level. It will also prove useful to all scientists starting to employ neutron scattering techniques in their research in condensed matter physics, as well as to all the scientists already active in this field.

The book starts with three chapters introducing general, theoretical and instrumental aspects of neutron scattering, followed by twelve chapters on the most important topics of condensed matter physics where neutron scattering has provided important information. Whenever possible, the topical chapters include illustrative experimental results which were chosen from the myriads of literature data according to their didactical suitability for a textbook. The results thus include both "historical" data from the pioneering time of neutron scattering and data of today's research. Exercises have been given at the ends of some topical chapters, which are intended to illustrate the concepts as well as to deepen the understanding of the text. The exercises thus form an essential part of the book. The readers are encouraged either to solve these problems or to look at the solutions summarized at the end of the respective chapter. In order to keep the topical chapters to a reasonable length, some mathematical and tabular material has been removed from the text and collected in several appendices.

The references given in each chapter are by no means complete in terms of a professional review, but they should merely be regarded as being representative for the particular subjects. In fact, we tried to keep the list of references short in order to allow a smooth reading of the text, and references are mainly used to indicate the source of the experimental results presented. Nevertheless, for each chapter we added some references in the section *Further Reading* for those who are interested to enter the subjects from a different point of view.

We profited a great deal from discussions with colleagues and students whose positive criticism was helpful in properly shaping the text. It is impossible to recall all the occasions when we received advice from colleagues during enlightening discussions on numerous issues. So we anonymously thank all those who provided assistance and gave permission to display figures from their publications.

Finally, the authors are indebted to World Scientific for the invaluable efforts towards a rapid and professional publication of the book.

Villigen, March 2009 *Albert Furrer, Joël Mesot, Thierry Strässle*

Thierry Strässle, Joël Mesot and Albert Furrer (from left to right).

About the authors

Albert Furrer received his Masters and Ph.D. degrees in experimental physics at ETH Zurich (Switzerland). He held post-doctoral and research scientist positions at the Risö National Laboratory (Denmark) and at the Oak Ridge National Laboratory (USA). In 1984 he became Head of the Laboratory for Neutron Scattering, a joint venture between ETH Zurich and the Paul Scherrer Institute (PSI) Villigen. His main research activities cover neutron scattering studies in the fields of magnetism and superconductivity. He is author and co-author of more than 400 scientific publications and editor of seven books. He received (together with H.U. Güdel) the Walter Hälg Prize 2005 of the European Neutron Scattering Association in recognition of pioneering neutron scattering investigations of magnetic molecular compounds and spin-dimer systems. At present he is Professor Emeritus at ETH Zurich and one of the Managing Directors at SwissNeutronics.

Joël Mesot received his Masters and Ph.D. degrees in experimental physics at ETH Zurich (Switzerland). From 1992 to 1997 he built the time-of-flight spectrometer FOCUS at the spallation neutron source SINQ of the Paul Scherrer Institute (PSI). He then spent two years at Argonne National Laboratory (USA) performing angle resolved photoemission spectroscopy (ARPES) experiments. From 2004 to 2008 he lead the Laboratory for Neutron Scattering, a joint venture between ETH Zurich and PSI Villigen. Since 2008 he has served as Director of PSI and occupies a chair in solid-state physics jointly at ETH Zurich and EPF Lausanne. His research interests are in the field of metal oxides with unusual electronic and magnetic properties. In 2002 he was awarded the Latsis prize of ETH Zurich and from 2003 to 2008 he was the main Editor of Neutron News.

Thierry Strässle received his Masters and Ph.D. degrees in experimental physics at ETH Zurich (Switzerland). He has acquired expertise in the field of high-pressure neutron scattering during his post-doctoral stay at the Université Pierre & Marie Curie, Paris (France). In 2005 he joined the Laboratory for Neutron Scattering at PSI Villigen, where he is principally responsible of the time-of-flight spectrometer FOCUS at the Swiss Spallation Neutron Source. His research activities include neutron studies on multiferroic materials, unconventional superconductors, magnetic clusters and molecular solids under high-pressure.

Contents

Chapter 1

Introduction

1.1 Why neutron scattering?

In order to understand the materials naturally occurring in our world and artificially produced by modern technologies, a detailed understanding of their properties is required on an atomic scale. This information is the basis for any kind of research in physics, chemistry, biology, and materials science. Among the various experimental methods neutron and x-ray (photon) scattering have become the key techniques of choice. Both techniques are highly complementary. The most relevant, unique character of slow neutrons, which can hardly be matched by any other experimental technique, can be summarized as follows:

- The neutron interacts with the atomic nucleus, and not with the electrons as photons do. This has important consequences: The response of neutrons from light atoms (e.g., hydrogen, oxygen) is much higher than for x-rays; neutrons can easily distinguish atoms of comparable atomic number; and finally, neutrons easily distinguish isotopes which allows, e.g., by deuteration of specific parts of macromolecules (or biological substances), to focus on specific aspects of their atomic arrangement. A comparison of the scattering strength of some atoms and isotopes for x-rays and neutrons is sketched in Fig. 1.1.

- For the same wavelength as hard x-rays the neutron energy is much lower and comparable to the energy of elementary excitations in matter. Therefore, neutrons do not only allow the determination of the static average chemical structure, but also the investigation of the dynamic properties of atomic arrangements which are directly related to the physical properties of materials.

- By virtue of its neutrality the neutron is rather weakly interacting with matter which has several important consequences:
  - (i) The neutron produces a very small disturbance of the sample's properties which can be treated as small fluctuations from the equilibrium state, thus linear-response theory is an excellent approximation to extract the scattering law from the experimental data on an absolute scale. This condition is not always fulfilled in experiments with x-rays.
  - (ii) The neutron has a large penetration depth and therefore the bulk properties of matter can be studied.
  - (iii) The large penetration depth of the neutron is beneficial for the investigation of materials under extreme conditions such as very low and very high temperatures, high pressures, high magnetic and electric fields, or several of these together; in such cases the studied sample is always surrounded by numerous shields which largely prevent the use of x-rays.
  - (iv) Due to the weak interaction there is almost no radiation damage to the objects under study (e.g., living biological objects).
- The neutron carries a magnetic moment which makes it an excellent probe for the determination of the static and dynamical magnetic properties of matter (magnetic ordering phenomena, magnetic excitations, spin fluctuations).

In summary, the neutron scattering technique covers an ideal wavevector-energy range $(\boldsymbol{Q}, \omega)$ for investigations of condensed matter as visualized in Fig. 1.2.

1.2 Basic properties of the neutron

The energies of free neutrons span many orders of magnitude. For neutron scattering we are only interested in slow neutrons. Table 1.1 gives a summary of terms commonly used to characterize different neutron energy regimes. The kinetic energy of slow neutrons with velocity $\boldsymbol{v}$ is given by

$$E = \frac{mv^2}{2}, \tag{1.1}$$

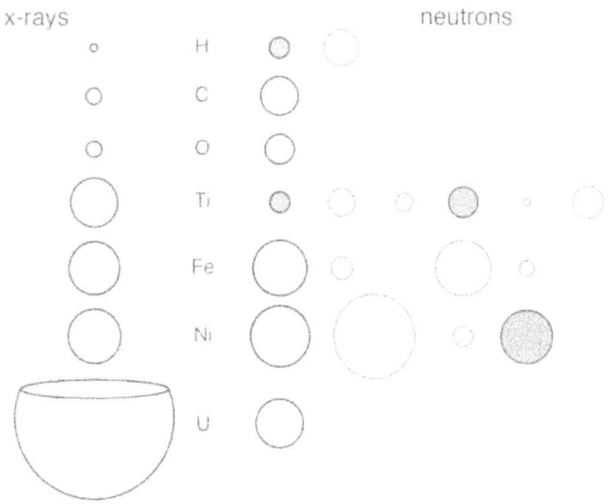

Fig. 1.1 Scattering strength of some atoms and isotopes for x-rays and neutrons. For neutrons only coherent scattering cross-sections are considered. Thick and thin lined circles refer to elements in their natural abundance and to selected isotopes, respectively. Isotopes with pronounced incoherent scattering are shown shaded.

Table 1.1 Approximate limits of neutron energy regimes classified by names.

Energy range	Classification nuclear physics	Classification neutron scattering	Energy range neutron scattering
		ultra cold	< 0.1 meV
		very cold	$0.1 \div 0.5$ meV
< 1 keV	slow	cold	$0.5 \div 5$ meV
		thermal	$5 \div 100$ meV
		epithermal or hot	$0.1 \div 1$ eV
		resonant	$1 \div 100$ eV
1 keV $\div$ 0.5 MeV	intermediate		
$0.5 \div 10$ MeV	fast		
$10 \div 50$ MeV	very fast		
$0.05 \div 10$ GeV	high energy or ultra fast		
> 10 GeV	relativistic		

where $m = 1.675 \cdot 10^{-27}$ kg is the mass of the neutron. The de Broglie wavelength λ of the neutron is defined by

$$\lambda = \frac{h}{mv}, \tag{1.2}$$

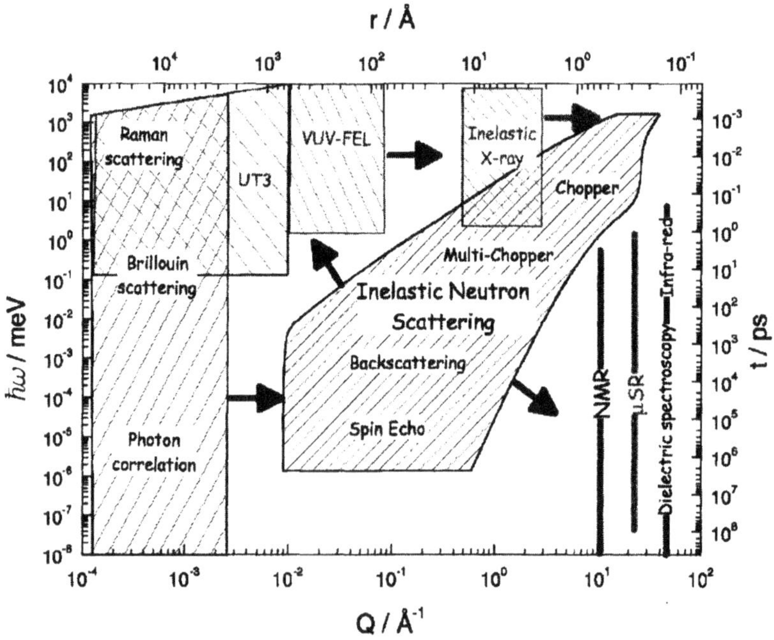

Fig. 1.2 Dynamical ranges $(\boldsymbol{r}, t)$ and $(\boldsymbol{Q}, \omega)$ in real and reciprocal space, respectively, for neutrons and some other probes used for investigations of condensed matter. The variables $\boldsymbol{Q}$ and ω are defined by Eqs (2.1) and (2.2), respectively (taken from *The ESS Project, Vol. II*, ed. by D. Richter (FZ Jülich, 2002), p.5-4)

where $h = 6.626 \cdot 10^{-34}$ J·s is the Planck constant. The wavevector $\boldsymbol{k}$ of the neutron has the magnitude

$$k = \frac{2\pi}{\lambda}, \tag{1.3}$$

its direction being that of its velocity $\boldsymbol{v}$. Eqs (1.1) - (1.3) define the momentum $\boldsymbol{p}$ and the energy E of the neutron:

$$\boldsymbol{p} = \hbar \boldsymbol{k}, \tag{1.4}$$

$$E = \frac{\hbar^2 k^2}{2m}, \tag{1.5}$$

with $\hbar = h/2\pi$. It is conventional to say that a neutron with energy E corresponds to a temperature T:

$$E = k_B T, \tag{1.6}$$

where $k_B = 1.381 \cdot 10^{-23}$ J/K is the Boltzmann constant. Combining Eqs (1.1) - (1.6) yields

$$E = \frac{h^2}{2m\lambda^2} = \frac{\hbar^2 k^2}{2m} = \frac{mv^2}{2} = k_B T. \tag{1.7}$$

By inserting the values of the elementary constants we arrive at the following relations between the energy, wavelength, wavevector, velocity, and temperature for thermal neutrons:

$$E = 81.81 \cdot \frac{1}{\lambda^2} = 2.072 \cdot k^2 = 5.227 \cdot v^2 = 0.08617 \cdot T, \tag{1.8}$$

where the units are meV for E, Å for λ, Å^{-1} for k, km/s for v, and K (Kelvin) for T. In neutron scattering the energies are usually quoted in meV. Another energy unit frequently used is terahertz (THz), and other spectroscopic techniques often use wavenumbers in units of cm^{-1}. We then have

$$1 \text{ meV} = 0.242 \text{ THz} = 8.07 \text{ cm}^{-1} = 11.6 \text{ K} = 17.3 \text{ T}, \tag{1.9}$$

where the conversions to temperature (K) and magnetic field (T=Tesla) are included for completeness.

Chapter 2

Basic Principles of Neutron Scattering

2.1 Aim of a neutron scattering experiment

The principal aim of a neutron scattering experiment is the determination of the probability that a neutron which is incident on the sample with wavevector $\boldsymbol{k}$ is scattered into the state with wavevector $\boldsymbol{k}'$. The intensity of the scattered neutrons is thus measured as a function of the momentum transfer

$$\hbar \boldsymbol{Q} = \hbar(\boldsymbol{k} - \boldsymbol{k}'), \tag{2.1}$$

where $\boldsymbol{Q}$ is known as the scattering vector, and the corresponding energy transfer is given by

$$\hbar\omega = \frac{\hbar^2}{2m}(k^2 - k'^2). \tag{2.2}$$

Eqs (2.1) and (2.2) describe the momentum and energy conservation of the neutron scattering process, respectively. The momentum conservation is schematically sketched in Fig. 2.1. For $k = k'$ we have from Eq. (2.2) $\hbar\omega = 0$, i.e., elastic scattering. Fig. 2.1(a) shows the particular situation

$$\boldsymbol{Q} = \boldsymbol{k} - \boldsymbol{k}' = \boldsymbol{\tau}, \tag{2.3}$$

which is just the condition known as Bragg's law (coherent elastic scattering, Eq. (4.8)). If $\boldsymbol{Q}$ does not coincide with a reciprocal lattice vector $\boldsymbol{\tau}$, we have inelastic neutron scattering (inelastic neutron scattering). For inelastic scattering as shown in Fig. 2.1(b) the scattering vector can be decomposed according to $\boldsymbol{Q} = \boldsymbol{\tau} + \boldsymbol{q}$, where $\boldsymbol{q}$ is the wavevector of an elementary excitation to be specified. Neutron scattering turns out to be the most precise experimental technique which is able to measure the dispersion relation $\hbar\omega(\boldsymbol{q})$ at any predetermined point in reciprocal space.

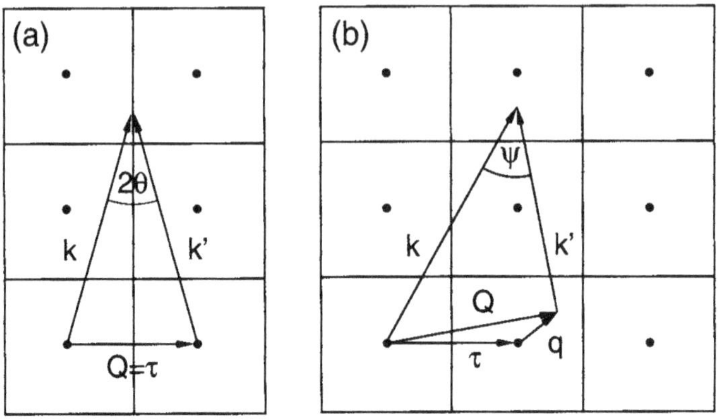

Fig. 2.1 Visualization of Eq. (2.1) in reciprocal space for elastic (a) and inelastic (b) neutron scattering. 2θ and Ψ denote the Bragg scattering angle and the general scattering angle, respectively. The lines indicate the boundaries of the Brillouin zone, and the dots denote the zone centers.

2.2 Neutron scattering cross-section

The neutron scattering cross-section corresponds to the number of neutrons scattered per second into a (small) solid angle $d\Omega$ with energy transfers between $\hbar\omega$ and $\hbar(\omega + d\omega)$, divided by the flux of the incident neutrons. Theoretical expressions for the cross section usually start from Fermi's Golden Rule

$$\frac{d^2\sigma}{d\Omega d\omega} = \left(\frac{m}{2\pi\hbar^2}\right)^2 \frac{k'}{k} \sum_{\lambda',\sigma'} \sum_{\lambda,\sigma} p_\lambda p_\sigma |\langle \boldsymbol{k}', \boldsymbol{\sigma}', \lambda' | \hat{U} | \boldsymbol{k}, \boldsymbol{\sigma}, \lambda \rangle|^2 \times \delta(\hbar\omega + E_\lambda - E_{\lambda'}) \tag{2.4}$$

where $|\lambda\rangle$ denotes the initial state of the scatterer, with energy E_λ and thermal population factor p_λ, and its final state is $|\lambda'\rangle$. $\boldsymbol{\sigma}$ and $\boldsymbol{\sigma}'$ are the spin states of the incoming and scattered neutrons, respectively, and p_σ is the polarization probability. The δ-function describes the law of energy conservation (see Appendix A for the definition and some useful properties of the δ-function). $\hat{U}$ is the interaction operator of the neutron with the sample which depends on the specific scattering process. E.g., neutron scattering from nuclei at fixed positions $\boldsymbol{R}_j$ is well approximated by the Fermi pseudopotential:

$$\hat{U}(\boldsymbol{r}) = \frac{2\pi\hbar^2}{m}\sum_j b_j\delta(\boldsymbol{r}-\hat{\boldsymbol{R}}_j) \tag{2.5}$$

where b_j is the scattering length. The magnitude of b_j is of the order 10^{-12} cm, i.e., for nuclear scattering the cross section Eq. (2.4) amounts to about 10^{-24} cm^2 (= 1 barn). In principle, the scattering length is a complex number; the real part describes the energy-independent scattering, and the imaginary part the energy-dependent absorption. The imaginary part is large, if the energy of the compound nucleus (= nucleus + neutron) is close to the energy of an excited nuclear state. However, only a few nuclei fulfill this property for thermal neutrons, the most prominent examples being ^{113}Cd and ^{157}Gd. In the following we only consider the real part of the scattering lengths which are listed in the Appendix B for selected nuclei.

Let us describe the incoming and outgoing neutrons by plane waves (neglecting the spin states $\boldsymbol{\sigma}$ and $\boldsymbol{\sigma}'$ which will be considered in Chaps 2.6 and 2.7):

$$|\boldsymbol{k}\rangle = e^{i\boldsymbol{k}\cdot\boldsymbol{r}}, \qquad |\boldsymbol{k}'\rangle = e^{i\boldsymbol{k}'\cdot\boldsymbol{r}}. \tag{2.6}$$

Inserting Eqs (2.5) and (2.6) into the matrix element of Eq. (2.4) yields

$$\langle\boldsymbol{k}',\lambda'|\hat{U}|\boldsymbol{k},\lambda\rangle = \left(\frac{2\pi\hbar^2}{m}\right)\langle\lambda'|\int \mathrm{d}\boldsymbol{r}\, e^{-i\boldsymbol{k}'\cdot\boldsymbol{r}}\sum_j b_j\delta(\boldsymbol{r}-\hat{\boldsymbol{R}}_j)e^{i\boldsymbol{k}\cdot\boldsymbol{r}}|\lambda\rangle. \tag{2.7}$$

With use of the transformation $\boldsymbol{R} = \boldsymbol{r} - \boldsymbol{R}_j$ and the definition $\boldsymbol{Q} = \boldsymbol{k} - \boldsymbol{k}'$ Eq. (2.7) is rewritten as

$$\langle\boldsymbol{k}',\lambda'|\hat{U}|\boldsymbol{k},\lambda\rangle = \left(\frac{2\pi\hbar^2}{m}\right)\langle\lambda'|\int \mathrm{d}\boldsymbol{R}\, e^{i\boldsymbol{Q}\cdot\boldsymbol{R}}\delta(\boldsymbol{R})\sum_j b_j e^{i\boldsymbol{Q}\cdot\hat{\boldsymbol{R}}_j}|\lambda\rangle, \tag{2.8}$$

which by application of Eq. (A.8) further reduces to

$$\langle\boldsymbol{k}',\lambda'|\hat{U}|\boldsymbol{k},\lambda\rangle = \left(\frac{2\pi\hbar^2}{m}\right)\langle\lambda'|\sum_j b_j e^{i\boldsymbol{Q}\cdot\hat{\boldsymbol{R}}_j}|\lambda\rangle. \tag{2.9}$$

Combining Eqs (2.4) and (2.9) and using Eq. (A.6) for the δ-function yields the cross section

$$\begin{aligned}\frac{\mathrm{d}^2\sigma}{\mathrm{d}\Omega\mathrm{d}\omega} &= \frac{k'}{k}\sum_\lambda p_\lambda\sum_{\lambda'}\sum_{j,j'} b_j b_{j'}\langle\lambda|e^{-i\boldsymbol{Q}\cdot\hat{\boldsymbol{R}}_{j'}}|\lambda'\rangle\langle\lambda'|e^{i\boldsymbol{Q}\cdot\hat{\boldsymbol{R}}_j}|\lambda\rangle \\ &\quad\times\frac{1}{2\pi\hbar}\int_{-\infty}^{\infty} e^{i(E_{\lambda'}-E_\lambda)t/\hbar}e^{-i\omega t}\mathrm{d}t \\ &= \frac{1}{2\pi\hbar}\frac{k'}{k}\sum_\lambda p_\lambda\sum_{\lambda'}\sum_{j,j'} b_j b_{j'}\int_{-\infty}^{\infty}\langle\lambda|e^{-i\boldsymbol{Q}\cdot\hat{\boldsymbol{R}}_{j'}}e^{iE_{\lambda'}t/\hbar}|\lambda'\rangle \\ &\quad\times\langle\lambda'|e^{i\boldsymbol{Q}\cdot\hat{\boldsymbol{R}}_j}e^{-iE_\lambda t/\hbar}|\lambda\rangle\, e^{-i\omega t}\mathrm{d}t.\end{aligned} \tag{2.10}$$

With the theorem for Hermitian operators $\hat{A}$, $\hat{B}$

$$\sum_{\lambda'} \langle\lambda|\hat{A}|\lambda'\rangle\langle\lambda'|\hat{B}|\lambda''\rangle = \langle\lambda|\hat{A}\hat{B}|\lambda''\rangle \tag{2.11}$$

and the eigenvalue equation for the Hamiltonian $\hat{H}$

$$e^{-\imath\hat{H}t/\hbar}|\lambda\rangle = e^{-\imath E_\lambda t/\hbar}|\lambda\rangle \tag{2.12}$$

we can rewrite the matrix elements of Eq. (2.10) as

$$\langle\lambda|e^{-\imath\boldsymbol{Q}\cdot\hat{\boldsymbol{R}}'_j}e^{\imath\hat{H}t/\hbar}e^{\imath\boldsymbol{Q}\cdot\hat{\boldsymbol{R}}_j}e^{-\imath\hat{H}t/\hbar}|\lambda\rangle. \tag{2.13}$$

Now we replace the Schrödinger operators $e^{\imath\boldsymbol{Q}\cdot\hat{\boldsymbol{R}}_j}$ by time-dependent Heisenberg operators:

$$e^{\imath\boldsymbol{Q}\cdot\hat{\boldsymbol{R}}_j(t)} = e^{\imath\hat{H}t/\hbar}e^{\imath\boldsymbol{Q}\cdot\hat{\boldsymbol{R}}_j}e^{-\imath\hat{H}t/\hbar}. \tag{2.14}$$

Likewise we define a Heisenberg operator for $t = 0$:

$$e^{\imath\boldsymbol{Q}\cdot\hat{\boldsymbol{R}}_j(0)} = e^{\imath\hat{H}t/\hbar}e^{\imath\boldsymbol{Q}\cdot\hat{\boldsymbol{R}}_j}e^{-\imath\hat{H}t/\hbar}\,|_{t=0}. \tag{2.15}$$

By making use of the expectation value of an operator $\hat{A}$:

$$\langle\hat{A}\rangle = \sum_{\lambda} p_\lambda\langle\lambda|\hat{A}|\lambda\rangle \tag{2.16}$$

we arrive at the final cross-section formula

$$\boxed{\frac{\mathrm{d}^2\sigma}{\mathrm{d}\Omega\mathrm{d}\omega} = \frac{k'}{k}\frac{1}{2\pi\hbar}\sum_{j,j'} b_j b_{j'}\int_{-\infty}^{\infty}\langle e^{-\imath\boldsymbol{Q}\cdot\hat{\boldsymbol{R}}_{j'}(0)}e^{\imath\boldsymbol{Q}\cdot\hat{\boldsymbol{R}}_j(t)}\rangle e^{-\imath\omega t}\mathrm{d}t} \tag{2.17}$$

2.3 Correlation functions

The operator part of Eq. (2.17) corresponds to the intermediate pair correlation function (also called intermediate scattering function):

$$I(\boldsymbol{Q},t) = \frac{1}{N}\sum_{j,j'}\langle e^{-\imath\boldsymbol{Q}\cdot\hat{\boldsymbol{R}}_{j'}(0)}e^{\imath\boldsymbol{Q}\cdot\hat{\boldsymbol{R}}_j(t)}\rangle \tag{2.18}$$

where N denotes the number of atoms in the system. Fourier transformation of Eq. (2.18) with respect to $\boldsymbol{Q}$ and t yields the space-time pair correlation function introduced by van Hove [van Hove (1954)]

$$G(\boldsymbol{r},t) = \frac{1}{(2\pi)^3}\int I(\boldsymbol{Q},t)e^{-\imath\boldsymbol{Q}\cdot\boldsymbol{r}}\mathrm{d}\boldsymbol{Q} \tag{2.19}$$

and the dynamical structure factor

$$S(\boldsymbol{Q},\omega) = \frac{1}{2\pi\hbar}\int I(\boldsymbol{Q},t)e^{-\imath\omega t}\mathrm{d}t, \tag{2.20}$$

respectively. $S(\boldsymbol{Q},\omega)$ is also called the scattering law, as it is directly related to the cross section Eq. (2.17).

We now want to elucidate the physical meaning of the pair correlation function $G(\boldsymbol{r},t)$. Combining Eqs (2.18) and (2.19) yields

$$\boxed{G(\boldsymbol{r},t) = \frac{1}{(2\pi)^3}\frac{1}{N}\int\sum_{j,j'}\langle e^{-\imath\boldsymbol{Q}\cdot\hat{\boldsymbol{R}}_{j'}(0)}e^{\imath\boldsymbol{Q}\cdot\hat{\boldsymbol{R}}_j(t)}\rangle e^{-\imath\boldsymbol{Q}\cdot\boldsymbol{r}}\mathrm{d}\boldsymbol{Q}} \tag{2.21}$$

It should be noted that the operators $\hat{\boldsymbol{R}}_{j'}(0)$ and $\hat{\boldsymbol{R}}_j(t)$ do not commute in general, thus the expectation value in Eq. (2.21) cannot be replaced by

$$\langle e^{-\imath\boldsymbol{Q}\cdot(\hat{\boldsymbol{R}}_{j'}(0)-\hat{\boldsymbol{R}}_j(t))}\rangle.$$

Applying Eq. (A.8):

$$e^{-\imath\boldsymbol{Q}\cdot\hat{\boldsymbol{R}}_{j'}(0)} = \int e^{-\imath\boldsymbol{Q}\cdot\boldsymbol{r}'}\delta\left(\boldsymbol{r}'-\hat{\boldsymbol{R}}_{j'}(0)\right)\mathrm{d}\boldsymbol{r}'$$

and regrouping some terms in Eq. (2.21) yields

$$\begin{aligned} G(\boldsymbol{r},t) = \frac{1}{(2\pi)^3}\frac{1}{N}\int\sum_{j,j'}\langle\delta\left(\boldsymbol{r}'-\hat{\boldsymbol{R}}_{j'}(0)\right) \\ \times\int e^{-\imath\boldsymbol{Q}\cdot(\boldsymbol{r}+\boldsymbol{r}'-\hat{\boldsymbol{R}}_j(t))}d\boldsymbol{Q}\rangle\mathrm{d}\boldsymbol{r}'. \end{aligned} \tag{2.22}$$

Application of Eq. (A.9):

$$\int e^{-\imath\boldsymbol{Q}\cdot(\boldsymbol{r}+\boldsymbol{r}'-\hat{\boldsymbol{R}}_j(t))}\mathrm{d}\boldsymbol{Q} = (2\pi)^3\delta\left(\boldsymbol{r}+\boldsymbol{r}'-\hat{\boldsymbol{R}}_j(t)\right)$$

yields finally

$$G(\boldsymbol{r},t) = \frac{1}{N}\sum_{j,j'}\int\langle\delta\left(\boldsymbol{r}'-\hat{\boldsymbol{R}}_{j'}(0)\right)\delta\left(\boldsymbol{r}'+\boldsymbol{r}-\hat{\boldsymbol{R}}_j(t)\right)\rangle\mathrm{d}\boldsymbol{r}'. \tag{2.23}$$

$G(\boldsymbol{r},t)$ describes the correlation between the atom j' at the time $t=0$ at the position $\boldsymbol{r}'$ and the atom j at a later time t at another position $\boldsymbol{r}'+\boldsymbol{r}$, i.e., the probability of having two atoms j and j' in a well defined spatial and temporal correlation. $G(\boldsymbol{r},t)$ may therefore be considered as the most general description of the statics and dynamics of condensed matter on an atomic scale.

$G(\boldsymbol{r},t)$ can be split into two parts with $j = j'$ and $j \neq j'$, describing the self correlation and the distinct pair correlation $G_s(\boldsymbol{r},t)$ and $G_d(\boldsymbol{r},t)$, respectively:

$$G_s(\boldsymbol{r},t) = \frac{1}{N}\sum_j \int \langle\delta\left(\boldsymbol{r}' - \hat{\boldsymbol{R}}_j(0)\right)\delta\left(\boldsymbol{r}' + \boldsymbol{r} - \hat{\boldsymbol{R}}_j(t)\right)\rangle \mathrm{d}\boldsymbol{r}', \tag{2.24}$$

$$G_d(\boldsymbol{r},t) = \frac{1}{N}\sum_{j\neq j'} \int \langle\delta\left(\boldsymbol{r}' - \hat{\boldsymbol{R}}_{j'}(0)\right)\delta\left(\boldsymbol{r}' + \boldsymbol{r} - \hat{\boldsymbol{R}}_j(t)\right)\rangle \mathrm{d}\boldsymbol{r}'. \tag{2.25}$$

2.4 Coherent and incoherent scattering

We now proceed to introduce the correlation functions defined in Chap. 2.3 into the cross-section formula Eq. (2.17). Hereby we have to take into account that the scattering lengths b_j of each element depend on both the isotope and the spin quantum number I of the nucleus. Since there is no correlation between the scattering lengths b_j and $b_{j'}$ of the nuclei, the corresponding sum in Eq. (2.17) has to be averaged over the sample volume:

$$\begin{aligned} &\text{for } j \neq j': \quad && \langle b_j b_{j'}\rangle = \langle b_j\rangle\langle b_{j'}\rangle = \langle b\rangle^2; \\ &\text{for } j = j': \quad && \langle b_j b_{j'}\rangle = \langle b_j^2\rangle = \langle b^2\rangle. \end{aligned} \tag{2.26}$$

Eq. (2.17) can then be written as

$$\frac{\mathrm{d}^2\sigma}{\mathrm{d}\Omega\mathrm{d}\omega} = N\frac{k'}{k}\int_{-\infty}^{\infty}\mathrm{d}t\, e^{-i\omega t}\int \mathrm{d}\boldsymbol{r}e^{i\boldsymbol{Q}\cdot\boldsymbol{r}}\left(\langle b^2\rangle G_s(\boldsymbol{r},t) + \langle b\rangle^2 G_d(\boldsymbol{r},t)\right).$$

With $G(\boldsymbol{r},t) = G_s(\boldsymbol{r},t) + G_d(\boldsymbol{r},t)$ we can eliminate $G_d(\boldsymbol{r},t)$:

$$\boxed{\begin{aligned}\frac{\mathrm{d}^2\sigma}{\mathrm{d}\Omega\mathrm{d}\omega} = N\frac{k'}{k}&\int_{-\infty}^{\infty}\mathrm{d}t\, e^{-i\omega t}\int \mathrm{d}\boldsymbol{r}e^{i\boldsymbol{Q}\cdot\boldsymbol{r}} \\ &\times\left(\langle b\rangle^2 G(\boldsymbol{r},t) + \left[\langle b^2\rangle - \langle b\rangle^2\right] G_s(\boldsymbol{r},t)\right)\end{aligned}} \tag{2.27}$$

The two terms in Eq. (2.27) are called coherent and incoherent neutron scattering cross sections:

$$\left(\frac{\mathrm{d}^2\sigma}{\mathrm{d}\Omega\mathrm{d}\omega}\right)_{\text{coh}} = N\frac{k'}{k}\langle b\rangle^2 S_{coh}(\boldsymbol{Q},\omega), \tag{2.28}$$

$$\left(\frac{\mathrm{d}^2\sigma}{\mathrm{d}\Omega\mathrm{d}\omega}\right)_{\text{inc}} = N\frac{k'}{k}\left[\langle b^2\rangle - \langle b\rangle^2\right] S_{inc}(\boldsymbol{Q},\omega). \tag{2.29}$$

For coherent scattering, the total cross section is given by the squared average of the sum of the scattering lengths:

$$\sigma_{\mathrm{coh}} = 4\pi \left(\frac{1}{N} \sum_j b_j \right)^2 = 4\pi \langle b \rangle^2, \tag{2.30}$$

whereas the total incoherent scattering has its origin in the disorder of the scattering lengths:

$$\sigma_{\mathrm{inc}} = 4\pi \left(\langle b^2 \rangle - \langle b \rangle^2 \right). \tag{2.31}$$

The average scattering lengths are defined by

$$\langle b \rangle = \sum_j p_j b_j, \quad \langle b^2 \rangle = \sum p_j b_j^2, \qquad \text{with} \quad \sum_j p_j = 1, \tag{2.32}$$

where p_j denotes the probability that a nucleus has the scattering length b_j. For the case of a nucleus with nuclear spin quantum number $I \neq 0$ the interacting system, i.e., nucleus plus neutron, can be either in an ortho state $|+\rangle$ or in a para state $|-\rangle$:

$\|+\rangle$ state	total spin quantum number: $I + 1/2$ degeneracy of the state: $2(I + 1/2) + 1 = 2I + 2$ scattering length: b^+
$\|-\rangle$ state	total spin quantum number: $I - 1/2$ degeneracy of the state: $2(I - 1/2) + 1 = 2I$ scattering length: b^-

For unpolarized neutrons, the two states have therefore different probabilities to occur:

$$p^+ = \frac{2I+2}{4I+2} = \frac{I+1}{2I+1}, \quad p^- = \frac{2I}{4I+2} = \frac{I}{2I+1}. \tag{2.33}$$

The average scattering lengths can then be calculated from Eq. (2.32).

Hydrogen with $I = 1/2$ is an excellent example for a strong incoherent scatterer, because the scattering length of the triplet state $|+\rangle$ ($b^+ = 1.085 \cdot 10^{-12}$ cm) and the singlet state $|-\rangle$ ($b^- = -4.750 \cdot 10^{-12}$ cm) are very different, leading to $\sigma_{inc} = 80.3$ barn $\gg \sigma_{coh} = 1.76$ barn. In contrast, the corresponding values for deuterium ($I = 1$) are $\sigma_{inc} = 2.05$ barn and $\sigma_{coh} = 5.59$ barn. Therefore it is possible to distinguish between coherent and incoherent scattering processes by deuteration of the sample which is one of the unique features of neutron scattering (see, e.g., Chaps 5.3 and 13.4).

2.5 Principle of detailed balance

In inelastic neutron scattering experiments the energy transfer $\hbar\omega$ defined by Eq. (2.2) may be positive and negative, corresponding to neutron energy-loss and energy-gain processes, respectively. It is therefore of interest to investigate the scattering law $S(\boldsymbol{Q},\omega)$ upon time reversal, i.e., $\omega \rightarrow -\omega$. For this purpose we introduce $S(\boldsymbol{Q},\omega)$ into Eq. (2.4) and assume the states $|\lambda\rangle$ to obey Boltzmann statistics:

$$S(\boldsymbol{Q},\omega) = \frac{1}{NZ}\sum_{\lambda,\lambda'} e^{-\frac{E_\lambda}{k_BT}} |\sum_j \langle\lambda'|e^{i\boldsymbol{Q}\cdot\hat{\boldsymbol{R}}_j}|\lambda\rangle|^2 \delta(\hbar\omega + E_\lambda - E_{\lambda'}), \quad (2.34)$$

where $Z = \sum_\lambda e^{-\frac{E_\lambda}{k_BT}}$ is the partition function. Time reversal implies the exchange of the states $|\lambda\rangle$ and $|\lambda'\rangle$:

$$\begin{aligned} S(-\boldsymbol{Q},-\omega) = & \frac{1}{NZ}\sum_{\lambda,\lambda'} e^{-\frac{E_{\lambda'}}{k_BT}} |\sum_j \langle\lambda|e^{-i\boldsymbol{Q}\cdot\hat{\boldsymbol{R}}_j}|\lambda'\rangle|^2 \\ & \times \ \delta(-\hbar\omega + E_{\lambda'} - E_\lambda). \end{aligned} \quad (2.35)$$

Inserting the relation (valid for Hermitian operators)

$$|\sum_j \langle\lambda|e^{-i\boldsymbol{Q}\cdot\hat{\boldsymbol{R}}_j}|\lambda'\rangle|^2 = |\sum_j \langle\lambda'|e^{i\boldsymbol{Q}\cdot\hat{\boldsymbol{R}}_j}|\lambda\rangle|^2 \quad (2.36)$$

and Eq. (A.2) into Eq. (2.35) yields

$$\begin{aligned} S(-\boldsymbol{Q},-\omega) &= \frac{1}{NZ}\sum_{\lambda,\lambda'} e^{-\frac{E_{\lambda'}}{k_BT}} |\sum_j \langle\lambda'|e^{i\boldsymbol{Q}\cdot\hat{\boldsymbol{R}}_j}|\lambda\rangle|^2 \\ & \quad \times \delta(\hbar\omega + E_\lambda - E_{\lambda'}) \\ &= \frac{1}{NZ}\sum_{\lambda,\lambda'} e^{-\frac{E_{\lambda'}-E_\lambda}{k_BT}} e^{-\frac{E_\lambda}{k_BT}} |\sum_j \langle\lambda'|e^{i\boldsymbol{Q}\cdot\hat{\boldsymbol{R}}_j}|\lambda\rangle|^2 \\ & \quad \times \delta(\hbar\omega + E_\lambda - E_{\lambda'}). \end{aligned} \quad (2.37)$$

With $\hbar\omega = E_{\lambda'} - E_\lambda$ and by comparison with Eq. (2.34) we obtain

$$\boxed{S(-\boldsymbol{Q},-\omega) = e^{-\frac{\hbar\omega}{k_BT}} S(\boldsymbol{Q},\omega)} \quad (2.38)$$

which is known as principle of detailed balance. The principle of detailed balance unambiguously relates the neutron energy-gain and energy-loss processes to each other, similar to other spectroscopies for which emission and absorption processes (often called Stokes and Anti-Stokes processes, respectively) are possible. This is schematically sketched in Fig. 2.2 and

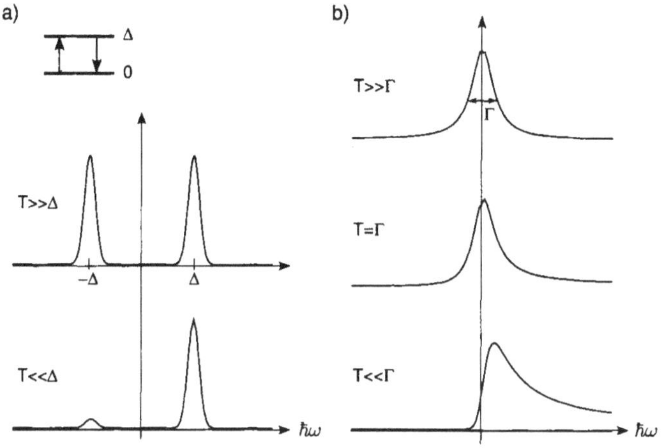

Fig. 2.2 Schematic sketch of the temperature dependence of (a) inelastic scattering and (b) quasielastic diffuse magnetic scattering for both energy-gain and energy-loss processes.

visualized in Fig. 2.3 for both inelastic scattering (observation of a transition in a two-level system with energy separation $\hbar\omega = \Delta$) and quasielastic scattering (observation of a broad distribution with half-width Γ at zero-energy transfer). In the latter case we see that the observation of an apparently inelastic line at $T \ll \Gamma$ reduces to quasielastic scattering by applying Eq. (2.38).

2.6 Magnetic scattering

We describe the operator $\hat{U}$ of Eq. (2.4) for the case of magnetic scattering by the interaction of a neutron with the magnetic field $\boldsymbol{H}$ associated with the sample:

$$\hat{U} = \hat{\boldsymbol{\mu}} \cdot \boldsymbol{H} = -\gamma \mu_N \hat{\boldsymbol{\sigma}} \cdot \boldsymbol{H}, \tag{2.39}$$

where $\hat{\boldsymbol{\mu}}$ is the magnetic moment operator of the neutron, $\gamma = -1.91$ the gyromagnetic ratio, $\mu_N = 5.05079 \cdot 10^{-27}$ J/T the nuclear magneton, and $\hat{\boldsymbol{\sigma}}$ a Pauli spin operator (see Appendix C). For a large class of magnetic compounds the magnetic field $\boldsymbol{H}$ used in Eq. (2.39) is generated by unpaired electrons. The magnetic field due to a single electron moving with velocity

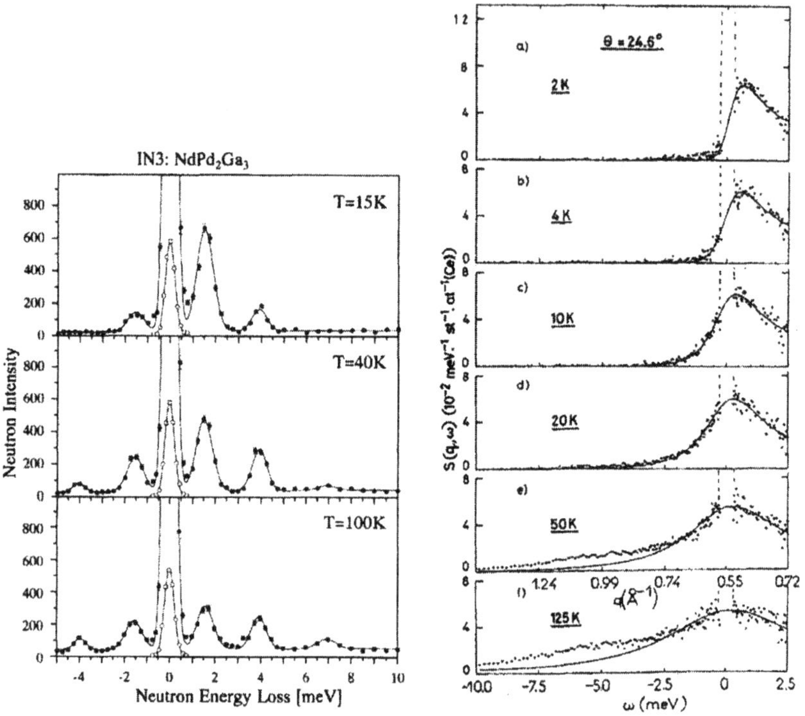

Fig. 2.3 Left: energy spectra of crystal-field transitions of Nd within $NdPd_2Ga_3$. Reprinted with kind permission from Springer Science+Business Media [Dönni *et al.* (1997)], Fig. 2, copyright 1997. The inelastic lines correspond to the transitions $\Gamma_7 \leftrightarrow \Gamma_9^{(1)}$ (± 1.5 meV), $\Gamma_9^{(1)} \leftrightarrow \Gamma_8^{(1)}$ (± 4 meV) and $\Gamma_8^{(1)} \rightarrow \Gamma_8^{(2)}$ (7 meV). Right: quasielastic diffuse magnetic neutron scattering of $CeAl_3$. Reprinted from [Murani *et al.* (1980)]. Copyright 1980, with permission from Elsevier. The lines represent Lorentzian fits to the data.

$\boldsymbol{v}_e$ is given by

$$\boldsymbol{H} = \nabla \wedge \left(\frac{\boldsymbol{\mu}_e \wedge \boldsymbol{R}}{|\boldsymbol{R}|^3} \right) - \frac{e}{c} \frac{\boldsymbol{v}_e \wedge \boldsymbol{R}}{|\boldsymbol{R}|^3}, \tag{2.40}$$

where $\boldsymbol{R}$ is the distance from the electron to the point at which the field is measured, $e = 1.602 \cdot 10^{-19}$ C the elementary charge, and $c = 2.99792 \cdot 10^8$ m/s the velocity of light. The magnetic moment operator of an electron is

$$\hat{\boldsymbol{\mu}}_e = -2\mu_B \hat{\boldsymbol{s}}, \tag{2.41}$$

where $\mu_B = 9.27402 \cdot 10^{-24}$ J/T is the Bohr magneton and $\hat{\boldsymbol{s}}$ the spin operator of the electron. The first term of Eq. (2.40) arises from the spin of the electron, and the second from its orbital motion.

The major task in the evaluation of the magnetic neutron cross-section is the calculation of the transition matrix element in Eq. (2.4). The corresponding procedure is quite involved, thus we refer to the Appendix D for details. For unpolarized neutrons, identical magnetic ions with localized electrons, and spin-only scattering the following master formula is obtained:

$$\boxed{\frac{d^2\sigma}{d\Omega d\omega} = (\gamma r_0)^2 \frac{k'}{k} F^2(\boldsymbol{Q}) e^{-2W(\boldsymbol{Q})} \sum_{\alpha,\beta} \left(\delta_{\alpha\beta} - \frac{Q_\alpha Q_\beta}{Q^2}\right) S^{\alpha\beta}(\boldsymbol{Q},\omega)} \tag{2.42}$$

where $S^{\alpha\beta}(\boldsymbol{Q},\omega)$ is the magnetic scattering function:

$$\boxed{\begin{aligned} S^{\alpha\beta}(\boldsymbol{Q},\omega) = \sum_{j,j'} e^{\imath \boldsymbol{Q}\cdot(\boldsymbol{R}_j - \boldsymbol{R}_{j'})} \sum_{\lambda,\lambda'} p_\lambda \langle\lambda|\hat{S}^\alpha_{j'}|\lambda'\rangle\langle\lambda'|\hat{S}^\beta_j|\lambda\rangle \\ \times \quad \delta(\hbar\omega + E_\lambda - E_{\lambda'}) \end{aligned}} \tag{2.43}$$

$F(\boldsymbol{Q})$ the dimensionless magnetic form factor defined as the Fourier transform of the normalized spin density associated with the magnetic ions, $e^{-2W(\boldsymbol{Q})}$ the Debye-Waller factor, and $\hat{S}^\alpha_j$ ($\alpha = x,y,z$) the spin operator of the jth ion at site $\boldsymbol{R}_j$. From the magnitude of r_0 we expect the magnetic neutron cross-section to be of the order 10^{-24} cm^2, i.e., similar to the size of the nuclear cross section (see Chap. 2.2).

The essential factor in Eq. (2.42) is the magnetic scattering function $S^{\alpha\beta}(\boldsymbol{Q},\omega)$ which will be discussed in more detail below. There are two further factors which govern the cross section for magnetic neutron scattering in a characteristic way: Firstly, the magnetic form factor $F(\boldsymbol{Q})$ which usually falls off with increasing modulus of the scattering vector $\boldsymbol{Q}$ [Freeman and Desclaux (1979); Brown (1999)]. Secondly, the polarization factor $(\delta_{\alpha\beta} - Q_\alpha Q_\beta/Q^2)$ tells us that neutrons can only couple to magnetic moments or spin fluctuations perpendicular to $\boldsymbol{Q}$ which unambiguously allows to determine moment directions or to distinguish between different polarizations of spin fluctuations.

Eq. (2.43) strictly applies to cases where the orbital angular momentum of the magnetic ions is either zero or quenched by the crystal field. A theoretical treatment of the scattering by ions with unquenched orbital moment has been given by Johnston [Johnston (1966)], however, the calculation is complicated, and we simply quote the result for $Q \to 0$. In this case the cross section measures the magnetization, $\boldsymbol{\mu} = -\mu_B(\boldsymbol{L} + 2\boldsymbol{S})$, i.e., a combination of spin and orbital moments that does not allow their separation. This clearly contrasts to magnetic scattering by x-rays [Lovesey and Collins (1996)]. For magnetic neutron scattering an approximate result can be obtained for modest values of Q. We replace the spin operator $\hat{S}_j^\alpha$ in Eq. (2.43) by

$$\hat{S}_j^\alpha = \frac{1}{2} g \hat{J}_j^\alpha \tag{2.44}$$

where

$$g = 1 + \frac{J(J+1) - L(L+1) + S(S+1)}{2J(J+1)} \tag{2.45}$$

is the Landé splitting factor and $\hat{J}_j^\alpha$ is an effective angular momentum operator (e.g., for rare-earth ions J is the total angular momentum quantum number resulting from the spin-orbit coupling which combines the spin and orbital angular momentum S and L, respectively).

Using the integral representation Eq. (A.6) of the δ-function,

$$\delta(\hbar\omega + E_\lambda - E_{\lambda'}) = \frac{1}{2\pi\hbar} \int_{-\infty}^{\infty} e^{i(E_{\lambda'} - E_\lambda)t/\hbar} e^{-i\omega t} dt,$$

the scattering function $S^{\alpha\beta}(\boldsymbol{Q}, \omega)$ (Eq. (2.43)) transforms into a physically transparent form:

$$\boxed{S^{\alpha\beta}(\boldsymbol{Q}, \omega) = \frac{1}{2\pi\hbar} \sum_{j,j'} \int_{-\infty}^{\infty} e^{i\boldsymbol{Q}\cdot(\boldsymbol{R}_j - \boldsymbol{R}_{j'})} \langle \hat{S}_j^\alpha(0) \hat{S}_{j'}^\beta(t) \rangle e^{-i\omega t} dt} \tag{2.46}$$

$\langle \hat{S}_{j'}^\alpha(0) \hat{S}_j^\beta(t) \rangle$ is the thermal average of the time-dependent spin operators. It corresponds to the van Hove pair correlation function (see Eq. (2.23)) and gives essentially the probability that, if the magnetic moment of the j'th ion at site $\boldsymbol{R}_j'$ has some specified (vector) value at time zero, then the moment of the jth ion at site $\boldsymbol{R}_j$ has some other specified value at time t. A neutron scattering experiment measures the Fourier transform of the pair correlation function in space and time, which is clearly just what is needed to describe a magnetic system on an atomic scale.

The van Hove representation of the cross section in terms of pair correlation functions is related to the fluctuation-dissipation theorem:

$$\boxed{S^{\alpha\beta}(\boldsymbol{Q}, \omega) = \frac{N\hbar}{\pi} \left(1 - e^{-\frac{\hbar\omega}{k_B T}}\right)^{-1} \mathrm{Im}\chi^{\alpha\beta}(\boldsymbol{Q}, \omega)} \tag{2.47}$$

where N is the total number of magnetic ions. Physically speaking, the neutron may be considered as a magnetic probe which effectively establishes a frequency- and wavevector-dependent magnetic field, $H^{\beta}(\boldsymbol{Q},\omega)$, in the scattering sample, and detects its response, $M^{\alpha}(\boldsymbol{Q},\omega)$, to this field by

$$M^{\alpha}(\boldsymbol{Q},\omega) = \chi^{\alpha\beta}(\boldsymbol{Q},\omega)H^{\beta}(\boldsymbol{Q},\omega), \tag{2.48}$$

where $\chi^{\alpha\beta}(\boldsymbol{Q},\omega)$ is the generalized magnetic susceptibility tensor. This is really the outstanding property of the neutron in a magnetic scattering measurement, and no other experimental technique is able to provide such detailed microscopic information about magnetic compounds.

2.7 Polarized neutrons

In the preceding sections we only considered the scattering of neutrons from one momentum state to another. However, the interaction of the magnetic moment of the neutron with the system under investigation involves also spin-dependent terms in the neutron scattering cross-section. These terms give rise to interesting polarization effects and provide additional details about the scattering system. Polarized neutrons are often used for determining magnetic correlations, for distinguishing between collective and single-particle excitations (i.e., for discriminating between coherent and incoherent scattering processes), and for high-resolution spectroscopy (neutron spin-echo).

The magnetic moment of the neutron can be expressed in terms of the Pauli spin operators $\hat{\boldsymbol{\sigma}}$ (see Appendix C). We denote the spin states of the neutron by $|+\rangle$ and $|-\rangle$ which correspond to the spin-up and spin-down states with eigenvalues +1 and -1 for the operator σ_z (i.e., we choose the z-axis as the polarization and quantization direction). If a beam of neutrons has a fraction f of neutrons in the $|+\rangle$ state, then the polarization of the beam is defined by a vector $\boldsymbol{P}$ in the z direction with magnitude

$$|\boldsymbol{P}| = 2f - 1. \tag{2.49}$$

For a completely polarized beam we have $|\boldsymbol{P}| = 1$ and for an unpolarized beam $\boldsymbol{P} = 0$.

The neutron scattering cross-section can now be split into four parts. The processes $|+\rangle \rightarrow |+\rangle$ and $|-\rangle \rightarrow |-\rangle$ involve no change of spin . The processes $|+\rangle \rightarrow |-\rangle$ and $|-\rangle \rightarrow |+\rangle$ involve a change of spin and are known as spin-flip processes. This is taken into account by considering the spin states $\boldsymbol{\sigma}$ and $\boldsymbol{\sigma}'$ in the matrix elements $\langle \boldsymbol{k}', \boldsymbol{\sigma}', \lambda'|\hat{U}|\boldsymbol{k}, \boldsymbol{\sigma}, \lambda\rangle$ of Fermi's Golden Rule Eq. (2.4). The interaction operator $\hat{U}$ then reads

$$\hat{U} = \hat{b} + A\boldsymbol{I} \cdot \hat{\boldsymbol{\sigma}} + B\boldsymbol{M} \cdot \hat{\boldsymbol{\sigma}}, \tag{2.50}$$

where $\hat{b}$ corresponds to the operator defined by Eq. (2.5) and the spin-dependent operators describe the interaction with the nuclear spins ($\boldsymbol{I}$) and with the magnetic moments of the electrons ($\boldsymbol{M}$) of the system. Because of the orthogonality between the momentum and spin states of the neutron the matrix element can be separated:

$$\langle \boldsymbol{k}', \boldsymbol{\sigma}', \lambda'|\hat{U}|\boldsymbol{k}, \boldsymbol{\sigma}, \lambda\rangle = \langle \boldsymbol{k}', \lambda'|\hat{U}|\boldsymbol{k}, \lambda\rangle \cdot \langle \boldsymbol{\sigma}'|\hat{U}|\boldsymbol{\sigma}\rangle. \tag{2.51}$$

Applying the operator $\hat{U}$ to the two spin states of the neutron yields

$$\begin{aligned}
\hat{U}|+\rangle &= (b + AI_z + BM_z)|+\rangle \\
&\quad + [A(I_x + \imath I_y) + B(M_x + \imath M_y)]\,|-\rangle, \\
\hat{U}|-\rangle &= (b - AI_z - BM_z)|-\rangle \\
&\quad + [A(I_x - \imath I_y) + B(M_x - \imath M_y)]\,|+\rangle.
\end{aligned} \tag{2.52}$$

The matrix elements of the four possible processes then read

$$\begin{aligned}
\langle +|\hat{U}|+\rangle &= b + AI_z + BM_z, \\
\langle -|\hat{U}|-\rangle &= b - AI_z - BM_z, \\
\langle -|\hat{U}|+\rangle &= A(I_x + \imath I_y) + B(M_x + \imath M_y), \\
\langle +|\hat{U}|-\rangle &= A(I_x - \imath I_y) + B(M_x - \imath M_y).
\end{aligned} \tag{2.53}$$

We recognize the important result that neutron scattering from magnetic materials involves a spin flip only when the magnetization direction is perpendicular to the neutron polarization $\boldsymbol{P}$; for parallel configuration no spin flip is observed. These features will be discussed in more detail in Chap. 7.

In Chap. 2.4 we have seen that the nuclear scattering lengths are dependent upon the nuclear spin quantum number I. We can therefore consider the right-hand side of Eq. (2.53) in the absence of electronic magnetic moments ($|\boldsymbol{M}| = 0$) as the scattering lengths for the four spin-state transitions. This opens useful opportunities to separate coherent and incoherent scattering processes as outlined below.

For coherent scattering we have to evaluate the squares of the matrix elements of Eq. (2.53) with $|\boldsymbol{M}| = 0$. As long as the directions of nuclear spins are randomly oriented we have $\langle I_x\rangle = \langle I_y\rangle = \langle I_z\rangle = 0$. Therefore

$$\begin{aligned}
\langle +|\hat{U}|+\rangle^2 &= \langle -|\hat{U}|-\rangle^2 = \langle b\rangle^2, \\
\langle -|\hat{U}|+\rangle^2 &= \langle +|\hat{U}|-\rangle^2 = 0,
\end{aligned} \tag{2.54}$$

i.e., nuclear coherent scattering does not involve a spin flip.

For incoherent scattering we have to evaluate the quantities $|\langle\pm|\hat{U}|\pm\rangle|^2 - \langle\pm|\hat{U}|\pm\rangle^2$. We have again $\langle I_x\rangle = \langle I_y\rangle = \langle I_z\rangle = 0$, but $\langle I_x^2\rangle = \langle I_y^2\rangle = \langle I_z^2\rangle = \frac{1}{3}I(I+1)$. We then find

$$\begin{aligned}
|\langle+|\hat{U}|+\rangle^2| - \langle+|\hat{U}|+\rangle^2 &= |\langle-|\hat{U}|-\rangle^2| - \langle-|\hat{U}|-\rangle^2, \\
&= \langle b^2\rangle - \langle b\rangle^2 + \frac{1}{3}A^2 I(I+1), \\
|\langle-|\hat{U}|+\rangle^2| - \langle-|\hat{U}|+\rangle^2 &= |\langle+|\hat{U}|-\rangle^2| + \langle+|\hat{U}|-\rangle^2 \\
&= \frac{2}{3}A^2 I(I+1). \qquad (2.55)
\end{aligned}$$

In case of a scattering system with only one isotope we have $\langle b\rangle^2 = \langle b^2\rangle$, thus according to Eq. (2.55) the cross section for non spin-flip is one half the cross section for spin-flip processes. In this way incoherent scattering due to nuclear spin can be extracted from experimental data.

2.8 Dynamical neutron scattering

In the preceding chapters we have assumed that the neutron is scattered at one particular position $\boldsymbol{R}$ in the system. This assumption (which is called the kinematical theory of neutron scattering) is justified as long as the scattering system has a small volume so that multiple scattering processes can be neglected. Whenever multiple scattering occurs (particularly in perfect single crystals), there is interference of the scattered neutron waves which weakens the outgoing neutron beam. In the kinematical theory this is taken into account by introducing so-called extinction coefficients (see Chap. 4.5) which describe the observed attenuation of the scattered neutron beam in an integral manner, but cannot describe the interference phenomena. The latter is dealt with by the dynamical theory of neutron scattering as outlined below.

We start from the Schrödinger equation for the neutron wavefunction ψ within the scattering system

$$\frac{\hbar^2}{2m}\nabla^2\psi + \left(E - \hat{U}(\boldsymbol{r})\right)\psi = 0 \qquad (2.56)$$

and match the outside and inside wavefunctions at the surface. We now insert into Eq. (2.56) the operator $\hat{U}$ according to Eq. (2.5) and the eigenvalue $E = \frac{\hbar^2}{2m}k_a^2$ corresponding to the kinetic energy of the neutron with wavevector $\boldsymbol{k}_a$ outside the system. Application of Eqs (A.10) - (A.12) then yields

$$\nabla^2\psi + k_a^2\psi = \frac{4\pi}{v_0}N\langle b\rangle\left(\sum_{\boldsymbol{\tau}} e^{-\imath\boldsymbol{\tau}\cdot\boldsymbol{r}}\right)\psi. \tag{2.57}$$

The solution of Eq. (2.57) is a Bloch function of the form

$$\psi = \sum_{\boldsymbol{\tau}} a_{\boldsymbol{\tau}} e^{\imath(\boldsymbol{k}_i-\boldsymbol{\tau})\cdot\boldsymbol{r}}, \tag{2.58}$$

with unknown amplitude $a_{\boldsymbol{\tau}}$ and wavevector $\boldsymbol{k}_i$ inside the system. Inserting ψ into the Schrödinger equation Eq. (2.57) and using the relation $\nabla^2 e^{\imath\boldsymbol{Q}\cdot\boldsymbol{r}} = -Q^2 e^{\imath\boldsymbol{Q}\cdot\boldsymbol{r}}$ yields

$$\sum_{\boldsymbol{\tau}} a_{\boldsymbol{\tau}}\left(\boldsymbol{k}_a^2 - (\boldsymbol{k}_i-\boldsymbol{\tau})^2\right)e^{\imath(\boldsymbol{k}_i-\boldsymbol{\tau})\cdot\boldsymbol{r}} = \frac{4\pi}{v_0}N\langle b\rangle\sum_{\boldsymbol{\tau}'} e^{-\imath\boldsymbol{\tau}'\cdot\boldsymbol{r}}\sum_{\boldsymbol{\tau}''} a_{\boldsymbol{\tau}''}e^{\imath(\boldsymbol{k}_i-\boldsymbol{\tau}'')\cdot\boldsymbol{r}}.$$

Equating terms in $e^{-\imath(\boldsymbol{k}_i-\boldsymbol{\tau})\cdot\boldsymbol{r}}$ gives

$$a_{\boldsymbol{\tau}}\left(\boldsymbol{k}_a^2 - (\boldsymbol{k}_i-\boldsymbol{\tau})^2\right) = \frac{4\pi}{v_0}N\langle b\rangle\sum_{\boldsymbol{\tau}'} a_{\boldsymbol{\tau}-\boldsymbol{\tau}'}. \tag{2.59}$$

We now consider two cases. First we investigate the small-angle scattering with no Bragg reflections, i.e., we set $\boldsymbol{\tau} = 0$ in Eq. (2.59). Since $(k_a - k_i)/k_a \ll 1$ we have $k_a + k_i \approx 2k_a$, so that

$$k_a - k_i = \frac{4\pi}{v_0}N\langle b\rangle\frac{1}{2k_a^2}. \tag{2.60}$$

In analogy to optics we define the refractive index n from Eq. (2.60):

$$n = \frac{k_i}{k_a} = 1 - \frac{4\pi}{v_0}N\langle b\rangle\frac{1}{2k_a^2}. \tag{2.61}$$

With the atomic density $\rho = N/v_0$ and the neutron wavelength λ defined in Eq. (1.3) we find

$$\boxed{n = 1 - \frac{1}{2\pi}\rho\lambda^2\langle b\rangle} \tag{2.62}$$

For all known compounds the refractive index is very close to $n = 1$. As a consequence, total reflection of neutrons occurs at very small angles

$$\gamma \leq \arccos(n) \tag{2.63}$$

of the order of minutes (e.g., for nickel we have $\gamma = 30'$ for $\lambda = 5$ Å).

Now we consider the case of scattering near a Bragg reflection, i.e., we have a single $\boldsymbol{\tau} \neq 0$. Then two equations result from Eq. (2.59), one for $\boldsymbol{\tau} = 0$ and another for $\boldsymbol{\tau} \neq 0$, which combine to

$$\left(\boldsymbol{k}_a^2 - \frac{4\pi}{v_0}N\langle b\rangle - \boldsymbol{k}_i^2\right)\left(\boldsymbol{k}_a^2 - \frac{4\pi}{v_0}N\langle b\rangle - (\boldsymbol{k}_i - \boldsymbol{\tau})^2\right) = \left(\frac{4\pi}{v_0}N\langle b\rangle\right)^2 \quad (2.64)$$

giving rise to two solutions $\boldsymbol{k}_{i_1}$ and $\boldsymbol{k}_{i_2}$ with very close moduli. The two neutron waves propagating inside the system interfere with each other, i.e., the intensity amplitude exhibits a sine modulation which was experimentally verified by Shull [Shull (1968)] and is known as Pendellösung.

2.9 Further reading

- E. Balcar and S. W. Lovesey, *Theory of magnetic neutron and photon scattering* (Clarendon Press, Oxford, 1989)
- P. Böni, in *Magnetic neutron scattering*, ed. by A. Furrer (World Scientific, Singapore, 1995), p. 27: *Polarized neutrons*
- T. Chatterji, in *Neutron scattering from magnetic materials*, ed. by T. Chatterji (Elsevier, Amsterdam, 2006), p. 1: *Magnetic neutron scattering*
- S. Chen and M. Kotlarchyk, *Interactions of photons and neutrons with matter* (World Scientific, Singapore, 2007)
- A. J. Dianoux and G. Lander, *ILL Neutron data booklet*, 2nd edition (Old City Publishing, Philadelphia, 2003)
- P. A. Egelstaff, *Thermal neutron scattering* (Academic Press, London, 1965)
- W. E. Fischer, in *Complementarity between neutron and synchrotron x-ray scattering*, ed. by A. Furrer (World Scientific, Singapore, 1998), p. 3: *Neutron and synchrotron x-ray scattering (the theoretical principles)*
- S. W. Lovesey, *Theory of neutron scattering from condensed matter*, Vol. 1 and 2 (Clarendon Press, Oxford, 1984)
- V. McLaine, C. L. Dunford and P. F. Rose, *Neutron cross sections*, Vol. 2 (Academic Press, Boston, 1988)
- P. C. H. Mitchell, S. F. Parker, A. J. Ramirez-Cuesta and J. Tomkinson, in *Vibrational spectroscopy with neutrons* (World Scientific, Singapore, 2005), p. 13: *The theory of inelastic neutron scattering spectroscopy*
- R. M. Moon, T. Riste and W. C. Koehler, Phys. Rev. B 181, 920 (1969): *Polarization analysis of thermal-neutron scattering*

- R. Nathans, C. G. Shull, G. Shirane and A. Andresen, J. Phys. Chem. Solids 10, 138, (1959): *The use of polarized neutrons in determining the magnetic scattering by iron and nickel*
- D. L. Price and K. Sköld, in *Methods of experimental physics*, Vol. 23, Part A, ed. by D. L. Price and K. Sköld (Academic Press, London, 1986), p. 1: *Introduction to neutron scattering*
- R. Scherm, Ann. Phys. 7, 349 (1972): *Fundamentals of neutron scattering by condensed matter*
- G. L. Squires, *Introduction to the theory of thermal neutron scattering* (Dover Publications, New York, 1996)

Chapter 3

Instrumentation

3.1 Neutron sources

3.1.1 *Historical evolution of neutron sources*

The quality and precision of neutron scattering experiments are primarily determined by the counting rate and, therefore, by the flux of the available neutrons, usually quoted in units of neutrons per square centimeter per second ($n{\cdot}cm^{-2}{\cdot}s^{-1}$). Although about half of our world is made up of neutrons, they are tightly bound deep inside the atomic nucleus and quite difficult to set free. Facilities that accomplish this task in a manner useful for scientific and technological applications are called neutron sources.

For all neutron sources we are dealing with nuclear reactions. Some nuclear reactions with a practical potential are listed in Table 3.1. Neutrons were observed for the first time in 1932 by Chadwick who used the interaction of α-particles from decay of natural polonium with beryllium, and subsequently neutron sources based on natural α-radiation were the basis of the earliest neutron physics research. Nuclear fission reactions were the next generation of sources, however, they were initially built for research in nuclear industry, and the ability to do neutron scattering was simply an unforeseen spin off. The development started in 1942 with a thermal neutron flux of 10^7 $n{\cdot}cm^{-2}{\cdot}s^{-1}$ (CP-1, USA) and had its climax in 1972 with the reactor at the Institute Laue-Langevin (ILL) in Grenoble (France), a dedicated neutron scattering facility with thermal neutron fluxes just exceeding 10^{15} $n{\cdot}cm^{-2}{\cdot}s^{-1}$.

Meanwhile, pulsed sources, starting with Alvarez's use of an rf-pulsed cyclotron, and in particular electron accelerators that produce neutrons through bremsstrahlung photoneutron reactions have been increasingly applied for slow-neutron research. In addition, repetitively pulsed reactors

Table 3.1 Neutron yields and deposited heat for some neutron producing reactions.

Reaction	Energy/Event	Yield (n/event)	Deposited heat (MeV/n)
T(d,n)	0.2 MeV	$8 \cdot 10^{-5}$ n/d	2500
W(e,n)	35 MeV	$1.7 \cdot 10^{-2}$ n/e	2000
^{9}Be(d,n)	15 MeV	$1.2 \cdot 10^{-2}$ n/d	1200
^{235}U(n,f)	fission	~ 1 n/fission	200
(T,d)	fusion	~ 1 n/fusion	3
Pb spallation	1 GeV	~ 20 n/p	23
^{238}U spallation	1 GeV	~ 40 n/p	50

have been developed. However, all these facilities could not really compete for instance with the overall performance of the ILL reactor. Today the most intense pulsed neutron sources are based on proton accelerators and the production of neutrons by spallation. Presently the world-wide leading facility is the spallation neutron source SNS at Oak Ridge (USA) which provides instantaneous thermal neutron fluxes of over 10^{17} n$\cdot$cm$^{-2}\cdot$s^{-1} with short pulse lengths of the order of 50 μs.

3.1.2 *Practical requirements for neutron sources*

The primary goal of a neutron source must be to release as many neutrons as possible in as small a volume as possible to achieve a high luminosity. Another important property of neutron sources is the heat deposition going along with the neutron release, since cooling problems are a limiting factor in practically all neutron source designs. In this respect fusion is by far the most optimal process for neutron production as can be seen from Table 3.1, followed by spallation and fission. Fusion may be the technique of neutron production in the far future, but today the two most commonly used reactions are thermal nuclear fission in ^{235}U and spallation by protons in the energy range around 1 GeV, thus we will restrict ourselves in the following to the latter two reactions.

3.1.3 *Fission sources*

Fission of the uranium isotope 235 by slow neutron capture has been the most frequently used reaction in neutron sources up to the present. The reaction can be made self-sustaining because it is exothermal and releases more neutrons per fission process than are needed to initiate the process.

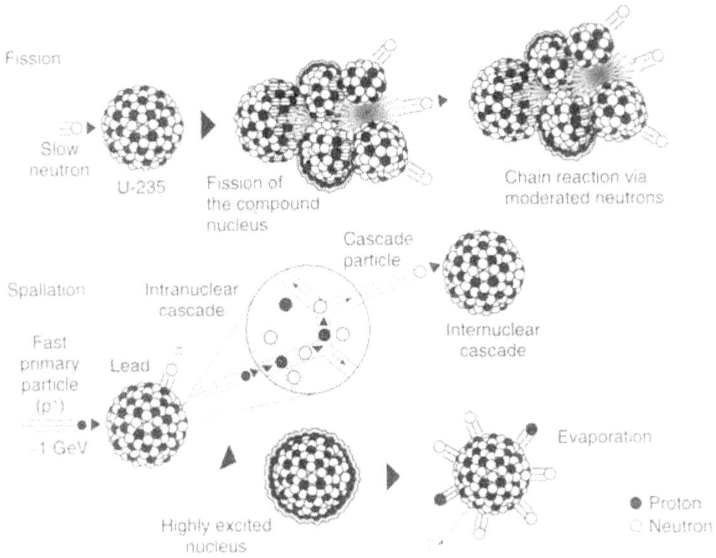

Fig. 3.1 Schematic representation of fission (upper part) and spallation (lower part).

If a slow neutron is captured by a fissionable nucleus, the resulting deformation can cause the nucleus to break into two fragments as visualized in Fig. 3.1. Very often a neutron is released directly during this process, but mostly the neutrons evaporate from the fragments. This is a very important feature, because a small fraction of these evaporation neutrons are released with a time delay of the order of seconds (up to minutes) and thus enable a critical arrangement to be run in a controlled fashion. The spectral distribution of the fission neutrons can be well described by a Maxwellian

$$n(E) = 2\sqrt{\frac{E}{\pi E_T^3}} \cdot e^{-E/E_T} \tag{3.1}$$

with a characteristic energy $E_T = 1.29$ MeV. Fission reactors produce a continuous flux of neutrons.

3.1.4 *Spallation sources*

The term spallation is applied to a sequence of events that take place, if target nuclei are bombarded with particles (e.g., protons) of a de Broglie wavelength $\lambda = h/\sqrt{2mE}$ which is shorter than the linear dimension of the nucleus. In this case collisions can take place with individual nuclides inside the nucleus and large amounts of energy are transferred to the nuclides which, in turn, can hit other nuclides in the same nucleus. The net effect of this intra-nuclear cascade is twofold (see Fig. 3.1): Firstly, energy is more or less evenly distributed over the nucleus leaving it in a highly excited state; secondly, energetic particles may leave the nucleus and carry the cascade on to the other nuclei (inter-nuclear cascade) or escape from the target. The excited nucleus left behind will start to evaporate neutrons (and to a lesser extent protons). The low-energy part of the spectrum of these evaporation neutrons is quite similar to the one resulting from fission (Eq. (3.1)), but as a consequence of the neutron escape during the intra-nuclear cascade the spectrum extends to energies up to that of the incident particles (i.e., up to 1 GeV). The release of spallation neutrons takes place within less than 10^{-15} s after the nucleus was hit, so that the time distribution of spallation neutrons is exclusively determined by the time distribution of the driving particle pulse.

Modern spallation sources are usually based on a linear accelerator (linac). The process starts at the front end of the linac with the creation of negatively charged hydrogen ions by powerful ion sources. As they are electrically charged these hydrogen ions can be accelerated in radio frequency structures with strong fields along the linac to kinetic energies in the GeV range (which is about 90% of the speed of light). When this highly energetic particle stream exits the linac, the hydrogen ions are stripped off their electrons by letting them pass through a thin carbon sieve, so that the hydrogen ions have become protons. The protons are then fed into a compressor ring which collects the protons from a large number of successive bunches fired out of the linac into a single very high intensity proton pulse. For this purpose an assembly of magnets bends each accelerated proton bunch into a circular orbit of such a diameter (of the order of 50-100 m) that the next bunch of protons arrives exactly when the previous has gone once around. In this way all these bunches pile up. After about 1000 revolutions sufficient intensity is accumulated, and the full proton pulse with a pulse length of about 1 μs is extracted and propelled towards the target which is usually a liquid metal (mercury or a lead-bismuth eutectic mixture), encased in special materials to take up a beam power of a few MW. This whole process – from the creation of the hydrogen ions to the arrival of the highly energetic protons at the target – should occur with a pulse repetition rate in the range 10-100 Hz in order to achieve an optimal neutron economy in scattering experiments.

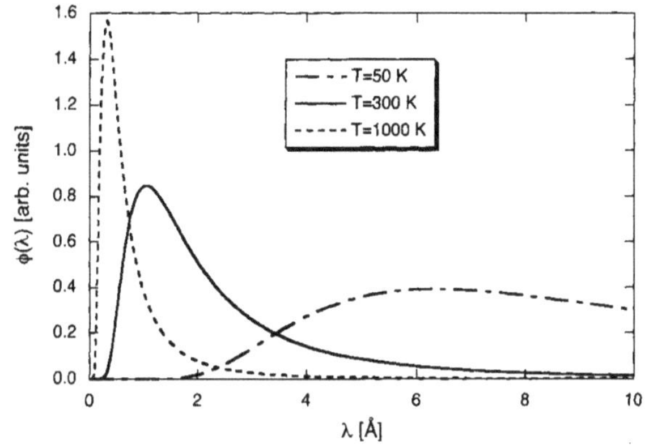

Fig. 3.2 Wavelength distribution of neutrons from cold (50 K), thermal (300 K), and hot (1000 K) moderators.

3.1.5 *Moderation of neutrons*

The energy spectrum of neutrons released from neutron sources is in the MeV range, whereas meV neutrons are required for scattering experiments in condensed matter research. Therefore an energy shift of several orders of magnitude is necessary, which is accomplished by collisions with the atoms of a moderator substance. The goal in the layout of a moderator is to create the highest possible flux of moderated neutrons either in the shortest possible time (pulsed neutron sources) or in the largest possible volume (continuous neutron sources). This can be achieved by using moderators made of light atoms such as H_2O and D_2O. The time for slowing down the neutrons is of the order of 10^{-6} s after which the neutrons are in thermal equilibrium with the moderator kept at a constant temperature T according to the Maxwellian distribution

$$\Phi(\lambda) \propto \frac{1}{\lambda^3} \exp\left(-\frac{h}{2k_B T m \lambda^2}\right), \tag{3.2}$$

where λ and m are the wavelength and the mass of the neutron, respectively. Moderators are usually kept at room temperature, and this is the reason why the corresponding neutrons are called thermal neutrons, with a maximum peak flux around the neutron wavelength $\lambda \approx 1$ Å, see Fig. 3.2.

Scattering experiments which require cold or hot neutrons (i.e., neutrons with wavelengths considerably different from $\lambda \approx 1$ Å) would experience a tremendous flux penalty when working with the thermal neutron spectrum. However, the Maxwellian energy spectrum of the neutrons can be shifted by inserting either a cold source (e.g., a vessel containing D_2 at $T \geq 20$ K) or a hot source (e.g., a graphite bloc heated up to $T \leq 2000$ K) into the moderator; typical spectral shifts for cold and hot neutrons are displayed in Fig. 3.2. By choosing the adequate neutron spectrum the scattering experiments can be optimally tailored to the particular experimental requirements.

3.2 Instrument components

3.2.1 *Beam tubes and static collimators*

Once neutrons have been produced and moderated in the neutron source, it is still necessary to transport the neutrons to the instrument positions. Since neutron sources radiate isotropically, they must be surrounded by heavy biological shielding. Beam tubes are inserted into the shielding to transport the neutron beam in a line of sight from the moderator surface to the instruments for neutron scattering. It is essential that these beam tubes are arranged tangentially to the core of the neutron source in order to reduce significantly unwanted radiation (e.g., fast neutrons). The beam tubes limit the orientation of the neutron wavevectors $\boldsymbol{k} = (k_x, k_y, k_z)$ in space as illustrated in Fig. 3.3a, with k_z being the component pointing along the axis of the beam tube. The solid angle $\mathrm{d}\Omega$ of a beam passing through a beam hole of length L and rectangular cross section $a \times b$ is $\mathrm{d}\Omega = ab/4\pi L^2$. For typical dimensions $L = 500$ cm and $a = b = 5$ cm we find $\mathrm{d}\Omega \approx 10^{-5}$, i.e., the neutron flux at the exit of the beam hole is drastically reduced as compared to the isotropic flux in the core of the neutron source. The horizontal and vertical angular divergences of the neutron beam at the exit are $\alpha = a/L$ and $\beta = b/L$, respectively, which are of the order of 1°. The angular spreads α and β can be improved by reducing the dimensions a and b of the beam hole, but this results immediately in a loss of neutron flux. This problem can be alleviated by subdividing the cross section into slits, using thin blades of a material which is opaque to neutrons. Such an arrangement with only vertical slits is called a Soller collimator (see Fig. 3.3b). It limits the divergence of the beam in the scattering plane from $\alpha = a/L$ to α_s, while leaving the divergence perpendicular to the scattering plane unchanged at $\beta = b/L$. This is permissible in most cases, since the divergence β affects the resolution in $\boldsymbol{Q}$ and ω only in second order.

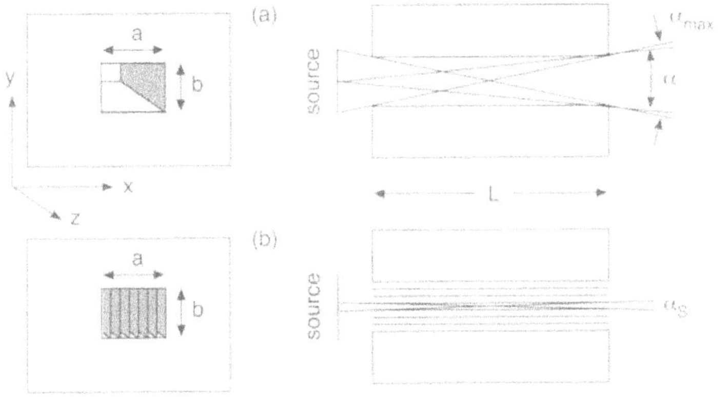

Fig. 3.3 Side views (left) and top views (right) of beam tube of rectangular cross section (a) and beam tube with Soller collimator (b), respectively.

3.2.2 *Neutron guides*

The neutron flux at an instrument positioned at a distance r from the neutron source suffers a drastic reduction according to the well-known r^{-2} law for isotropic radiation. This loss in luminosity can be largely eliminated by using neutron guides. A neutron guide works via total reflection of neutrons from the smooth walls of the guide material. This occurs for scattering angles less than the critical angle γ_c

$$\gamma_c = \lambda\sqrt{\frac{\rho b}{\pi}} \tag{3.3}$$

defined by Eq. (2.63) Among common materials, nickel is the best choice with critical angles θ_c [°]= $\lambda \cdot 10^{-1}$ [Å]. The angular acceptance of a neutron guide can be dramatically increased by reflection from so-called supermirrors which consist of a sequence of layers of variable thickness with alternating high positive (e.g. Ni) and negative (e.g. Ti) scattering length density. The multilayers represent an artificial one-dimensional lattice, thus Bragg reflections occur at appropriate scattering vectors $\boldsymbol{Q}$ with modulus (from Eq. (4.8))

$$Q = \frac{4\pi}{\lambda}\sin(\theta). \tag{3.4}$$

Supermirrors exploit this property and provide a regime of continuous Bragg reflection from a depth-graded multilayer as schematically shown in Fig. 3.4. Supermirrors can be characterized by a number m defining the

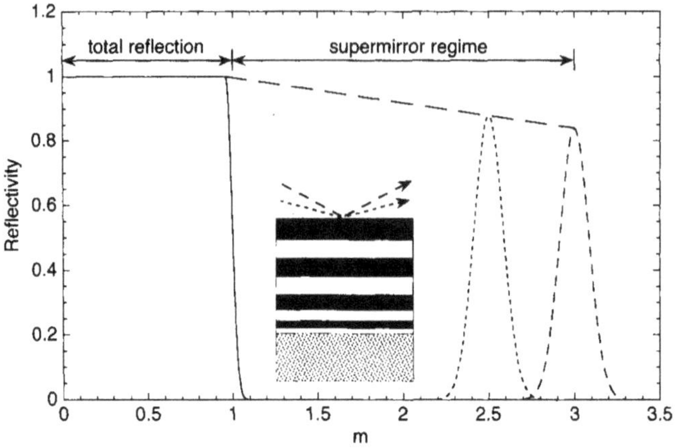

Fig. 3.4 Schematic sketch illustrating the concepts for neutron reflection of a supermirror.

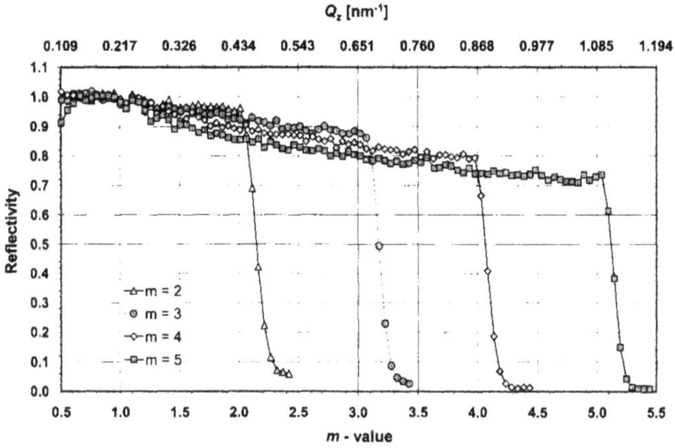

Fig. 3.5 Measured neutron reflectivities for Ni/Ti supermirrors with $m = 2, 3, 4, 5$ (by courtesy of SwissNeutronics AG, CH-5313 Klingnau).

increase of γ_c compared to nickel. At present $m = 5$ can be routinely achieved as demonstrated in Fig. 3.5. The use of supermirrors results in an enormous flux increase at the instrument position as compared to a conventional beam tube. Neutron guides can be up to 100 m long, and the neutron loss from the entrance to the exit of the guide is usually less than 2% per 10 m.

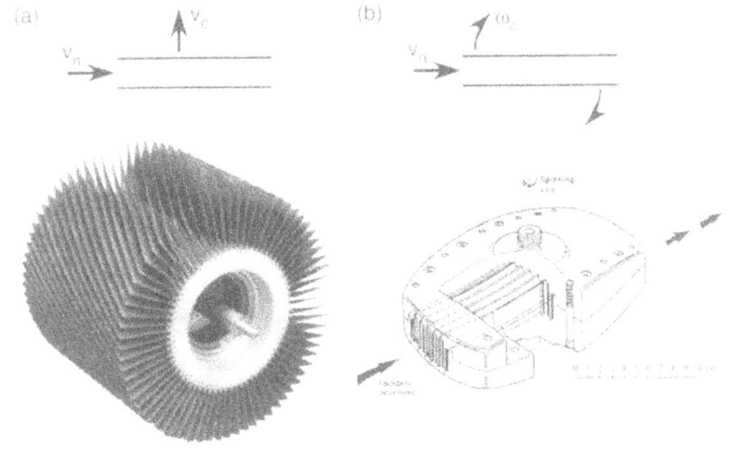

Fig. 3.6 Monochromatization of neutrons by moving a collimator (a) in translational or (b) in rotational motion.

3.2.3 *Time-of-flight monochromators*

A static Soller collimator has no effect on the magnitude k_z of the wavevector of the transmitted neutrons, i.e., it does not "monochromate". This can be achieved by moving the collimator at right angles to its direction of transmission in the scattering plane. This movement can be translational (Fig. 3.6a) or rotational (Fig. 3.6b). In practice, case (a) is accomplished by arranging many slits on a drum whose axis is parallel to the neutron beam, while case (b) is accomplished by slits in a drum whose axis of rotation is perpendicular to the neutron beam. In both cases the slits (coated with a material that is opaque to neutrons) are curved to match the flight path of the neutrons in the moving frame of reference for optimum transmission. Case (a) allows a continuous transmission of neutrons of the desired velocity range; it is called a mechanical velocity selector. Case (b), on the other hand, chops the beam into pulses; it is called a Fermi chopper.

A monochromatic effect through flight time is also possible by having two disks which are opaque for neutrons with a transmitting slit rotating at a certain distance from each other. To define the opening time more precisely, normally a stationary slit is placed near a disk chopper (Fig. 3.7) or a second disk rotating in opposite direction close to the first one can be used.

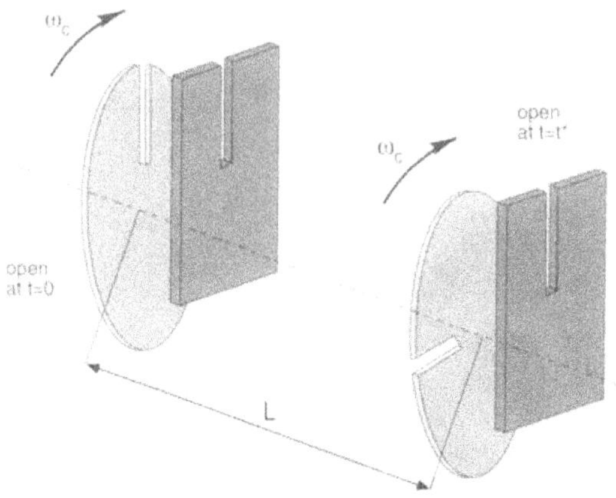

Fig. 3.7 Monochromating effect of two disk choppers separated by a distance L. The first disk only chops the beam, whereas a limited velocity band is selected by the phase angle between the two disks.

3.2.4 *Monochromator (and analyzer) crystals*

While the transmission through a collimator limits and defines the direction of the neutron wavevector $\boldsymbol{k}$ in space, Bragg reflection from a single crystal changes the direction of $\boldsymbol{k}$. This is illustrated in Fig. 3.8. A single crystal is placed in a polychromatic neutron beam which propagates along a fixed direction denoted by DO. The crystal is oriented such that the angle between the reciprocal lattice vector $\boldsymbol{\tau}_{hkl}$ and the incoming neutrons is φ. For an arbitrary wavevector $\boldsymbol{k}$ (marked by AO) there is no Bragg scattering; only when the condition $\tau = 2k\cos\varphi$ is fulfilled, Bragg scattering occurs (i.e. for BO). This is the standard method of producing a beam of monochromatic neutrons, which is also called Laue method.

Since the multiples $\boldsymbol{\tau}_{2h,2k,2l}$, $\boldsymbol{\tau}_{3h,3k,3l}$,... of a particular reciprocal lattice vector $\boldsymbol{\tau}_{h,k,l}$ are always parallel to the latter, the Bragg condition is also fulfilled for the multiples of $\boldsymbol{k}$ as illustrated in Fig. 3.8. Therefore the Laue method does not provide purely monochromatic neutrons of wavevector $\boldsymbol{k}$, but also higher-order neutrons with wavevectors $2\boldsymbol{k}$, $3\boldsymbol{k}$,..., i.e., neutrons of half, third,... of the targeted wavelength. Nevertheless, there are several tools to avoid higher-order Bragg scattering by choosing appropriate filters (see Chap. 3.2.5) or single crystals with vanishing $2nd$ order Bragg reflections (see Chap. 4.2).

The setup of a single crystal monochromator system usually involves col-

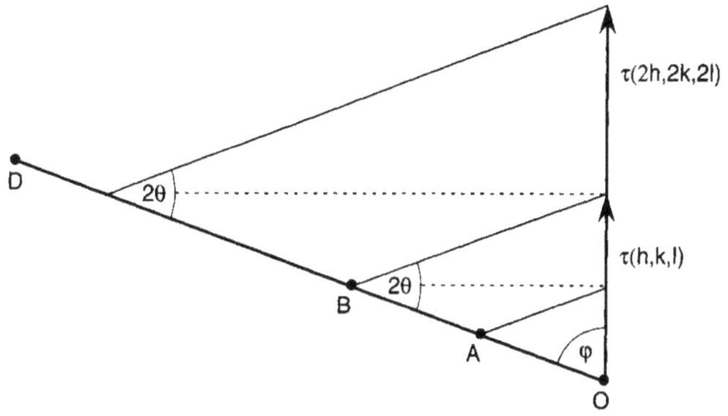

Fig. 3.8 Bragg reflection of a single crystal placed in a polychromatic neutron beam.

limators for both the incoming polychromatic and outgoing monochromatic neutron beam defining the respective directions. The resulting $\boldsymbol{k}$ resolution of the monochromatic neutron beam is therefore affected by both the mosaicity η of the crystal and the angular resolution α of the collimators:

$$\mathrm{d}k_z \approx k_z\sqrt{\eta^2+\alpha^2}\cot(\theta_B)\mathrm{d}\theta_B, \tag{3.5}$$

where θ_B is the Bragg angle.

The single crystal setup described above can also be used to analyze the wavevector $\boldsymbol{k}'$ of neutrons scattered from the sample; in this case the system is called analyzer crystal.

Many crystal spectrometers are equipped with focusing monochromator systems in order to focus the neutrons at the sample position. Concepts for fixed vertical and horizontal focusing were first reported by Riste [Riste (1970)] and Scherm et al. [Scherm *et al.* (1977)], respectively. Bührer et al. [Bührer *et al.* (1981)] introduced a flexible focusing system with individual adjustment of all the single crystals arranged in an $m \times n$ matrix. In such a setup the central crystal is fixed, whereas the adjacent crystals in the m columns and in the n rows are symmetrically turned around the vertical and horizontal axes by $\pm\delta_h$, $\pm 2\delta_h$, ... and by $\pm\delta_v$, $\pm 2\delta_v$, ..., respectively. Since the focusing conditions are energy-dependent, the turn angles $\delta_{h,v}$ have to be permanently adjusted, e.g., in a constant-Q scan of a triple-axis spectrometer. Modern focusing monochromator systems are able to focus an initial neutron beam with a typical cross section of 20×20 cm^2 down to an area of about 5×2 cm^2, resulting in an order of magnitude increase of the flux density at the sample position, see Fig. 3.9.

Fig. 3.9 Doubly focusing monochromator system consisting of a 13×9 matrix of single crystals (by courtesy of SwissNeutronics AG, CH-5313 Klingnau).

If the crystal of a monochromator setup is rotating around an axis perpendicular to the scattering plane, the condition of Bragg reflection is only fulfilled twice per revolution, resulting in the emission of short monochromatic neutron pulses. Such an arrangement is called a rotating crystal monochromator, which serves as a valuable alternative to the mechanical time-of-flight monochromators discussed in Chap. 3.2.3. Since in this case the reflecting lattice planes of the crystal are moving, the neutrons experience the Doppler effect, i.e., the wavenumber of the scattered neutrons is shifted by

$$\Delta k(r) = \frac{2\pi \nu m r}{\hbar} \sin \theta_B, \tag{3.6}$$

where ν is the frequency of the spinning crystal, and r the distance from the rotation axis. The net effect over the beam cross section from a rotating crystal monochromator is that a wider energy band is deflected than from a crystal at rest and that, depending on the sense of rotation of the crystal, the beam is either focussed or defocussed in time.

3.2.5 *Neutron beam filters*

As discussed in the preceding chapter, the beam scattered from a crystal monochromator not only contains the wavevector $\boldsymbol{k}$, but may also be contaminated by higher order reflections $2\boldsymbol{k}$, $3\boldsymbol{k}$,..., depending on the structure factor of these reflections. In order to eliminate or at least substantially reduce these unwanted reflections, material is placed in the beam path which has good transmission properties for the desired wavevector and a large cross section for the unwanted ones. Strong energy-dependent variation in neutron cross-section occurs either in the case of resonant absorption or in polycrystals whenever the diameter of the Ewald sphere coincides with the magnitude of a reciprocal lattice vector.

The most prominent examples for polycrystal filters are beryllium and (BeO) which have a Bragg cutoff at 5 meV (4 Å) and 3.8 meV (4.6 Å), respectively. Neutrons with energies above the cutoff energy are completely scattered out of the beam, whereas the low-energy neutrons pass the filter almost unaffected, apart from losses due to phonon scattering. The latter are usually minimized by cooling the filter down to the temperature of liquid nitrogen (77 K). Be and BeO are ideal filters for experiments with cold neutrons.

An example for an absorption filter is ^{239}Pu which has a strong resonant absorption at 300 meV, thus ^{239}Pu is a reasonably good second-order filter for thermal neutrons with energies of about 75 meV.

A very unusual case is pyrolytic graphite (PG) which has reasonably good single crystal properties along the c-axis. However, since the planes perpendicular to the c-axis are randomly oriented with respect to each other, the Bragg points of these planes can be considered to be Bragg rings about the $(00l)$ axis. For certain values of the energy of the incoming neutrons, the Ewald sphere will intersect one or more rings. The Bragg condition will thus be satisfied for several reflections simultaneously. For these energies, part of the neutron beam will be scattered out of the incoming beam direction. This effect is rather pronounced for neutrons in the energy range 13-15 meV with wavelength $\lambda \approx 2.44$ Å and $\lambda \approx 2.34$ Å. While neutrons with these wavelengths λ will pass through the filter almost unattenuated, there are drastic reductions of neutrons with wavelengths 2λ and 3λ due to $2nd$ and $3rd$ order Bragg scattering as shown in Fig. 3.10.

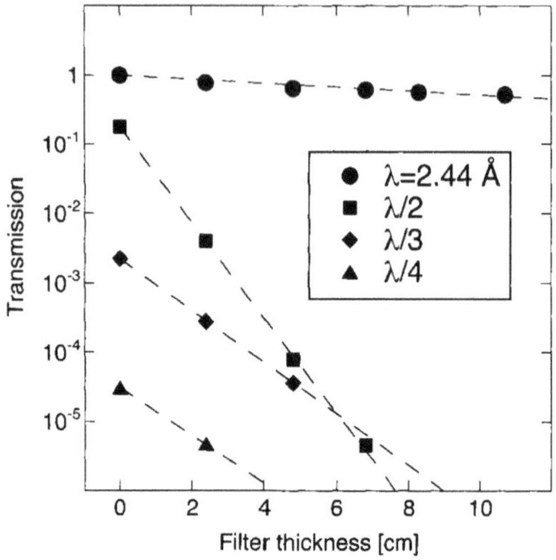

Fig. 3.10 Attenuation of neutrons resulting from higher-order Bragg reflections as a function of the thickness of a PG filter [Hälg (1981)].

3.2.6 *Spin polarizers (and spin analyzers)*

Most neutron experiments are concerned only with specific changes of the momentum and the energy of the scattered neutrons. However, since the neutron has a spin, the use of polarized neutrons offers additional information as discussed in Chap. 2.7. Here we mention the techniques to produce polarized neutrons (polarizers); the same techniques can also be used to analyze the spin state of the neutrons after the scattering process (analyzers). There are three types of polarizing devices:

(1) Bragg reflection from a magnetic single crystal, where the interference of the magnetic and nuclear part of the scattering amplitudes is constructive for one spin state and destructive for the opposite spin state. The most frequently used crystal polarizer is the Cu_2MnAl Heusler crystal, whose $(1,1,1)$ Bragg reflection produces a polarization of about 95%.

(2) Total reflection from mirrors or supermirrors (see Chap. 3.2.2) with a magnetic layered structure based on constructive and destructive interference as in case (1). Here the best polarizing systems are supermirrors made of Fe/Si layers with a degree of polarization above

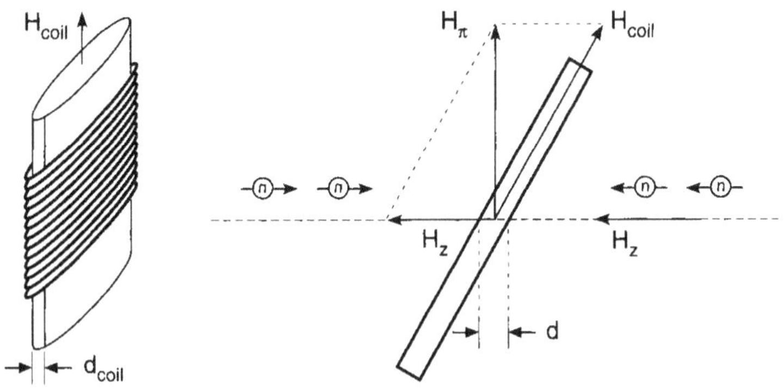

Fig. 3.11 Schematic illustration (side and top view) of a neutron π-flipper (Mezei coil).

95%.

(3) Filters that produce by absorption or extinction a polarized neutron beam in transmission. ^{3}He filters are of particular importance, since they exhibit high absorption for neutrons having their spins antiparallel aligned to the spins of the ^{3}He nuclei. The neutron transmission and the degree of polarization can be optimized to the neutron energies of interest by varying the gas pressure. So far beam polarizations up to 80% have been achieved.

3.2.7 *Guide fields and spin flippers*

Once polarized neutrons have been produced, a magnetic field defining a quantization axis has to be used to maintain the polarization of the neutron beam. The neutron spins will align either parallel or antiparallel to the field direction. The strength of the guide field has to be weak, so that the sample magnetization is not significantly affected, but sufficiently stronger than the earth magnetic field.

An important device for working with polarized neutrons is a spin flipper, see Fig. 3.11. One can use the Larmor precession in a coil to turn the neutron spin. The field inside the coil is perpendicular to both the polarization and the flight direction of the neutron. The spin flipper has to be placed inside a guide field to avoid depolarization, but the strength of the guide field has to be compensated by appropriately positioning the spin flipper. Neutrons with velocity v passing through the coil (with thickness d) experience a sudden field change and perform a precession. A rotation

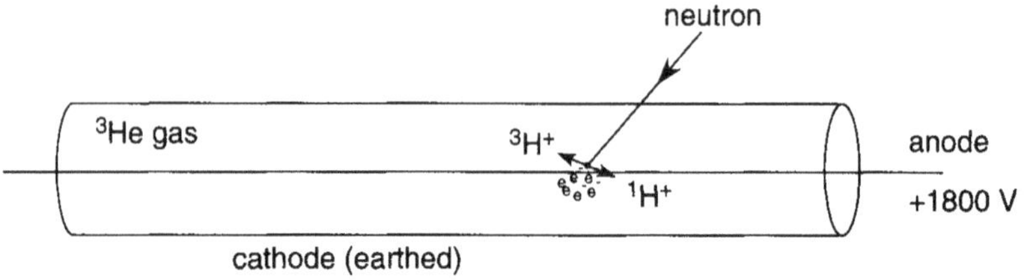

Fig. 3.12 Schematic sketch of a ^{3}He gas tube.

of 180° (π-flip) is realized under the condition

$$\pi = \gamma_L \cdot \frac{d \cdot H_\pi}{v}, \tag{3.7}$$

where $\gamma_L = 2.916$ kHz/Oe is the Larmor constant and H_π the resulting field. By this technique any type of spin-flip can be achieved (e.g., a $\pi/2$–flipper).

A less frequently used alternative is a guide field whose field direction changes slowly along the neutron flight path. The neutron spin will then follow adiabatically the field direction as long as the field turn is much smaller than the corresponding Larmor precession.

3.2.8 *Detectors*

The detection of neutrons is based on the measurement of electric currents. Thermal neutrons with their small energy and without an electric charge can only be measured indirectly following nuclear reactions with target atoms which emit either ionizing radiations or ionizing particles. The most important reactions for neutron detection are:

$$^1_0\text{n} + ^3_2\text{He} \rightarrow ^3_1\text{H} + ^1_1\text{H} + 0.77\ \text{MeV}$$

$$^1_0\text{n} + ^6_3\text{Li} \rightarrow ^3_1\text{H} + ^4_2\text{He} + 4.77\ \text{MeV}$$

$$^1_0\text{n} + ^{10}_5\text{B} \rightarrow \begin{cases} ^4_2\text{He} + ^7_3\text{Li} + \gamma\ (0.48\ \text{MeV}) + 2.3\ \text{MeV} & (93\%) \\ ^4_2He + ^7_3\text{Li} + 2.79\ \text{MeV} & (7\%) \end{cases}$$

^{3}He and ^{10}BF$_3$ are used in gas tubes, whereas ^{6}Li is used in scintillator detectors.

A He gas tube is schematically shown in Fig. 3.12. The steel tube, filled with ^{3}He at a pressure of 5 – 10 bar, constitutes the cathode. A high voltage (about 1800 Volt) anode is spanned along the cylinder axis.

The electrons produced by the ionizing particles of the nuclear reaction are accelerated towards the anode, causing further ionization and an avalanche effect. The gas amplification with a gain factor up to 10^5 is proportional to the amount of primarily produced charges, therefore this type of gas tube is called proportional counter; it is also known as Geiger counter. Due to the high gain factor single neutrons are easily detectable. ^{3}He tubes have the advantage that γ-rays can be easily discriminated and that they are insensitive to magnetic fields. On the other hand, they have a rather slow response time, thus they are not suitable for high count-rate applications.

Geiger counters can also be used as position sensitive detectors. To measure the position along a single tube, the central anode is replaced by a resistive wire. The charge produced by the ionization process travels to both ends of the wire. From the ratio of the charges collected at both ends one can determine the position, where the neutron had its impact, usually with a precision of about 10 mm. An alternative is to have a large number of thin anode wires assembled in a plane and mounted between two plates which are composed of separate cathode strips; in this case a spatial resolution of 1 mm can be achieved.

Scintillator detectors were developed for high count-rate applications. These are solid state detectors using polycrystalline ZnS doped with ^{6}Li. The neutron impact produces a flash of light which is recorded by a photomultiplier tube, either directly or by optical means such as lenses, fibers or mirrors. Since the scintillator material is much denser than a gas, the photoactive layer is rather thin (about 1 mm), resulting in an excellent spatial resolution. The local count rate is up to several MHz/mm^2. However, scintillators are sensitive to light, γ-rays, and magnetic fields, thus appropriate shieldings are required.

All neutron instruments need beam monitors to measure the incident flux. Since the incident flux must not be attenuated, monitors are necessarily insensitive. Monitors can be either a scintillator or a low pressure ^{3}He tube.

3.3 Neutron instruments

3.3.1 *Introductory remarks*

The instrument components described in the previous chapter can be combined in a variety of ways to obtain the desired properties of a neutron scattering instrument, depending on the kind of science in question. The power and breadth of neutron scattering is based on the enormous range over which the variables $\boldsymbol{Q}$ and ω of the scattering law $S(\boldsymbol{Q},\omega)$ introduced in Chap. 2 can be varied: roughly six orders of magnitude in Q and twelve orders of magnitude in ω can be covered, but each Q,ω-range has its specific instrumental realization. Among the myriads of instrument solutions achieved to date we will concentrate here on the most important instrument types.

A useful classification of neutron scattering experiments is in terms of elastic and inelastic scattering. For elastic scattering the neutron does not suffer an energy transfer, thus there is no need to analyze the energy of the scattered neutrons. The corresponding instruments are called diffractometers; they also include instruments for small-angle scattering (SANS) and reflectometers. For inelastic scattering the variety lies in the energy range accessible by the particular instruments, which include time-of-flight and triple-axis spectrometers with energy resolutions of 10 μeV to 10 meV as well as backscattering and spin-echo spectrometers with energy resolutions of 1 μeV and below.

3.3.2 *Powder diffractometers*

For structure determination by neutron diffraction the scattering function $S(\boldsymbol{Q},\omega)$ is reduced to the special case without energy transfer ($\hbar\omega = 0$) and hence the scattering intensity is only depending on the scattering vector $\boldsymbol{Q}$. For structure investigations of powder samples there are two principally different diffraction techniques: the angular-dispersive method (ADP) and the energy-dispersive method (EDP) which are used at steady-state and pulsed neutron sources, respectively.

In the ADP measurement the sample is irradiated by a monochromatic neutron beam. To each lattice spacing d_{hkl} belongs a Bragg angle θ_{hkl} which gives rise to an elastic signal at the scattering angle $2\theta_{hkl}$. Rather than moving a single detector around the sample, the powder diffractometers are equipped with banks of individual detectors and/or position sensitive detectors to cover a wide range of scattering angles simultaneously. A schematic layout of a multicounter powder diffractometer is shown in Fig. 3.13. The angular resolution of a powder diffraction diagram is largely defined by Eq. (3.5), thus large monochromator angles are favorable to realize small line widths in order to avoid an overlap of neighboring Bragg reflections.

In the case of the EDP method the incoming neutrons are pulsed and

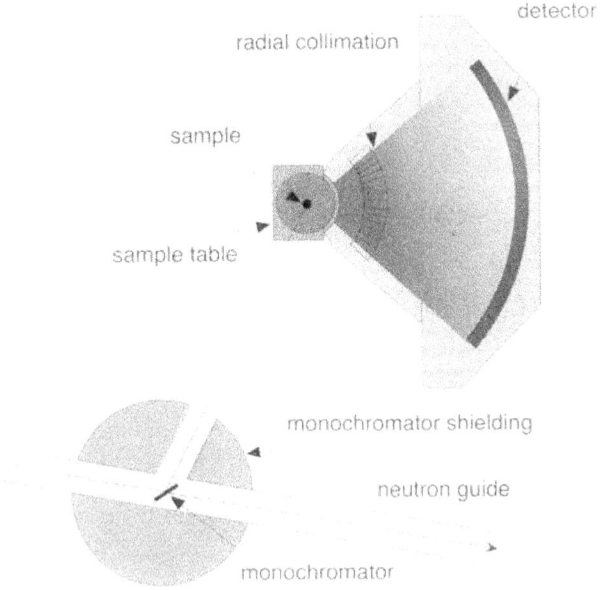

Fig. 3.13 Schematic layout of an angular-dispersive powder diffractometer equipped with a multicounter detector.

polychromatic, but the scattering angle is fixed. Each d_{hkl} value corresponds to a different flight time and hence wavelength of the neutron. The resolution $\Delta t/t$ of the time-of-flight analysis depends on the wavelength-dependent pulse structure of the moderator, on the length of the flight path, and on the scattering angle; thus, long flight paths of 50-100 m as well as the backscattering geometry are usually realized. If a short pulsed neutron source is available, the pulse width Δt in the slowing-down regime is inversely proportional to the neutron velocity, which leads to practically constant resolution over the whole diffraction range, which is one of the main advantages of the EDP method.

3.3.3 *Single-crystal diffractometers*

If single crystals are used to investigate the crystal structure, more information can be obtained as compared to powder diffraction, since in addition to the magnitude and structure factor of the reciprocal lattice vectors $\tau_{\rm hkl}$ also their orientation in space can be determined. To fulfill the Bragg con-

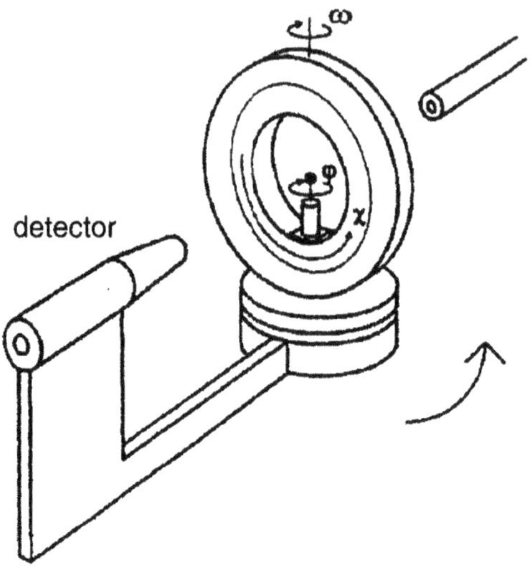

Fig. 3.14 Schematic sketch of a Eulerian cradle (after [Heger (2000)]).

dition for all $\tau_{\rm hkl}$, single crystal diffractometers are equipped with a special goniometer consisting of three independent rotations which is called a Eulerian cradle. It has in addition to the ω-axis ($\perp$ to the scattering plane) two further rotation axes χ and ϕ, which are perpendicular to each other (see Fig. 3.14). The χ-axis is also perpendicular to the ω-axis. Together with the rotation axis of the detector (parallel to the ω-axis) this mechanical unit is called 4-circle goniometer, leading to the name 4-circle diffractometer for this type of instrument. Since for a particular crystal setting only a few reciprocal lattice vectors $\tau_{\rm hkl}$ simultaneously fulfill the Bragg condition, it is sufficient to have not more than three detector arms. The use of position-sensitive detectors allows to determine the intensity distribution in two dimensions.

Besides the 4-circle diffractometer there is a second type called Laue diffractometer sketched in Fig. 3.15. The single crystal is mounted in a fixed position, and the scattered neutrons are usually recorded by an image plate detector placed around the sample. When the incoming neutron beam is parallel to a high-symmetry direction of the crystal, the Laue pattern also has high symmetry. For instance for cubic crystals, an incoming beam either parallel to one of the unit cell edges $\langle 001 \rangle$ or parallel to the body

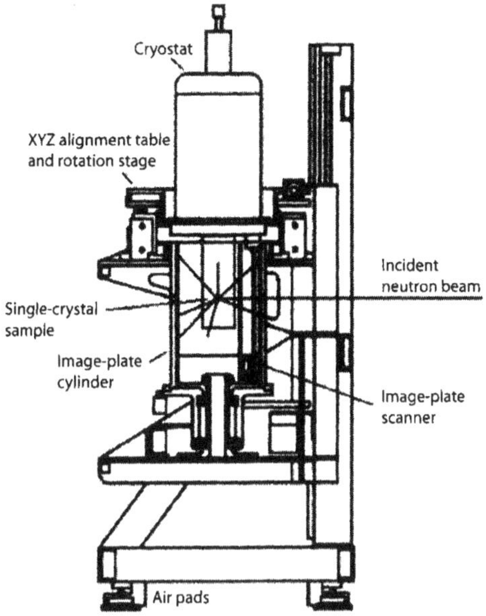

Fig. 3.15 Schematic sketch of a Laue diffractometer with a cylindrical image plate detector around the sample position. Reprinted from [McIntyre *et al.* (2006)]. Copyright 2006, with permission from Elsevier.

diagonal $\langle 111 \rangle$ of the unit cell produces a Laue pattern with either 4-fold or 3 fold symmetry, respectively.

3.3.4 *Small-angle scattering instruments*

Conventional diffractometers discussed in the previous chapters deliver structural information on the arrangements of atoms in condensed matter. If the systems under study are mesoscopic objects such as macromolecules with sizes exceeding 100 Å, the scattering angles 2θ resulting from Bragg's law are extremely small. Dedicated instruments were designed for this purpose which are called small-angle scattering instruments (SANS).

The principle layout of a SANS instrument is depicted in Fig. 3.16. A broad wavelength band of cold neutrons with $\delta\lambda/\lambda \approx 10$ % is delivered by a mechanical velocity selector (see Fig. 3.6). After passing two apertures the neutrons irradiate the sample, and the neutrons scattered under small scattering angles are counted by a two-dimensional detector (with a diameter of typically 1 m), whereas the primary beam is absorbed in the beam

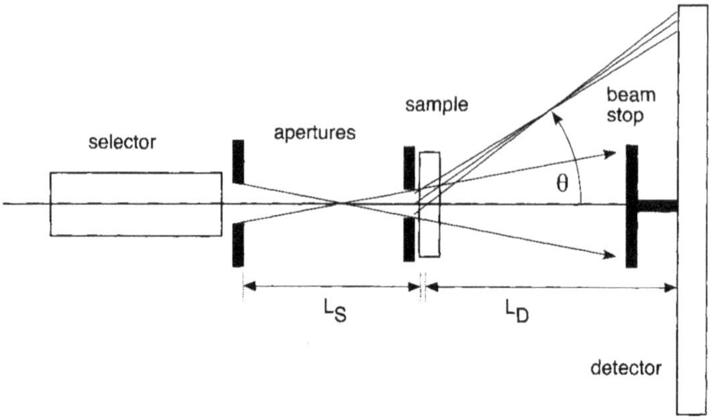

Fig. 3.16 Principle design of a pin-hole SANS instrument (after [Schwahn (2000)]).

stop in front of the detector. The resolution function of this experiment is given as [Schwahn (2000)]

$$\langle \delta Q^2 \rangle = \frac{k^2}{12}\left(\left(\frac{d_D}{L_D}\right)^2 + \left(\frac{d_E}{L_S}\right)^2 + d_S^2\left(\frac{1}{L_S} + \frac{1}{L_D}\right)^2 + \theta^2\left(\frac{\delta\lambda}{\lambda}\right)^2\right), \quad (3.8)$$

with the distances L_i (up to 20 m) and apertures d_i (adjusted to the sample size d_S) as indicated in Fig. 3.16. The requested Q-range as well as the resolution δQ is usually adjusted primarily by the distance L_D. The accessible Q values range from about 0.5 $Å^{-1}$ to 10^{-3} $Å^{-1}$.

3.3.5 *Reflectometers*

Reflectometers serve for investigations of surface properties, thin films, multilayers, and interfaces by means of total reflection of neutrons. A reflectometer measures the reflected intensity as a function of the scattering vector $\boldsymbol{Q}$ perpendicular to the reflecting surface. This can be achieved either by using a monochromatic neutron beam and scanning a large range of incident angles, or by using the broad-band neutron time-of-flight method with fixed scattering angle. On a pulsed neutron source the natural way to make measurements is the white beam time-of-flight method, whereas on continuous neutron sources the use of a monochromatic beam is more

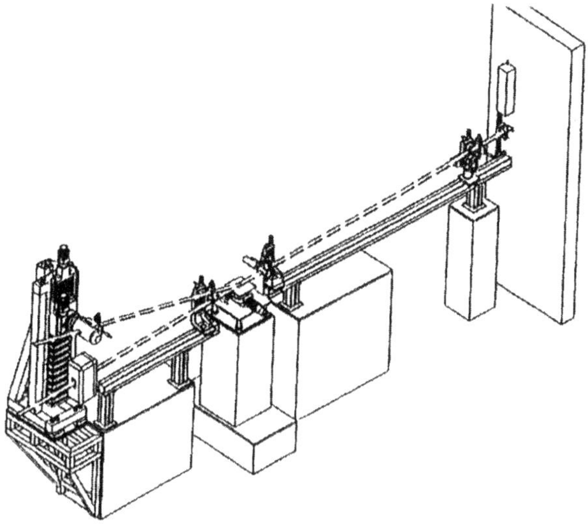

Fig. 3.17 Schematic diagram of a reflectometer showing the collimation slits (defining the beam size), frame overlap suppression mirrors, sample position, and detectors. Reprinted with permission from [Penfold *et al.* (1994)]). Copyright 1994, WorldScientific.

appropriate. In the case of the time-of-flight method the fixed sample geometry ensures constant sample illumination, and essentially constant Q resolution over the available Q range. Some reflectometers can also be operated in the polarized neutron mode for the study of magnetic multilayers and thin films; the polarization of the incident neutron beam can be achieved by inserting magnetic supermirrors (see Chap. 3.2.6).

A reflectometer is usually designed to have the scattering process perpendicular to the horizontal plane in order to allow investigations of liquid surfaces, see Fig. 3.17. The scattering angles γ for total reflection are extremely small, see Eq. (2.63), so both long flight paths of the order of 2 m and narrow well collimated beams with typical dimensions of 40 mm width and around 1 mm height are required. Frame overlap suppression can be achieved by additional choppers or nickel coated silicon mirrors, set at an angle to reflect out of the main beam direction unwanted neutrons. For the neutron detection either simple ^{3}He gas tubes or position-sensitive detectors are used.

The measured intensities I are usually reduced to reflectivity R on an absolute scale as a function of the scattering vector Q as follows:

$$R(Q) = f\frac{(I_d(Q) - B_d)\epsilon_m(Q)}{(I_m(Q) - B_m)\epsilon_d(Q)}, \tag{3.9}$$

where the subscripts d and m denote the detector and the beam monitor (placed immediately before the sample), respectively, $\epsilon_{d,m}$ the appropriate Q-dependent efficiencies, $B_{d,m}$ the associated background, and f an experimentally determined scale factor. The absolute scaling is obtained by normalization to the straight-through beam intensity by reference to the region of total reflection, or by normalization to a standard surface such as D_2O. The Q-resolution is given as

$$\frac{(\Delta Q)^2}{Q^2} = \frac{(\Delta x)^2}{x^2} + \frac{(\Delta\gamma)^2}{\gamma^2}, \tag{3.10}$$

where $x = t$ for the time-of-flight method and $x = \lambda$ for the monochromatic beam technique. Since $\Delta x/x \ll \Delta\gamma/\gamma$, the Q-resolution is dominated by $\Delta\gamma/\gamma$, and so is essentially constant in the time-of-flight method.

3.3.6 *Time-of-flight spectrometers*

Since neutrons commonly used in scattering experiments have velocities v of the order of a few hundred to a few thousand m/s, their energy can be conveniently determined by measuring their flight time t over a distance L of a few meters. Owing to the fact that neutrons are detected by nuclear reactions (see Chap. 3.2.8), they can only be detected once. This means that their starting time at a certain position must be defined by pulsing the beam. It also means that only either the incoming wavevector $\boldsymbol{k}$ or the scattered wavevector $\boldsymbol{k}'$ can be measured by time-of-flight techniques; the other one must therefore be enforced by the monochromatization techniques discussed in Chaps 3.2.3 and 3.2.4. Depending on whether $\boldsymbol{k}'$ or $\boldsymbol{k}$ is measured by time-of-flight, the method is called direct time-of-flight or inverted time-of-flight, respectively. Fig. 3.18 shows the space-time diagrams and the momentum space representations of the two techniques, while Fig. 3.19 gives a schematic overview on possible instrumental realizations. A special case are inverted time-of-flight spectrometers using neutron beam filters acting as a coarse monochromator; here preferentially filters with a Bragg cutoff at low energies such as Be are used (see Chap. 3.2.5) in order to obtain reasonably good resolution.

Time-of-flight spectrometers are usually equipped with large detector banks, covering scattering angles in the range $0 \leq 2\theta \leq \pi$. The intensities

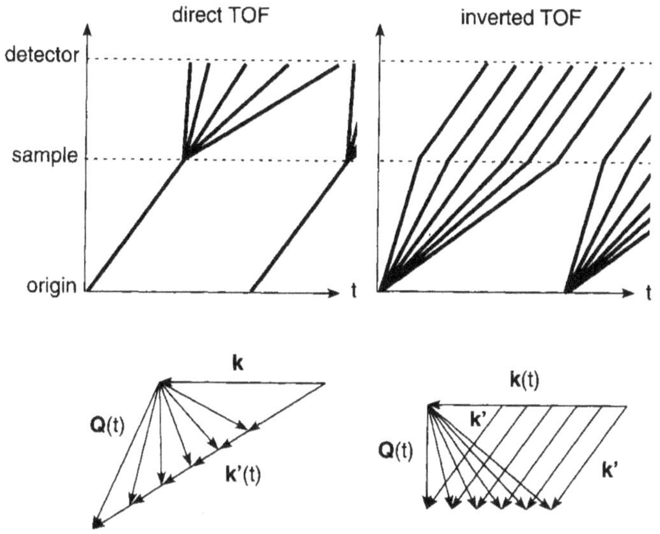

Fig. 3.18 Phase-space diagrams (top) and momentum representations (bottom) of direct and inverted time-of-flight techniques.

collected at the detectors represent the raw data of the form $I(2\theta, t)$, so that a transformation to the scattering law $S(\boldsymbol{Q}, \omega)$ is necessary. The flight time of the neutrons with wavevector $\boldsymbol{k}$ and $\boldsymbol{k}'$ are $t_0 = L/v$ and $t = L/v'$, respectively, thus using Eqs (2.1) and (2.2) yields

$$\omega(t) = \frac{m}{2\hbar} L^2 \frac{t^2 - t_0^2}{t^2 t_0^2} \tag{3.11}$$

and

$$Q = \frac{m}{\hbar} L \sqrt{\frac{t^2 + t_0^2 - 2t_0 t \cos(2\theta)}{t_0^2 t^2}} \tag{3.12}$$

The nonlinear mapping from time to energy given by Eq. (3.11) causes a strongly varying energy width of the (equidistant) time channels. From the derivative $d\omega/dt \propto 1/t^3$ and including the term $k'/k = t_0/t$ of the cross-section formula Eq. (2.4) we find the relation

$$S(\boldsymbol{Q}, \omega) \propto I(2\theta, t) \cdot t^4. \tag{3.13}$$

The factor t^4 in Eq. (3.13) causes a significant intensity enhancement of the early arriving neutrons, which is intimately connected with a corresponding loss of energy resolution. Eq. (3.13) tells us that the raw data $I(2\theta, t)$ are considerably distorted and should not be used for a detailed data interpretation.

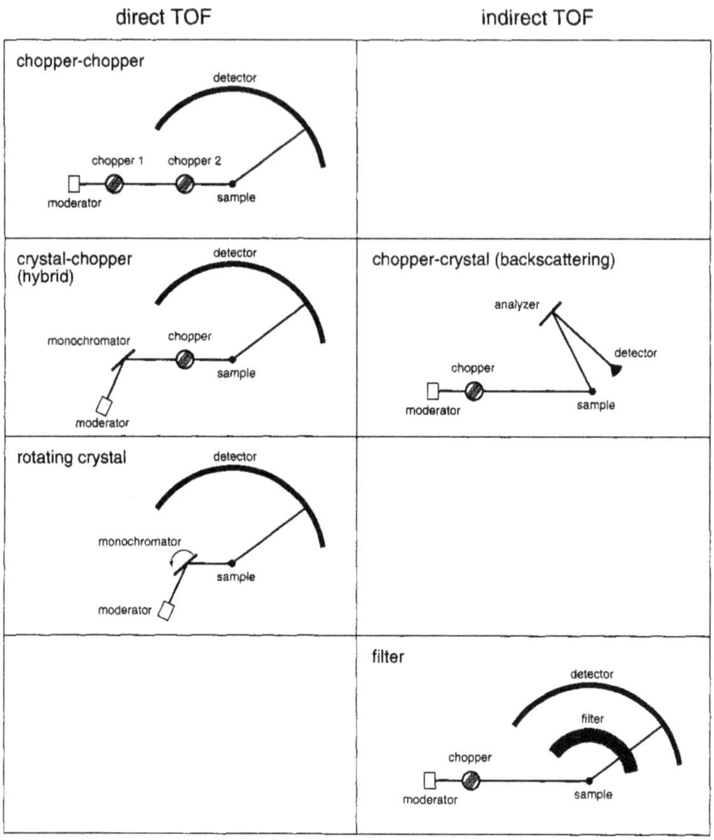

Fig. 3.19 Instrumental realizations of time-of-flight spectrometers.

3.3.7 *Triple-axis spectrometers*

Triple-axis spectrometers are the most versatile instruments for inelastic neutron scattering experiments, because they allow the controlled access to the variables $\boldsymbol{Q}$ and ω of the scattering law $S(\boldsymbol{Q}, \omega)$. As sketched in Fig. 3.20, an incident beam of neutrons with a well defined wavevector $\boldsymbol{k}$ is selected from the white spectrum of the neutron source by the monochromator crystal (first axis). The monochromatic beam is scattered from the sample (second axis). The intensity of the scattered beam with wavevector $\boldsymbol{k}'$ is reflected by the analyzer crystal (third axis) onto the neutron detector, thereby defining the energy transfer $\hbar\omega$ as well.

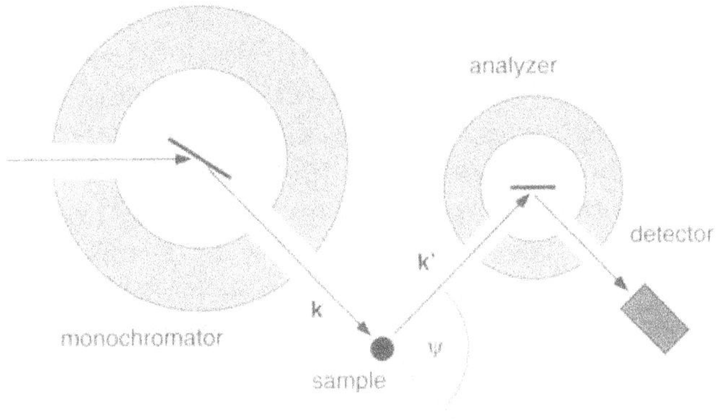

Fig. 3.20 Basic layout of a triple-axis spectrometer.

The outstanding advantage of a triple-axis spectrometer is that the data can be taken at any pre-determined point in reciprocal space (constant-$\boldsymbol{Q}$ scan) or for a fixed energy transfer $\hbar\omega$ (constant-E scan) along a particular line in reciprocal space, so that measurements of the dispersion relations in single crystals can be performed in a controlled manner (see, e.g., Chap. 5.2.3). Of course, general scans in $\boldsymbol{Q}$ and ω are also possible. Since on a triple-axis spectrometer all quantities $(\boldsymbol{k}, \boldsymbol{k}', \psi)$ defining the momentum and energy transfer (Eqs (2.1) and (2.2) and Fig. 2.1) are variable, there are many possibilities to measure the intensity at a point in $\boldsymbol{Q}$,ω space. The most commonly used option is a scan with $\boldsymbol{k}'$ fixed, because the conversion from count rate to cross section needs minimal corrections.

The resolution of a triple-axis spectrometer depends on the phase space volume transmitted by the monochromator and analyzer crystal in the scattering plane as well as on the slope of the dispersion of the measured excitation. The triple-axis spectrometer allows different configurations of the neutron path in terms of either clockwise or anticlockwise turn angles at each axis in order to achieve optimum resolution characteristics. The general features of the resolution function including analytic expressions and guides for its experimental determination were given by Cooper and Nathans [Cooper and Nathans (1967)] and Dorner [Dorner (1972)].

3.3.8 *Backscattering spectrometers*

If one wishes to obtain an extremely high energy resolution in a reflection from a single crystal, the scattering angle 2θ must be π (or very close to π) according to Eq. (3.5). This property is taken advantage of in the analyzer part of very high-resolution spectrometers in order to achieve a very precise determination of $\boldsymbol{k}'$. In this case large areas of nearly perfect crystals are arranged on spherical supports around the sample position. Neutrons coming from the sample and fulfilling the backscattering condition are therefore reflected back to a set of detectors arranged near the sample. The analyzer system thus works at a fixed energy. In order to measure an energy transfer in the sample, the energy of the incoming neutrons must be varied with similarly high resolution. This can be achieved in various ways:

(a) by very good resolution of time-of-flight monochromators (choppers) producing the incident beam which, however, requires flight paths from the pulsing device to the sample of the order of 100 m;
(b) by using a monochromator which is also designed to work in backscattering and whose lattice constant can be varied in small intervals as a function of time, e.g., by varying its temperature;
(c) by using a backscattering monochromator mounted on a velocity drive and using the Doppler effect to produce incremental changes Δk according to Eq. (3.6).

The energy resolution that can be obtained on a backscattering spectrometer is in the μeV-range. This is about the ultimate resolution that can be achieved with spectrometers based on crystal monochromators and time-of-flight devices.

3.3.9 *Spin-echo spectrometers*

The spin-echo spectrometer is based on the precession of the neutron spin in a magnetic field. Since a neutron spin performs a large number of revolutions while travelling in a magnetic field of useful length, a change in energy is best analyzed by reversing the spin direction after the scattering process and detecting the difference in spin orientation after having passed the second precession field. A schematic drawing of such an instrument is shown in Fig. 3.21.

Longitudinally polarized neutrons with velocity v and velocity distribution $f(v)$ enter the spectrometer from the left. In the first $\pi/2$-flipper the spin is rotated such that on exit it is orthogonal to the longitudinal magnetic field H of the precession path, thereby providing a well defined starting polarization $P = 1$. Upon traversing the first precession coil of

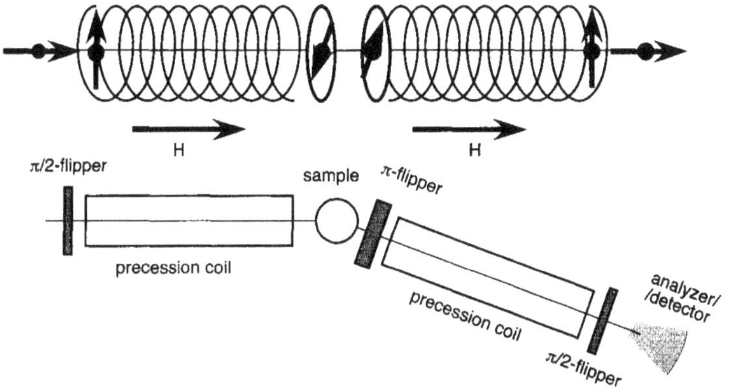

Fig. 3.21 Schematic layout of a spin-echo spectrometer.

length L, the total precession angle is given by (see Eq. (3.7))

$$\phi = \gamma_L \cdot \frac{L \cdot H}{v}, \tag{3.14}$$

Due to the velocity spread the beam suffers a depolarization in the precession coil, resulting in an average polarization

$$P = \langle \cos \phi \rangle = \int f(v) \cos(\phi(v))\, \mathrm{d}v. \tag{3.15}$$

However, the depolarization can be reverted, if the beam traverses a second, identical coil with opposite direction of precession. To reverse the direction of precession, a π-flipper is placed between the two precession coils. Finally, at the end of the second precession coil, the neutron spin is flipped back into the original longitudinal direction to allow the analysis of its polarization. A fully polarized beam will only be observed if the two precession coils are exactly identical and no energy change occurs at the sample.

If the neutron suffers an energy change $\hbar\omega$ at the sample and therefore a velocity change Δv, the spin orientation after the second precession coil will be off by an angle

$$\Delta\phi \approx \gamma_L \frac{L \cdot H}{v^2} \Delta v. \tag{3.16}$$

Thus, in a spin-echo spectrometer the energy transfer is not measured by defining $\boldsymbol{k}$ and $\boldsymbol{k}'$ very precisely as, e.g., in a backscattering spectrometer (see Chap. 3.3.8), but rather directly by the difference in neutron velocity before and after the scattering through Eq. (3.16). Therefore a relatively broad neutron wavelength band of the order of $\delta\lambda/\lambda \sim 10^{-1}$ can be used, whereas in backscattering $\delta\lambda/\lambda \lesssim 10^{-4}$ is generally required. This decoupling of the energy transfer from the neutron energy is an important concept, which provides outstanding energy resolutions down to the neV range.

3.4 Sample environment

Many interesting physical properties of condensed matter emerge at non-ambient conditions, i.e., at temperatures different from room temperature or under the influence of an externally applied magnetic field H, pressure P or electric field E. Furthermore, the variation of these thermodynamic variables allows probing the underlying interaction potential V of the system. Effects upon variation of T, H, P and E are a crucial benchmark to theoretical models describing the system under investigation.

Devices enabling measurements at non-ambient conditions over a vast range of thermodynamic variables are hence of utmost importance to any experimental method and to a large amount define its power and versatility.

Neutron scattering techniques are in a privileged position here. Owing to the fact that the neutron is an electrically neutral particle the penetration depth into engineering materials is large. Entrance and exit windows the neutron has to pass can thus be realized relatively easily compared to other similar experimental techniques like, e.g., x-ray scattering. Due to the fact that isotopes with very large neutron absorption are dispersed non-uniformly over the periodic table, a large choice of suitable materials for collimation of the incoming and scattered beam exists. Thus, e.g., windows are often made of aluminum (Al) with a very small absorption and almost negligible incoherent cross-section (see Appendix B), other materials with good transmission properties include, e.g., C and Al_2O_3. For collimation purposes materials with large neutron absorption are used, these include: Cd, Gd, B, Li.

3.4.1 *Temperature*

Variation of temperature is required for most neutron scattering studies. Low temperatures are required for the study of quantum mechanical properties of matter and in order to elucidate the corresponding ground-state

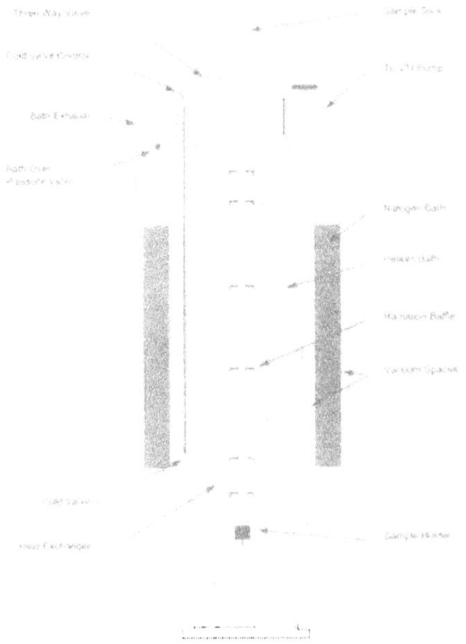

Fig. 3.22 Basic layout of a ^{4}He cryostat ("orange ILL-type cryostat") used for neutron scattering studies (by courtesy of AS Scientific Products Ltd., Abingdon, U.K.).

of the system. At elevated temperatures thermally excited states are populated and the respective properties consequently mixed. On the contrary, for the study of dynamical properties, e.g., diffusion processes, the system under investigation has to be thermally excited.

Temperatures down to 1.8 K are realized in bath ^{4}He cryostats. The sample is in contact with a few 10 mbar He exchange gas in the sample chamber, thus temperature gradients are usually negligible. Specific to cryostats, like for other sample environment used for neutron scattering is the consequent use of thin walled aluminum as the material for the sample chamber, heat screen and vessel for the insulation vacuum. Figure 3.22 shows a scheme of the orange ILL-type cryostat widely used at various neutron centers worldwide. Lower temperatures to $\sim$ 300 mK can be realized by evaporating ^{3}He (^{3}He cryostat) and to $\sim$ 10 mK by using a mixture of ^{3}He and ^{4}He (dilution cryostat). In both cases the sample chamber is evacuated and good thermal contact must be granted between the sample and the cold-finger the sample is mounted on. Vibrations and thermal radiation must further be kept to a minimum in order to reach lowest possible temperatures.

Cryogenic free cooling remove or reduce the dependency on supply of liquid He and hence enjoy increasing popularity. Cooling in these devices results from slow compression and fast decompression of He gas circulating in a closed system. Correspondingly they are referred to as CCR (closed-cycle refrigerator) devices. Temperatures of 4 K can be realized routinely, and in combination with, e.g., a Joule/Thompson stage, 1.5 K can be reached cryogenic free. These devices can also be combined with ^{3}He bath and ^{3}He/^{4}He dilution inserts to reach $\sim$ 250 mK and $\sim$ 50 mK, respectively. A major drawback of CCR devices consists of the vibrations inherent to the compression and decompression of the gas.

High temperatures can be realized with resistive furnaces where the sample placed in a crucible is in good thermal contact with a hot spot of the furnace but otherwise in vacuum that serves as thermal insulation to the outer vessel. The material of the crucible has to be chosen carefully in order to avoid chemical reaction with the sample. Common resistive furnaces can easily reach temperatures up to $\sim$ 1600 K.

Levitation methods can be employed if direct contact with the sample must be avoided, e.g., in the case of chemically reacting samples or where the purity of the sample has to be maintained (e.g., for the study of supercooled melts). In this case a sample of a few $\sim 10 \ldots 100$ mm^3 volume is elevated by an electromagnetic or electrostatic field, acoustic wave or gas flow around the sample. Temperatures up to $\sim$ 3000 K have been realized either by microwave absorption or heating with a laser.

It should be pointed out that devices covering continuously a large temperature range from a few 10 K to temperatures up to 1000°C are gaining importance in materials science research. Here modified cryostats (cryofurnaces) as well as modified CCR devices find broad applications.

3.4.2 *Magnetic field*

An externally applied magnetic field $H_{\rm ex}$ acts directly on the magnetic properties of the system under investigation. Thus external magnetic fields are used (i) to benchmark theoretical models of the system under investigation of which the response to $H_{\rm ex}$ can be calculated, (ii) to allow the differentiation of magnetic scattering from nuclear scattering, or (iii) to alter the ground-state of the system via level-crossing with a neighboring excited state leading to a field-induced quantum phase-transition. Magnetic fields of up to 16 T can be realized with superconducting coils. These need cooling by liquid He. Since magnetic fields are most often applied at low temperatures the magnet and cryostat are usually combined to a single device (cryomagnet). ^{3}He or dilution inserts allow for the measurement down to sub-K in cryomagnets. Magnetic fields in excess of 16 T can only be achieved with substantial technical efforts and usually require dedicated fixed infrastructures (i.e., can be operated only on dedicated beamlines).

For neutron scattering magnets with either field direction perpendicular (vertical) or parallel (horizontal) to the scattering plane are used. The devices are optimized to offer largest possible vertical and horizontal aperture with as thin as possible wall thicknesses of the entrance and exit windows. For vertical magnets the two coils are split by a cylindrical Al spacer thus allowing for a 360° horizontal aperture except for a small region needed for the feedthrough of cables connecting the two coils. For homogeneous fields over a large volume, the two coils have to be split sufficiently and consequently higher currents must be used for a given field strength. Magnets and cryomagnets are optimized for minimum stray fields outside the magnet, since forces on ferromagnetic objects, disturbance of electronics, ^{3}He spin filters and neighboring instruments must be avoided. Active shielding of a magnet can be achieved by a pair of coils with opposite polarity compensating the dipole moment at the outside of the device.

3.4.3 *Pressure*

Application of pressure allows the variation of interatomic (intermolecular) distances and hence probing the underlying interaction potential. Owing to a bulk modulus B of the few 10 and 100 GPa, several kbar (0.1 GPa) to several GPa are needed to alter atomic distances by as little as 1 % in, e.g., molecular solids and typical oxides, respectively. In the absence of a structural phase transition the volume V decreases linearly for pressures $P \lll B$, whereas for higher P the volume dependence is given, e.g., by the third-order Birch-Murnaghan equation of states [Poirier (2000)]

$$P = \frac{3}{2}B\left(x^{7/3} - x^{5/3}\right)\left(1 - \frac{3}{4}(4 - B')\left(x^{2/3} - 1\right)\right), \tag{3.17}$$

where $x = V/V_0$ denotes the change of volume compared to ambient pressure and B' is a dimensionless parameter with $B' \approx 4$ for many solids. The force F required for the generation of pressure on an area A is directly given by $F = P \cdot A$. Hence it becomes evident that forces of the order of a few tons are required to generate pressures of ~ 1 GPa on samples of a few mm diameter. In order to keep the required forces attainable, highest pressures can only be realized at the expense of sample volume. Contrary to, e.g., synchrotron techniques, neutron scattering here suffers from the limited brilliance of neutron sources. Pressures attainable for neutron techniques are hence limited to a few 10 GPa maximum. Beneficial, however, is the large penetration power of neutrons allowing easier transmission through thick walled pressure vessels. Devices allowing studies under pressure are termed pressure cells and we here review the most commonly used cells for neutron scattering.

Gas pressure-cells In these cells the sample is filled into a closed pressure vessel. A capillary attached to the vessel and connected to a gas compressor allows the increase of gas pressure in the vessel and hence onto the sample. The gas acts as the pressure transmitting medium. Common gases used include helium (down to cryogenic temperatures), argon and nitrogen. Advantages of this type of cell are that (i) the pressure can be changed *in-situ* (i.e. during the experiment on the instrument at, e.g., non-ambient temperatures), (ii) no addition of a pressure-transmitting medium is required (the sample is under truly hydrostatic conditions), (iii) the pressure acting on the sample is equal to the applied gas pressure that can be easily measured and controlled, and (iv) relatively large sample volumes of a few cm^3 can be compressed. Due to their cylindrical geometry and high elastic energy saved in the compressed gas these cells are limited to $\sim$ 1 GPa working pressure.

Piston-cylinder clamp cells Similar to gas pressure-cells the sample is filled in a cylindrically shaped vessel. However the pressure is applied by means of a piston acting force onto the sample directly or onto the sample immersed in a pressure-transmitting medium (Fig. 3.23). The force on the piston is usually applied *ex-situ* under a press and then clamped by screws. This type of cell is the most used one for pressures below $\sim$ 2 GPa. The effective pressure acting on the sample can only be estimated roughly from $P = F/A$ since large frictional forces between piston and cylinder may be present and since pressure may change as a function of temperature due to thermal expansion of the cell, sample and pressure-transmitting medium. Pressure is thus best determined by measurement of the lattice parameter of the sample via its equation of state or, if not known, by addition of a suitable pressure calibrant of which the equation of state is known (e.g., NaCl, CsCl, Pb, ...).

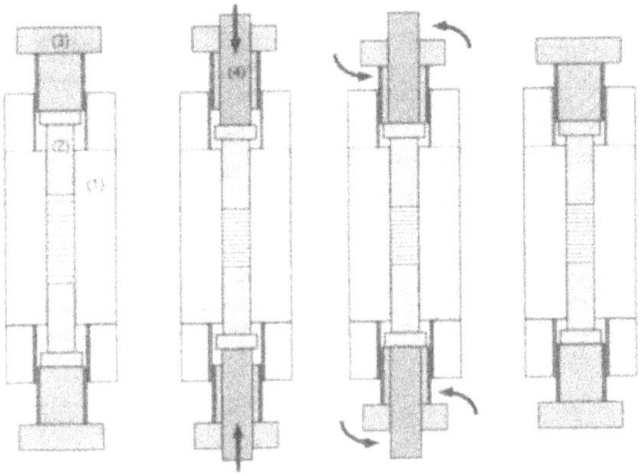

Fig. 3.23 Schematic use of a clamp pressure cell: the sample is loaded into the body of the cell (1), auxiliary pistons (4) are used to exert force onto the inner pistons (2), subsequently the inner pistons (2) are clamped with the screws (3) prior to release of the force and auxiliary pistons.

For both of the above cells the sample is placed in a cylinder subject to internal pressure. It can be shown that the maximum pressure before burst of a cylinder is about equal to the yield strength Y of the cylinder material irrespective of the ratio of the outer R_a and inner radius R_i for $R_a/R_i > 3$. By the technique of frettage the maximum reachable pressure can be further increased to about 2-3 times Y. Commonly used materials for pressure cells include steel and various alloys, including CuBe and TiZr all having Y in the range of 1-1.5 GPa, thus effectively limiting the use of cylindrical pressure cells to about 2-3 GPa.

Opposed-anvil cells To achieve pressures in excess of 3 GPa the cylindrical cell geometry has to be abandoned and instead the technique of opposed-anvils has to be used. The most known of this type of pressure cell is the diamond-anvil cell (DAC) where the sample (and pressure transmitting medium) is placed in a hole drilled into a thin metallic disk (gasket). Compression of this disk by two opposed anvils results in large frictional forces between the gasket and the anvils and to an enormously enhanced effective yield strength of the gasketing material due to radial forces aiming to collapse the hole in the gasket. If restricted to very small sample volumes (a few 10 μm^3) DAC can achieve pressures in the range of several Mbar (100 GPa). The limited brilliance of today's neutron sources require samples of a few mm^3 and thus limits

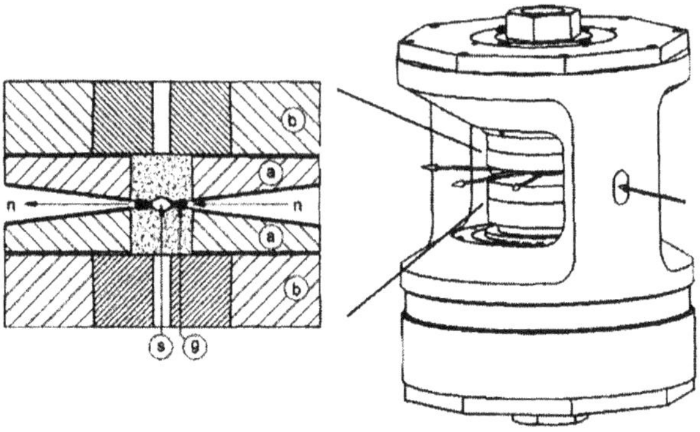

Fig. 3.24 Basic layout of a toroidal opposed-anvil setup used for neutron scattering under pressure (Paris-Edinburgh pressure cell). Left: the sample (s) is placed in between two anvils (a) and laterally confined by the gasket (g). Right: press (ram) acting force on the sample-anvil assembly (left) via backing seats (b). n denote the incoming and scattered neutron beam. Reprinted with permission from [Klotz *et al.* (2005b)]. Copyright 2005 by the American Institute of Physics.

pressures by the dimension of corresponding force-generating devices. Large-volume opposed anvil cells take the advantage that the attainable pressure in the center of a sphere is doubled ($P = 2F/A$). Anvils of this type hence have a hemispherical groove at the center for the sample and toroidal grooves for the gaskets. The latter allow for an additional stabilization of the gasket and correspondingly larger anvil-anvil distance. The anvils are typically produced of tungsten-carbide (WC), boron-nitride (BN) or sintered diamond. A schematic drawing of a large-volume opposed-anvil pressure cell is shown in Fig. 3.24. The force onto the anvils is generated by a hydraulically or pneumatically driven ram and allows the *in-situ* change of pressure. The pressure on the sample is usually determined via the equation of state of the sample or additional pressure calibrant.

Materials for the building of pressure cells (or gasketing material) should exhibit little absorption of the neutron beam and high yield strength. Alloys of aluminum fulfill this condition and can be used for cylindrical cells up to 1.2 GPa. For higher pressures hardened steel or special alloys must be used (e.g., NiCrAl alloy for cylinder cells up to 2.5 GPa). CuBe widely used for high-pressure applications has the disadvantage of getting activated and exhibits moderate transmission due to small amounts of Co present in the CuBe alloy. In the case of powder diffraction, Bragg scattering from the cell is undesired and TiZr alloy can be used. Ti exhibits a negative scattering length (see Appendix B). Combined with Zr an alloy of effective zero scattering length can be produced (often referred to as zero-matrix alloy). This alloy shows no coherent Bragg scattering at the disadvantage of relatively large incoherent scattering due to Ti giving rise to an elevated flat background in the diffraction pattern.

3.5 Further reading

- G. S. Bauer, in *Neutron scattering*, ed. by A. Furrer (Proc. 93-01, ISSN 1019-6447, PSI Villigen, 1993), p. 331: *Neutron sources*
- G. S. Bauer and W. Bührer, in *Neutron scattering*, ed. by A. Furrer (Proc. 93-01, ISSN 1019-6447, PSI Villigen, 1993), p. 1: *Instruments for neutron scattering*
- P. Böni, in *Complementarity between neutron and synchrotron x-ray scattering*, ed. by A. Furrer (World Scientific, Singapore, 1998), p. 305: *Neutron beam optics*
- W. Bührer, in *Introduction to neutron scattering*, ed. by A. Furrer (Proc. 96-01, ISSN 1019-6447, PSI Villigen, 1996), p. 33: *Instruments*
- J. M. Carpenter and W. B. Yelon, in *Methods of experimental physics*, Vol. 23, Part A, ed. by D. L. Price and K. Sköld (Academic Press, London, 1986), p. 99: *Neutron sources*
- A. J. Dianoux and G. Lander, *ILL Neutron data booklet*, 2nd edition (Old City Publishing, Philadelphia, 2003)
- A. Furrer, in *Encyclopedia of condensed matter physics*, ed. by G. F. Bassani, G. L. Liedl and P. Wyder (Elsevier, Amsterdam, 2005), p. 69: *Neutron sources*
- F. Mezei, in *Encyclopedia of condensed matter physics*, ed. by G. F. Bassani, G. L. Liedl and P. Wyder (Elsevier, Amsterdam, 2005), p. 76: *History of neutrons and neutron scattering*
- F. Mezei, in *Neutron spin echo*, ed. by F. Mezei (Springer, Berlin, 1980), p. 3: *The neutron spin echo method*
- P. C. H. Mitchell, S. F. Parker, A. J. Ramirez-Cuesta and J. Tomkinson, in *Vibrational spectroscopy with neutrons* (World Scientific, Singapore, 2005), p. 67: *Instrumentation and experimental methods*
- G. Shirane, S. M. Shapiro and J. M. Tranquada, *Neutron scattering with a triple-axis spectrometer* (Cambridge University Press, Cambridge, 2002)
- C. G. Windsor, in *Methods of experimental physics*, Vol. 23, Part A, ed.

by D. L. Price and K. Sköld (Academic Press, London, 1986), p. 197: *Experimental techniques*

Chapter 4

Structure Determination

4.1 Cross section

The technique used for structure determination is known as neutron diffraction or elastic neutron scattering. We start from the basic cross-section formula Eq. (2.17):

$$\frac{\mathrm{d}^2\sigma}{\mathrm{d}\Omega\mathrm{d}\omega} = \frac{k'}{k}\frac{1}{2\pi\hbar}\sum_{j,j'} b_j b_{j'} \int_{-\infty}^{\infty} \langle e^{-\imath \boldsymbol{Q}\cdot\hat{\boldsymbol{R}}_{j'}(0)} e^{\imath \boldsymbol{Q}\cdot\hat{\boldsymbol{R}}_{j}(t)}\rangle e^{-\imath\omega t}\mathrm{d}t. \tag{4.1}$$

Since the aim of a structure study is to determine the positions of all the atoms in a unit cell at thermal equilibrium, we can neglect the time dependence of the operators $\hat{\boldsymbol{R}}_j(t)$. The time integration can easily be performed by using Eq. (A.6):

$$\frac{1}{2\pi\hbar}\int_{-\infty}^{\infty} e^{-\imath\omega t}\mathrm{d}t = \frac{1}{\hbar}\delta(\omega) = \delta(\hbar\omega), \tag{4.2}$$

which means that the scattering is elastic, since $\hbar\omega = 0$, thus $k = k'$. We can transform the double-differential cross section to a single-differential one by integrating over $\hbar\omega$ (see Eq. (A.1)):

$$\frac{\mathrm{d}\sigma}{\mathrm{d}\Omega} = \int_{-\infty}^{\infty}\left(\frac{\mathrm{d}^2\sigma}{\mathrm{d}\Omega\mathrm{d}\omega}\right)\mathrm{d}(\hbar\omega) = \sum_{j,j'} b_j b_{j'} \langle e^{-\imath \boldsymbol{Q}\cdot\hat{\boldsymbol{R}}_{j'}} e^{\imath \boldsymbol{Q}\cdot\hat{\boldsymbol{R}}_{j}}\rangle. \tag{4.3}$$

To simplify matters, we drop the operator formalism, since the atomic positions $\boldsymbol{R}_j$ are fixed (this simplification is mathematically not quite correct as mentioned below). By using the expressions Eqs (2.28) and (2.29) we arrive at the following elastic neutron cross-sections:

$$\left(\frac{\mathrm{d}\sigma}{\mathrm{d}\Omega}\right)_{\text{coh}} = \langle b\rangle^2 \sum_{j,j'} e^{-\imath \boldsymbol{Q}\cdot(\boldsymbol{R}_{j'}-\boldsymbol{R}_j)}, \tag{4.4}$$

$$\left(\frac{\mathrm{d}\sigma}{\mathrm{d}\Omega}\right)_{\text{inc}} = \left(\langle b^2\rangle - \langle b\rangle^2\right)\sum_{j=j'} e^{-\imath \boldsymbol{Q}\cdot(\boldsymbol{R}_{j'}-\boldsymbol{R}_j)} = N\left(\langle b^2\rangle - \langle b\rangle^2\right). \tag{4.5}$$

The incoherent elastic scattering is isotropic and yields a constant background. On the other hand, the coherent elastic scattering provides information about the mutual arrangement of the atoms due to the phase factor, thus we concentrate in the following on coherent scattering only. We simplify Eq. (4.4) by the substitution $\boldsymbol{r} = \boldsymbol{R}_j - \boldsymbol{R}_{j'}$:

$$\frac{\mathrm{d}\sigma}{\mathrm{d}\Omega} = N_0 \langle b \rangle^2 \sum_{\boldsymbol{r}} e^{\imath \boldsymbol{Q}\cdot\boldsymbol{r}}, \tag{4.6}$$

where N_0 denotes the number of unit cells. Applying Eq. (A.10) yields the coherent elastic neutron cross-section for a Bravais lattice:

$$\frac{\mathrm{d}\sigma}{\mathrm{d}\Omega} = N_0 \frac{(2\pi)^3}{v_0} \langle b \rangle^2 \sum_{\boldsymbol{\tau}} \delta(\boldsymbol{Q} - \boldsymbol{\tau}), \tag{4.7}$$

where $\boldsymbol{\tau}$ denotes a reciprocal lattice vector (see Appendix E). The δ-function implies that scattering only occurs for $\boldsymbol{Q} = \boldsymbol{\tau}$. According to Appendix E the reciprocal lattice vectors $\boldsymbol{\tau}$ are perpendicular to the corresponding reflection planes indexed by the Miller triple (h, k, l) with mutual distance $d_{hkl} = 2\pi/|\tau_{hkl}|$. From this and Fig. 2.1 follows immediately Bragg's law

$$\boxed{\lambda = 2d \sin\theta} \tag{4.8}$$

where 2θ denotes the scattering angle.

For systems with more than one atom per unit cell we define the position vector $\boldsymbol{R}$ of the atoms as follows:

$$\boldsymbol{R} = \boldsymbol{l}_j + \boldsymbol{d}_\alpha, \tag{4.9}$$

with $\boldsymbol{l}_j$ and $\boldsymbol{d}_\alpha$ being the position vectors of the jth unit cell and of the αth atom in the unit cell, respectively. Combining Eqs (4.4) and (4.9) yields

$$\frac{\mathrm{d}\sigma}{\mathrm{d}\Omega} = \sum_{j,j'} e^{\imath \boldsymbol{Q}\cdot(\boldsymbol{l}_j - \boldsymbol{l}_{j'})} \sum_{\alpha,\alpha'} b_\alpha b_{\alpha'} e^{\imath \boldsymbol{Q}\cdot(\boldsymbol{d}_\alpha - \boldsymbol{d}_{\alpha'})}. \tag{4.10}$$

Setting $\boldsymbol{d} = \boldsymbol{d}_\alpha - \boldsymbol{d}_{\alpha'}$, using the relation

$$\sum_{\alpha,\alpha'} b_\alpha b_{\alpha'} e^{\imath \boldsymbol{Q}\cdot(\boldsymbol{d}_\alpha - \boldsymbol{d}_{\alpha'})} = |\sum_{\boldsymbol{d}} b_{\boldsymbol{d}} e^{\imath \boldsymbol{Q}\cdot\boldsymbol{d}}|^2, \tag{4.11}$$

and applying Eq. (A.10) yields

$$\frac{\mathrm{d}\sigma}{\mathrm{d}\Omega} = N_0 \frac{(2\pi)^3}{v_0} |\sum_{\boldsymbol{d}} e^{\imath \boldsymbol{Q}\cdot\boldsymbol{d}}|^2 \sum_{\boldsymbol{\tau}} \delta(\boldsymbol{Q} - \boldsymbol{\tau}). \tag{4.12}$$

The δ-function implies that $\boldsymbol{Q} = \boldsymbol{\tau}$, thus the general coherent elastic neutron cross-section reads

$$\frac{d\sigma}{d\Omega} = N_0 \frac{(2\pi)^3}{v_0} \sum_{\boldsymbol{\tau}} |S_{\boldsymbol{\tau}}|^2 \delta(\boldsymbol{Q} - \boldsymbol{\tau}) \tag{4.13}$$

with

$$\boxed{S_{\boldsymbol{\tau}} = \sum_{\boldsymbol{d}} b_{\boldsymbol{d}} e^{i\boldsymbol{\tau}\cdot\boldsymbol{d}}} \tag{4.14}$$

where $S_{\boldsymbol{\tau}}$ is the so-called structure factor. Equation (4.13), however, is not complete due to the neglect of the operator character of the atomic positions $\boldsymbol{R}_j$ in Eq. (4.4), thereby losing the Debye-Waller factor $e^{-2W(\boldsymbol{Q})}$. A correct treatment will be carried out in Chap. 5. The Debye-Waller factor describes the mean-squared displacements $\langle u^2 \rangle$ of the atoms from their equilibrium position; for a Bravais crystal of cubic symmetry the following result is obtained [Squires (1996)]:

$$2W(\boldsymbol{Q}) = 2W(Q) = \frac{1}{3} Q^2 \langle u^2 \rangle. \tag{4.15}$$

Introducing Eq. (4.15) into Eq. (4.13) yields the final expression for the coherent elastic neutron cross-section:

$$\boxed{\frac{d\sigma}{d\Omega} = N_0 \frac{(2\pi)^3}{v_0} e^{-2W(\boldsymbol{Q})} \sum_{\boldsymbol{\tau}} |S_{\boldsymbol{\tau}}|^2 \, \delta(\boldsymbol{Q} - \boldsymbol{\tau})} \tag{4.16}$$

The results to be expected from a neutron diffraction experiment are threefold: (i) Information on the size and the form (i.e., the symmetry) of the unit cell by examining the scattering angles 2θ at which Bragg reflections occur; (ii) information on the location of the atoms within the unit cell by analyzing the intensities of the Bragg reflections through the structure factor $S_{\boldsymbol{\tau}}$; (iii) information on the atomic displacements $\langle u \rangle$ by studying the $\boldsymbol{Q}$-dependence of the Debye-Waller factor.

4.2 Examples of structure factors

Copper

Copper crystallizes in a face-centered cubic lattice (lattice parameter a) with four Cu atoms located at

$\boldsymbol{d}_1 = a(0,0,0)$, $\boldsymbol{d}_2 = a(1/2,1/2,0)$, $\boldsymbol{d}_3 = a(1/2,0,1/2)$, $\boldsymbol{d}_4 = a(0,1/2,1/2)$.

With $\boldsymbol{\tau}_{hkl} = 2\pi/a \cdot (h,k,l)$ we derive below some structure factors S_{hkl} from Eq. (4.14):

$$\begin{aligned}
S_{100} &= b_{\mathrm{Cu}} \cdot (1 + e^{\imath\pi} + e^{\imath\pi} + 1) = 0 \\
S_{200} &= b_{\mathrm{Cu}} \cdot (1 + e^{\imath 2\pi} + e^{\imath 2\pi} + 1) = 4\, b_{\mathrm{Cu}} \\
S_{111} &= b_{\mathrm{Cu}} \cdot (1 + e^{\imath 2\pi} + e^{\imath 2\pi} + e^{\imath 2\pi}) = 4\, b_{\mathrm{Cu}} \\
S_{222} &= b_{\mathrm{Cu}} \cdot (1 + e^{\imath 4\pi} + e^{\imath 4\pi} + e^{\imath 4\pi}) = 4\, b_{\mathrm{Cu}}
\end{aligned}$$

The non-vanishing structure factors constitute the selection rules for the occurrence of symmetry-specific Bragg scattering.

Diamond

Diamond crystallizes also in a face-centered cubic lattice, however, there are eight C atoms per unit cell located at $\boldsymbol{d}_1$, $\boldsymbol{d}_2$, $\boldsymbol{d}_3$, $\boldsymbol{d}_4$ as for Cu and additionally at

$$\boldsymbol{d}_5 = a(\frac{1}{4},\frac{1}{4},\frac{1}{4}),\ \boldsymbol{d}_6 = a(\frac{3}{4},\frac{3}{4},\frac{1}{4}),\ \boldsymbol{d}_7 = a(\frac{3}{4},\frac{1}{4},\frac{3}{4}),\ \boldsymbol{d}_8 = a(\frac{1}{4},\frac{3}{4},\frac{3}{4}).$$

Below we derive some structure factors:

$$\begin{aligned}
S_{111} &= b_{\mathrm{C}} \cdot (4 + e^{\imath\frac{3}{2}\pi} + 3e^{\imath\frac{7}{2}\pi}) = 4(1-\imath)b_{\mathrm{C}} \\
|S_{111}|^2 &= 16(1-\imath)(1+\imath)b_{\mathrm{C}}^2 = 32 b_{\mathrm{C}}^2 \\
S_{222} &= b_{\mathrm{C}} \cdot (4 + e^{\imath 3\pi} + 3e^{\imath 7\pi}) = (4-4)b_{\mathrm{C}} = 0
\end{aligned}$$

We realize that the Bragg reflection $(2,2,2)$, which is the $2nd$ order of the Bragg reflection $(1,1,1)$, vanishes. This effect is often used for Si and Ge monochromators to avoid $2nd$ order contamination of the incoming neutron beam (see Chap. 3.2.4).

4.3 Polycrystalline materials

For instrumental details of powder diffractometers we refer to Chap. 3.3.2. A polycrystalline (powder) sample always contains some crystallites which are properly oriented to fulfill the Bragg reflection condition (Eq. (4.8)). The measurement procedure is equivalent to the Debye-Scherrer method known in x-ray diffraction. Powder neutron diffraction experiments provide a rather quick insight into the structure of the investigated system as long as its structural details are not too complex. In particular, diffraction patterns measured for systems with low symmetry and/or large lattice constants often lack in instrumental resolution, i.e., the Bragg peaks partially overlap. Moreover, systems with a large number (typically 50) of atoms per unit cell usually cannot be analyzed unambiguously.

Examples of powder diffraction patterns for cubic face-centered NaH and NaD, obtained at the very early stage of neutron scattering, are shown in Fig. 4.1. For an explanation of the drastically different features of the diffraction patterns upon isotope substitution H→D we refer to the Exercise No. 4.2. Due to instrumental resolution effects the Bragg reflections do not have the shape of δ-functions as predicted in the cross-section formula Eq. (4.16), but rather the shape of Gaussian lines with linewidth Γ. The analysis of the observed intensities I_i is usually based on the following expression:

$$I_i = a + 2b\theta_i + m_{\mathrm{hkl}}|S_{\mathrm{hkl}}|^2 L(\theta_i) \exp\left(-4\ln 2\left(\frac{2\theta_i - 2\theta_{\mathrm{hkl}}}{\Gamma_i}\right)^2\right). \qquad (4.17)$$

a and b are the parameters defining a linear background, m_{hkl} denotes the multiplicity of the Bragg reflection (h,k,l) (e.g., for cubic symmetry we have $m_{\mathrm{h00}} = 6$, $m_{\mathrm{hh0}} = 12$, $m_{\mathrm{hhh}} = 8$, etc.), $L(\theta) = 1/(\sin\theta \sin 2\theta)$ is the so-called Lorentz factor which is tremendously enhanced at small scattering angles θ (see Exercise No. 4.1), and the linewidth Γ varies according to the empirical expression

$$\Gamma(2\theta) = \sqrt{U\tan^2\theta + V\tan\theta + W}, \qquad (4.18)$$

where the parameters U, V, and W depend on both the mosaicity of the monochromator crystal and on the angular divergences of the flight path from source to detector [Caglioti *et al.* (1958)]. The minimum of the linewidth usually occurs for $\theta \approx \theta_m$, where θ_m corresponds to the Bragg angle of the monochromator crystal.

In modern treatments of powder diffraction patterns the Rietveld method is commonly used [Rietveld (1969)], which is based on the minimization of the weighted sum of the squared difference between the observed and calculated intensities I_i^{obs} and I_i^{cal}, respectively:

$$\chi^2 = \sum_i w_i \left(I_i^{\mathrm{obs}} - I_i^{\mathrm{cal}}\right)^2, \qquad (4.19)$$

where $w_i = 1/\sigma^2(I_i^{\mathrm{obs}})$ is the inverse of the variance associated with the ith observation. The analysis of powder diffraction patterns and the corresponding refinement of the structural parameters is most conveniently carried out with the aid of computer programs (such as, e.g., FULLPROF

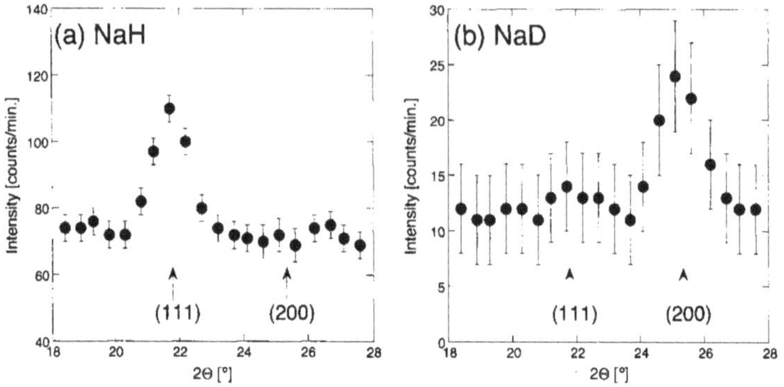

Fig. 4.1 Neutron diffraction patterns obtained for NaH and NaD at room temperature (after [Shull *et al.* (1948)]).

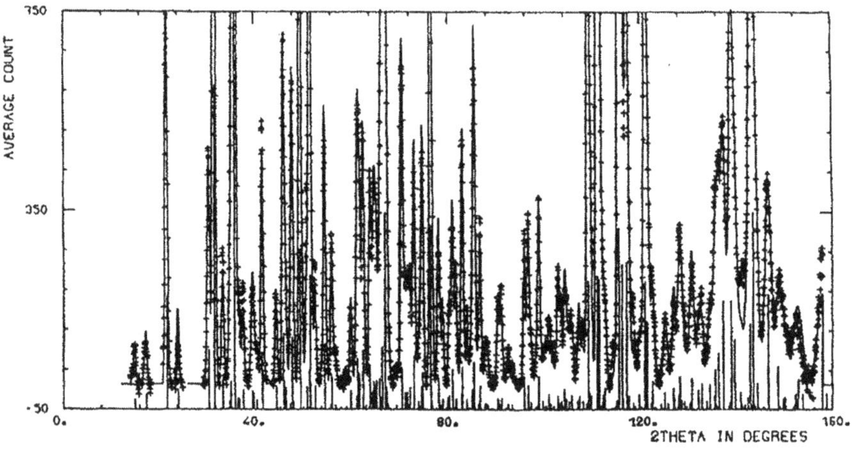

Fig. 4.2 Neutron diffraction pattern of deuterated terephthalic acid observed at $T = 2$ K. The positions of the Bragg reflections are marked on the base line. Reprinted with kind permission from [Meier (1984)].

[Rodriguez-Carvajal (1993)], GSAS [Larson and Von Dreele (2000)]) including internal tables for scattering lengths and - for magnetic structures - magnetic form factors.

Figure 4.2 presents the results of a low-temperature neutron diffraction measurement performed for a deuterated sample of terephthalic acid with chemical composition DOOC-C_6D_4-COOD which crystallizes in a triclinic

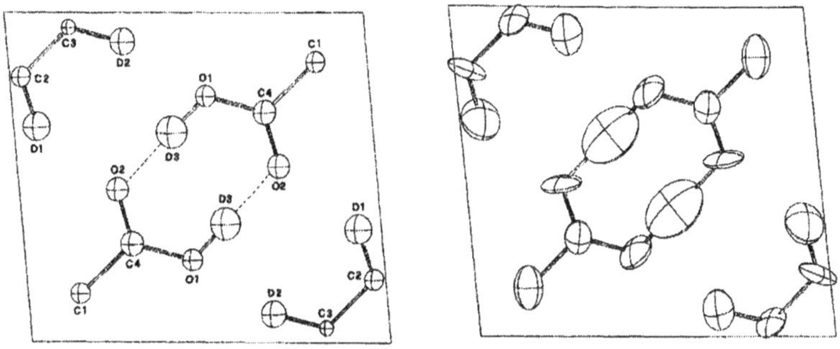

Fig. 4.3 Crystal structure of terephthalic acid at $T = 2$ K (left) and 300 K (right). The thermal motions of the atoms around their equilibrium positions are indicated by ellipsoids.

structure. The low symmetry as well as the large number of atoms in the unit cell required the use of a high-resolution powder diffractometer in order to be able to resolve as many Bragg lines as possible.

The specific interest in this study was the structural characterization of the hydrogen bonds as a function of temperature. Figure 4.3 displays the resulting crystal structures at $T = 2$ K and 300 K refined by the Rietveld method. While at low temperature the D_3 atoms are well localized within the carboxylic group COOD, the room temperature data place the D_3 atom almost in the middle of the O-D-O bond. Moreover, the D_3 atoms experience a huge thermal motion around their equilibrium positions, which supports the observed enhancement of the protonic jump activity within the O$-$D$\cdots$O bond with increasing temperature, see Chap. 15.2.

4.4 Single crystals

In neutron diffraction experiments on single crystals the problem of overlapping Bragg peaks does not exist, since only a single Bragg reflection can be observed for a particular experimental setting $\boldsymbol{Q} \| \boldsymbol{\tau}$. In analogy to x-ray diffraction there are two methods of single-crystal investigations, the rotating-crystal method and the Laue method.

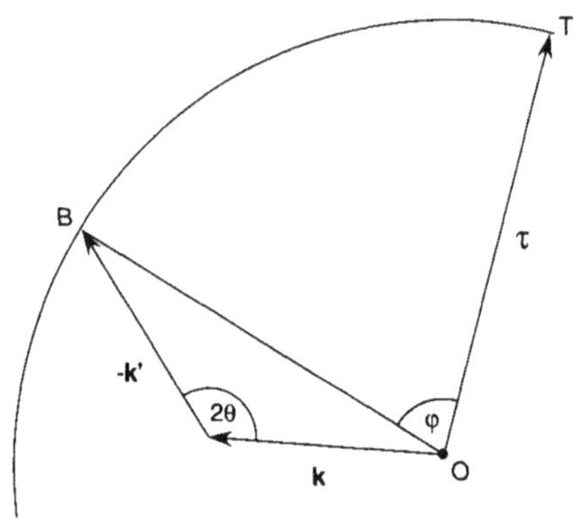

Fig. 4.4 Schematic sketch of the rotating-crystal method in single-crystal diffraction.

4.4.1 *Rotating crystal method*

The single crystal is placed in a monochromatic neutron beam with wavevector $\boldsymbol{k}$. The angle between $\boldsymbol{k}$ and the reciprocal lattice vector $\boldsymbol{\tau}$ is φ as shown in Fig. 4.4. When the crystal is rotated, i.e., the angle φ is varied, the endpoint of $\boldsymbol{\tau}$ moves on a circle with radius τ which is called Ewald's circle. The Bragg condition Eq. (4.8) is fulfilled when the endpoint of $\boldsymbol{\tau}$ passes through B. The diffraction pattern obtained by varying the angle ψ is known as a rocking curve. The aim of a single-crystal diffraction experiment is to record a sufficiently large number of Bragg reflections to allow a reliable refinement of the structure parameters. For a computer-controlled automatic data collection a detailed knowledge of the crystal lattice is needed. Therefore, a single-crystal diffraction experiment starts by a systematic search of reflections by varying χ and φ, with the restriction $\omega = \theta$ (for a definition of the Eulerian angles χ, φ, and ω see Chap. 3.3.3). From the accurate angular positions of typically twenty indexed reflections the lattice constants and the orientation matrix of the single crystal are determined. The measurement of integrated intensities is then performed by means of $\omega/2\theta$-scans, where 2θ is the detector angle.

4.4.2 *Laue method*

The principle of the Laue method is outlined in Chap. 3.3.3. It is a high-performance technique especially well suited to small crystals, rapid chemical crystallography (e.g., variation of bond lengths as a function of temperature), reciprocal space surveys, identification of twinning and incommensu-

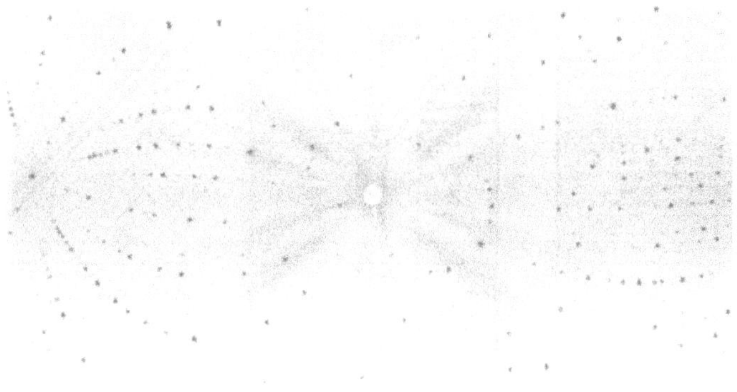

Fig. 4.5 Difference (10 K - 2 K) Laue diffraction pattern of $FeTaO_6$. Reprinted with permission from [Chung *et al.* (2004)]. Copyright 2004, Institute of Physics Publishing Ltd.

rability, studies of structural (and magnetic) phase transitions, etc. In the polychromatic neutron mode, it can provide a ten- to hundred-fold gain in efficiency over a conventional monochromatic diffractometer, however, without reaching the precision of the latter.

The capability to survey large volumes in reciprocal space is illustrated by the study of the magnetic ordering in the mineral $FeTaO_6$ [Chung *et al.* (2004)]. The difference between the Laue diffraction patterns just above and below the Néel temperature $T_N = 8$ K is shown in Fig. 4.5. The faint cross of radial streaks about the central hole (which allows passage of the transmitted neutron beam) reveals with striking clarity the two-dimensional antiferromagnetic ordering of the Fe moments which manifests itself as rods of scattering along c^* in various zones of reflections.

4.5 Extinction and absorption

If a monochromatic neutron beam falls on a single crystal oriented in the reflection condition for Bragg scattering, the incident beam will be diffracted by successive lattice planes and consequently will gradually diminish in intensity as it penetrates into the crystal, as shown schematically in Fig. 4.6a. The whole of the crystal will therefore not experience the same incident intensity, which – particularly for a highly perfect crystal – may even be exhausted in the near-surface region. The attenuation due to diffraction is termed primary extinction. However, most crystals contain imperfections such as dislocations that break the perfect regions up into a mosaic of small

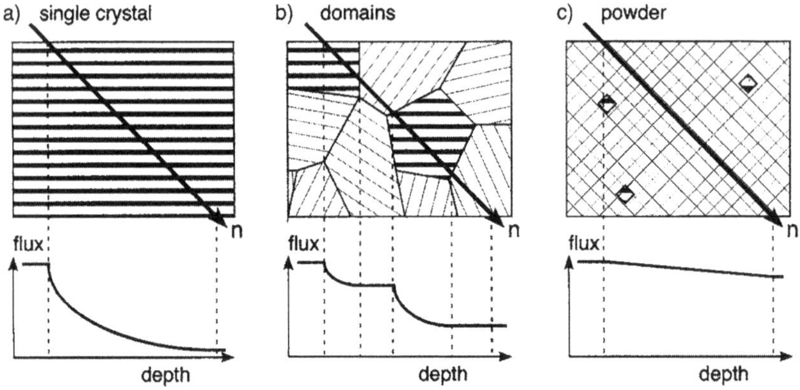

Fig. 4.6 Schematic sketch showing the loss of transmitted neutron flux due to primary (left) and secondary extinction (middle). Only the shaded grains satisfy the Bragg condition. A powder sample (right) is least affected by extinction.

crystallites at small angles to each other, making primary extinction small (see Fig. 4.6b). Indeed, a mosaic structure is important for single-crystal monochromators discussed in Chap. 3.2.4.

Primary extinction is of no concern for polycrystalline samples as visualized in Fig. 4.6c. As a result of the random distribution of the crystallites, the probability for the incident beam to hit crystallites with the same orientation to allow Bragg scattering is small, and accordingly the attenuation is small. For this case – as well as for the mosaic structure of a single crystal – the term secondary extinction is used.

The intensity of the incident neutron beam is also attenuated by absorption. After transmission through a sample with thickness d, the neutron intensity is reduced to

$$I_d = I_0 e^{-\mu d}, \tag{4.20}$$

where I_0 is the initial neutron intensity and μ the linear attenuation factor due to absorption (as well as due to all other cross sections that remove neutrons from the beam). An attenuation length $d_\mu = \mu^{-1}$ can be defined as the distance over which the incident neutron flux declines by a factor $1/e$. In principle, the attenuation length can be calculated from the element-specific cross sections listed in the Appendix B, but it is usually determined experimentally by a simple transmission experiment because of the differences between the theoretical and real densities of a (polycrystalline) sample. The attenuation effect depends on the geometrical shape of the sample (Eq. (4.20) applies for a plate); for cylindrical geometry, the attenuation factors were tabulated in [Weber (1967)].

Both the extinction and absorption parameters have to be introduced as correction factors into the cross-section formulae for the analysis of the observed intensities. While extinction is dominating for diffraction experiments on single crystals (small volume $\approx 1\ \text{mm}^3$), absorption is important for the study of polycrystalline samples (large volume $\approx 1\ \text{cm}^3$). The effects of extinction and absorption are usually included in procedures (such as the Rietveld method) that analyze neutron diffraction data.

4.6 Characterization of residual stress

Every mechanical component is stressed to some extent, even when it is not experiencing a load. This is due to various stages in its manufacture like rolling or machining, heat treatment or welding. These residual stresses remain inside the material, influencing its characteristics, and thus the performance and lifetime of the component. In applications where safety is a priority, it is important to know the distribution and size of stresses in each component.

Neutrons provide a unique tool for determining residual stresses deep inside matter. The principle is quite simple: compressive stresses reduce the spacing between atoms in the material, while tensile stresses stretch them. Since neutron diffraction can measure atomic distances, it can probe the stresses in an object. This neutron strain imaging technique which does not destroy the material, can map stresses with a spatial resolution of typically a cubic millimeter. The high penetration power of neutrons allows measurements to be carried out to several centimeters deep in steel.

In principle, residual stress measurements can be performed with use of a conventional powder diffractometer (see Chap. 3.3.2) by defining the gauge volume through an appropriate choice of the size of the apertures before and after the sample as illustrated in Fig. 4.7. In practice, specialized instruments are used for this purpose [Hutchings *et al.* (2005)]. In the simplest case, a single detector is scanned through the cone of the (hkl) Bragg reflection to give maximum intensity at the Bragg angle θ_{hkl} which defines the lattice spacing d_{hkl} through Bragg's law Eq. (4.8). The average elastic lattice macrostrain ϵ_{hkl} in the volume sampled is then given by the

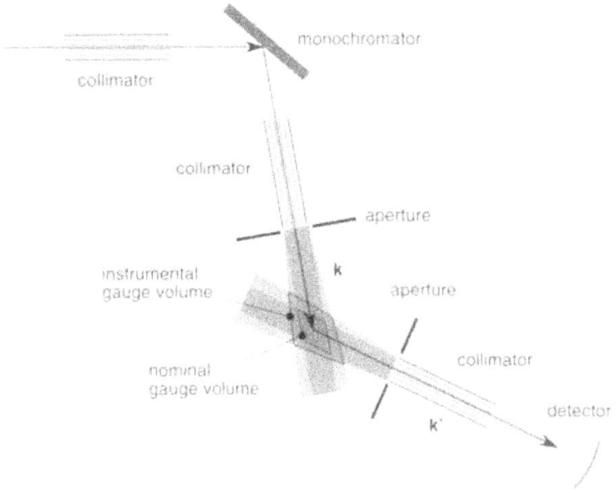

Fig. 4.7 Schematic illustration of the section of the instrumental gauge volume in the scattering plane.

relation

$$\epsilon_{\rm hkl} = \frac{d_{\rm hkl} - d^0_{\rm hkl}}{d^0_{\rm hkl}} = -\left(\cot\theta^0_{\rm hkl}\right)\left(\theta_{\rm hkl} - \theta^0_{\rm hkl}\right), \tag{4.21}$$

where $d^0_{\rm hkl}$ is the lattice spacing of a stress-free sample of the same material composition, and $\theta^0_{\rm hkl}$ the corresponding Bragg scattering angle. The direction in which the lattice strain $\epsilon_{\rm hkl}$ is measured is parallel to the scattering vector $\boldsymbol{Q}$. In order to determine the strain in different directions, the sample must be rotated accurately about the center of the gauge volume.

An example of a residual stress experiment is shown in Fig. 4.8. Of interest was the strain field around the end of an Al bead on a plate TIG weld which was mapped over an area of 40×20 mm^2 with a 1 mm^2 beam size [Owen *et al.* (2003)]. The experimental data are compared with the results of finite element models. Neutron diffraction turns out to be a valuable means of testing and improving the reliability of such models.

4.7 Further reading

- G. E. Bacon, *Neutron diffraction* (Clarendon Press, Oxford, 1975)
- H. Dachs, *Neutron diffraction*, Topics in current physics, Vol. 6 (Springer, Berlin, 1978)

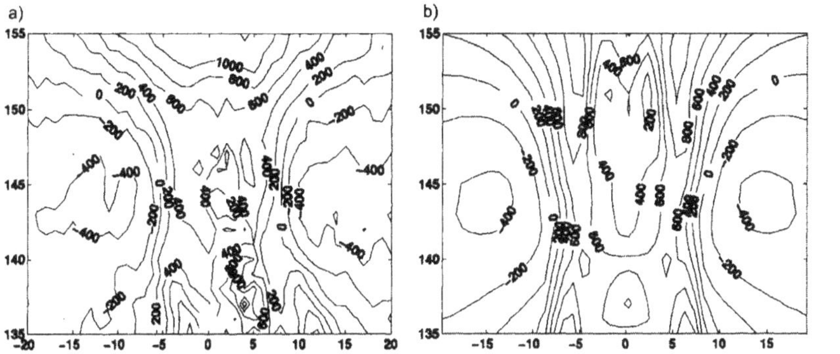

Fig. 4.8 Contour maps of the longitudinal residual strain component in a 40×20 mm^2 region of an Al bead in units of 10^{-6}. (a) Results derived from diffraction data. (b) Comparison with results from finite element calculations. Reprinted from [Owen *et al.* (2003)]. Copyright 2003, with permission from Elsevier.

- G. Heger, in *Introduction to neutron scattering*, ed. by A. Furrer (Proc. 96-01, ISSN 1019-6447, PSI Villigen, 1996), p. 52: *Neutron diffraction*
- G. Heger, in *Neutron scattering*, ed. by A. Furrer (Proc. 93-01, ISSN 1019-6447, PSI Villigen, 1993), p. 97: *Neutron diffraction from single crystals*
- M. T. Hutchings, P. J. Withers, T. M. Holden and T. Lorentzen, *Introduction to the characterization of residual stress by neutron diffraction* (Taylor and Francis, Boca Raton, 2005)
- L. Liang, R. Rinaldi and H. Schober (Eds.), *Neutron applications in earth, energy and environmental sciences* (Springer, Berlin, 2009)
- J. Rodriguez-Carvajal, in *Neutron scattering*, ed. by A. Furrer (Proc. 93-01, ISSN 1019-6447, PSI Villigen, 1993), p. 73: *Neutron diffraction from polycrystalline materials*
- V. F. Sears, *Neutron optics* (Oxford University Press, Oxford, 1989)
- B. T. M. Willis (Ed.), *Thermal neutron diffraction* (Oxford University Press, Oxford, 1970)
- C. C. Wilson, *Single crystal neutron diffraction from molecular materials* (World Scientific, Singapore, 2000)

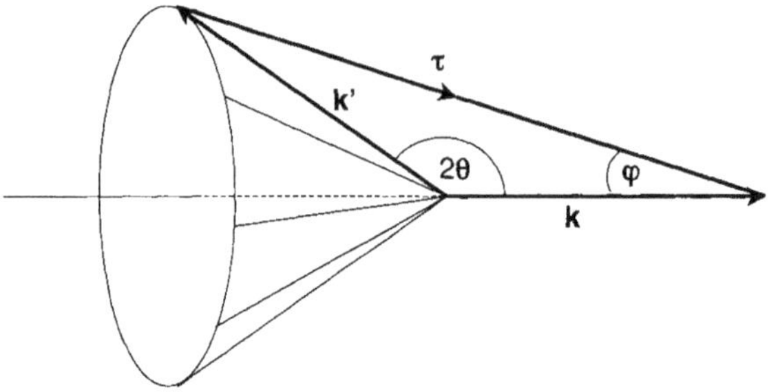

Fig. 4.9 Debye-Scherrer cone for Bragg scattering from polycrystalline materials.

4.8 Exercises

Exercise No. 4.1

Derive the Lorentz factor $L(\theta) = 1/(\sin\theta \sin 2\theta)$ introduced in Eq. (4.17). The origin of the Lorentz factor is twofold. (i) The statistical distribution of the crystallites in a polycrystalline sample has to be considered. (ii) The detector covers only part of the Debye-Scherrer cone which describes the Bragg scattering from polycrystalline materials. As sketched in Fig. 4.9, the wavevector $\boldsymbol{k}'$ of the scattered neutrons lies on a cone, known as Debye-Scherrer cone, where the axis of the cone is along the wavevector $\boldsymbol{k}$ of the incoming neutrons, and θ is the Bragg angle.

Exercise No. 4.2

Verify the diffraction patterns of Fig. 4.1. The atomic positions of NaH and NaD in the fcc unit cell are defined by

$$\boldsymbol{d}_{\mathrm{Na}} = a(0,0,0),\ a(1/2,1/2,0),\ a(1/2,0,1/2),\ a(0,1/2,1/2);$$
$$\boldsymbol{d}_{\mathrm{H/D}} = a(1/2,0,0),\ a(0,1/2,0),\ a(0,0,1/2),\ a(1/2,1/2,1/2).$$

Exercise No. 4.3

Neutron scattering has provided essential information for the understanding of high-temperature superconducting materials (see Chap. 11). The first contribution from neutron scattering concerned the location of the oxygen atoms in the compound $YBa_2Cu_3O_{6+x}$. For $x = 0$ two possible tetragonal structures were initially proposed as shown in Fig. 4.10. x-ray powder data

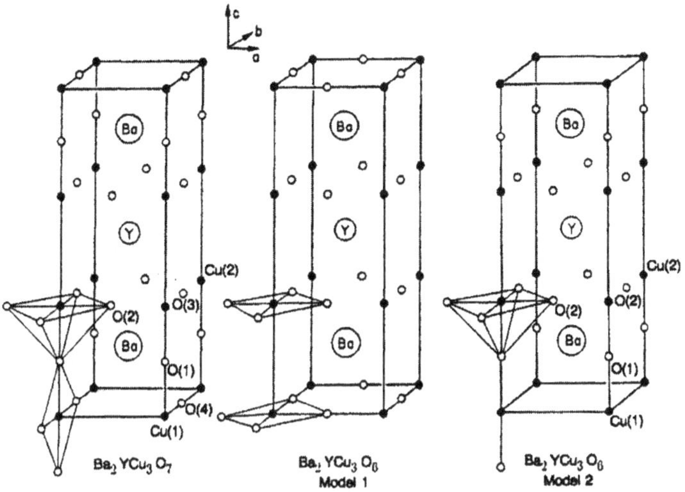

Fig. 4.10 Schematic unit cells for $YBa_2Cu_3O_{6+x}$. For the tetragonal compound with $x = 0$ two possible models were initially proposed from the analysis of x-ray data. Reprinted from [Santoro *et al.* (1987)]. Copyright 1987, with permission from Elsevier.

cannot easily distinguish between the two models due to the weak scattering strength of oxygen, but the model 1 could unambiguously be discarded from the analysis of the data obtained by powder neutron diffraction [Santoro *et al.* (1987)].

(a) Determine the allowed Bragg reflections (h, k, l) for $YBa_2Cu_3O_6$ which crystallizes in the space group P4/mmm with lattice parameters $a = b = 3.854$ Å and $c = 11.818$ Å. The atomic positions in the tetragonal unit cell are as follows:

$$\begin{aligned}
\text{Y}: &\ (1/2, 1/2, 1/2) \\
\text{Ba}: &\ (1/2, 1/2, z)\,, (1/2, 1/2, 1 - z)\,, z = 0.1944 \\
\text{Cu(1)}: &\ (0, 0, 0) \\
\text{Cu(2)}: &\ (0, 0, z)\,, (0, 0, 1 - z)\,, z = 0.3602 \\
\text{O(1)}: &\ (0, 0, z)\,, (0, 0, 1 - z)\,, z = 0.1511 \\
\text{O(2)}: &\ (1/2, 0, z)\,, (0, 1/2, z), (1/2, 0, 1 - z), (0, 1/2, 1 - z), z = 0.3791 \\
\text{O(4)}: &\ (1/2, 0, 0), (0, 1/2, 0)
\end{aligned}$$

(b) Calculate the intensities of the Bragg reflections $(0, 0, 1)$, $(0, 0, 2)$, $(0, 0, 3)$, and $(1, 0, 0)$ for both tetragonal models based on Eq. (4.17)

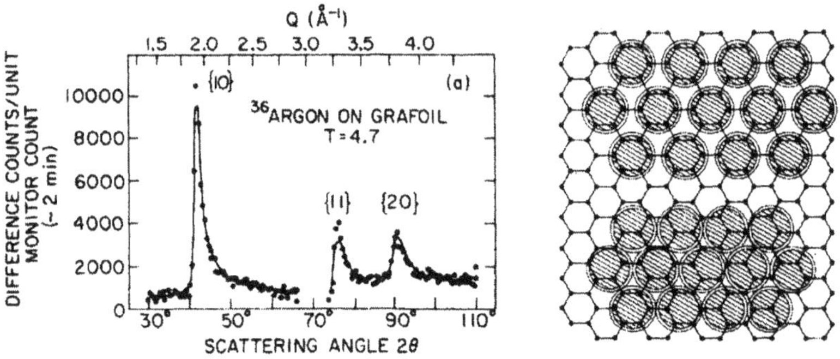

Fig. 4.11 (a) Diffraction pattern from a two-dimensional Ar monolayer (absorbed on graphite) showing the Bragg reflections (1,0), (1,1), and (2,0). (b) Schematic representation of a commensurate (top) and incommensurate (bottom) Ar monolayer phase. Reprinted with permission from [Taub *et al.* (1977)]. Copyright 1977 by the American Physical Society.

with $\lambda = 1.7$ Å. A comparison of the results demonstrates the power of neutron powder diffraction to determine the correct positions of the oxygen ions. The scattering lengths are listed in the Appendix B.

(c) By increasing the oxygen content x, superconductivity with $T_c \approx 60$ K sets in for $x \approx 0.4$. At $x \approx 0.7$ the critical temperature is stepwise enhanced to $T_c \approx 90$ K. The position of the oxygen ion O(1) was found to significantly move along the z-axis according to the two-plateau structure of T_c [Cava *et al.* (1990)]. Work out which Bragg reflections are best suited to provide information on the position of the oxygen ion O(1).

Exercise No. 4.4

Diffraction of neutrons from argon monolayers absorbed on graphite basal planes indicated that an ordered, two-dimensional triangular argon lattice is formed at low temperatures [Taub *et al.* (1977)], see Fig. 4.11a. There are two possible configurations for the argon monolayers, either commensurate with the graphite lattice or incommensurate corresponding to the closest packing as shown in Fig. 4.11b.

(a) Explain the asymmetric sawtooth profiles of the Bragg peaks in Fig. 4.11a.

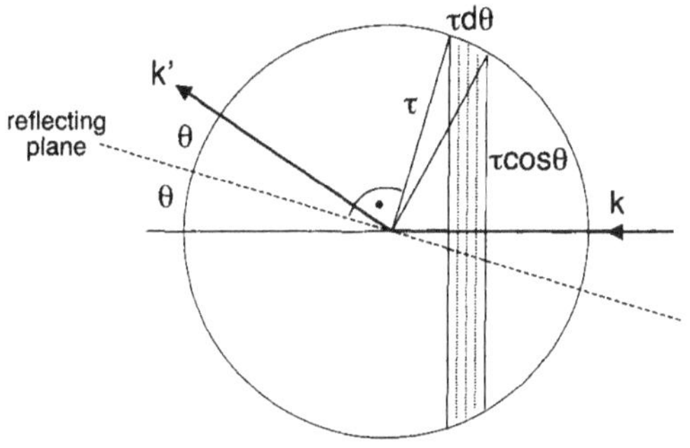

Fig. 4.12 Sketch showing the fraction of crystallites satisfying the Bragg condition.

(b) Determine from the observed Bragg reflections displayed in Fig. 4.11a whether the Ar monolayer is commensurate or incommensurate with the graphite lattice. The C atoms have a nearest-neighbor distance $a_C = 2.46$ Å in the hexagonal plane of graphite.

4.9 Solutions

Exercise No. 4.1

We consider first the statistical distribution of the crystallites in a polycrystalline sample. The fraction of microcrystals oriented to fulfill Bragg's law (Eq. (4.8)) can be obtained by considering Fig. 4.12. All crystallites with reciprocal lattice vectors lying in the dotted surface area of a sphere with radius τ contribute to the scattering. The active surface area amounts to $2\pi\tau^2 \cos\theta \, d\theta$, thus the fraction of properly oriented microcrystals is $2\pi\tau^2 \cos\theta d\theta / 4\pi\tau^2 = \cos\theta d\theta/2$. The total scattering for the Debye-Scherrer cone is then given by

$$\sigma_{\text{cone}} \propto \int_0^{\pi/2} \delta\left(k' - k\right) \cos\theta d\theta, \tag{4.22}$$

where $\delta\left(k' - k\right)$ confines the integration to elastic scattering. Using the geometry sketched in Fig. 4.9 we find

$$k'^2 - k^2 = \tau^2 - 2\tau k \cos\phi = \tau^2 - 2\tau k \sin\theta = (k' + k)(k' - k), \tag{4.23}$$

where we used the relation $\theta = \pi/2 - \phi$. Setting $k' \approx k$ yields

$$k' - k = \frac{1}{2k}\left(\tau^2 - 2\tau k \sin\theta\right). \tag{4.24}$$

Table 4.1 Squared structure factors of NaH and NaD.

$\lvert S_{\rm hkl}\rvert^2$	NaH	NaD
$\lvert S_{111}\rvert^2$	$8.69 \cdot 10^{-24}$ cm^2	$1.48 \cdot 10^{-24}$ cm^2
$\lvert S_{200}\rvert^2$	$0.002 \cdot 10^{-24}$ cm^2	$17.11 \cdot 10^{-24}$ cm^2

Combining Eqs (4.22) and (4.24) yields

$$\sigma_{\rm cone} \propto \int_0^{\pi/2} \delta\left(\tau^2 - 2\tau k \sin\theta\right) \cos\theta {\rm d}\theta. \tag{4.25}$$

We solve the integral in Eq. (4.25) by the substitution $x = 2\tau k \sin\theta$:

$$\sigma_{\rm cone} \propto \int \delta\left(\tau^2 - x\right) \frac{1}{2\tau k} {\rm d}x = \frac{1}{2\tau k}. \tag{4.26}$$

Setting $\tau = 2k \sin\theta$ from Bragg's law we find

$$\sigma_{\rm cone} \propto \frac{1}{\sin\theta}. \tag{4.27}$$

If the neutron detector with diameter d is at a distance r from the sample, it intercepts a fraction $q = d/2\pi r \sin(2\theta)$ of the neutrons in the cone. Multiplying $\sigma_{\rm cone}$ of Eq. (4.27) with the θ-dependent term of q yields the final result for the Lorentz factor:

$$L(\theta) = \frac{1}{\sin(\theta)\sin(2\theta)}. \tag{4.28}$$

Exercise No. 4.2

We calculate the structure factors for the $(1,1,1)$ and $(2,0,0)$ Bragg reflections according to Eq. (4.14) with $\tau_{\rm hkl} = (h,k,l)$ as follows:

$$\begin{aligned} S_{111} &= b_{\rm Na} e^{i\tau_{111}\cdot \boldsymbol{d}_{\rm Na}} + b_{\rm H/D} e^{i\tau_{111}\cdot \boldsymbol{d}_{\rm H/D}} = \\ &= b_{\rm Na}(1 + 3e^{2\pi i}) + 4b_{\rm H/D} e^{\pi i} = 4(b_{\rm Na} - b_{\rm H/D}), \\ S_{200} &= 2b_{\rm Na}(1 + e^{2\pi i}) + 2b_{\rm H/D}(1 + e^{2\pi i}) = 4(b_{\rm Na} + b_{\rm H/D}). \end{aligned}$$

Inserting the scattering lengths from the Appendix B yields the results listed in Table 4.1. Due to the same sizes but different signs of $b_{\rm Na}$ and $b_{\rm H}$ the intensity of the (111) reflection of NaH almost vanishes. Fig. 4.1 also demonstrates the importance of the isotope substitution H→D to reduce the background which is mainly due to incoherent scattering ($\sigma_i(\rm H) \gg \sigma_i(\rm D)$).

Table 4.2 Calculated intensities for models 1 and 2 based on Eq. (4.17).

(h,k,l)	$\theta[°]$	$L(\theta)$	m	model 1 $\lvert S_{\rm hkl}\rvert^2$	I [10^3 fm^2]	model 2 $\lvert S_{\rm hkl}\rvert^2$	I[10^3 fm^2]
$(0,0,1)$	4.12	96.9	2	135	26.2	272	52.7
$(0,0,2)$	8.27	24.5	2	312	15.3	5.32	0.3
$(0,0,3)$	12.46	11.0	2	985	21.7	75.1	1.7
$(1,0,0),(0,1,0)$	12.74	10.5	2	27.7	0.6	285	6.0

Exercise No. 4.3

(a) According to the *International Tables for Crystallography* (Kluwer, Dordrecht, 2002, p. 431), all Bragg reflections (h,k,l) are allowed.

(b) The results listed in Table 4.2 exhibit drastically different intensities for models 1 and 2, thus neutron powder diffraction allows an unambiguous determination of the oxygen positions. As shown in Fig. 4.13, the calculations of model 2 are nicely confirmed by diffraction data taken for the isostructural compound $TmBa_2Cu_3O_{6.11}$ (for Tm the scattering length is $b_c = 7.07$ fm, which is very close to $b_c = 7.75$ fm for Y).

(c) The variation of $|S_{\rm hkl}|^2$ as a function of the parameter z of the O(1) position is shown in Fig. 4.14. The reflections $(1,0,0)$ and $(0,1,0)$ are unaffected by z, the reflection $(0,0,2)$ has an extremely weak intensity, and the reflection $(0,0,3)$ has a parabolic shape around the z values of interest, thus only the reflection $(0,0,1)$ is suited for a precise determination of the parameter z.

Exercise No. 4.4

(a) The reciprocal lattice of a two-dimensional crystal consists of an ordered array of rods aligned normal to the scattering plane. In diffraction experiments there is for each Miller index pair (h,k) a minimum value of the scattering angle $2\theta = 2\theta_B$ where θ_B corresponds to the Bragg angle defined in Eq. (4.8). Depending on the size of the array, there will be a continuous distribution of diffracted intensity for scattering angles greater than $2\theta_B$. This produces a characteristic "sawtooth" line shape with a sharply rising leading edge on the low-angle side, followed by a trailing edge extending to larger scattering angles. The maximum of the diffracted intensity $I_{\rm hk}$ occurs at the scattering angle $2\theta_B$. For

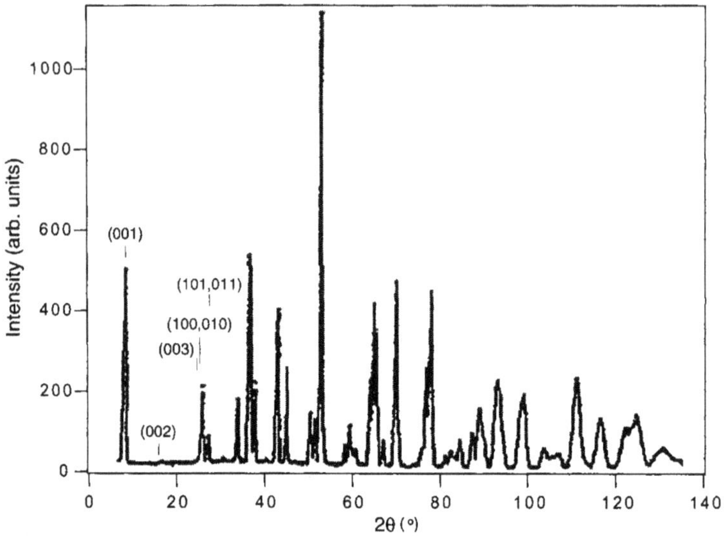

Fig. 4.13 Neutron diffraction pattern measured for $TmBa_2Cu_3O_{6.11}$ at 10 K with $\lambda = 1.7$ Å (after [Guillaume *et al.* (1994)]. The $(0,0,2)$ reflection disappears in the background due to its weak intensity predicted in Table 4.2.

a detailed calculation of the line shape we refer to the basic article by Warren [Warren (1941)] who derived the following result:

$$I_{\mathrm{hk}} = \frac{m_{\mathrm{hk}}|S_{\mathrm{hk}}|^2 L(\theta)}{\sin\theta\sqrt{\sin^2\theta - \sin^2\theta_B}}, \tag{4.29}$$

where the same notation is used as in Eq. (4.17).

(b) Following the expressions given in Appendix E, the reciprocal lattice vectors of a two-dimensional hexagonal lattice are given by

$$\tau_1 = \frac{2\pi a}{f_0}(\sqrt{3}/2, 1/2), \qquad \tau_2 = \frac{2\pi a}{f_0}(0,1), \qquad f_0 = \frac{\sqrt{3}}{2}a^2, \tag{4.30}$$

where a is the nearest-neighbor distance. By identifying the peaks observed at $Q = 1.9$, 3.3 and 3.8 Å^{-1} as the $(1,0)$, $(1,1)$ and $(2,0)$ Bragg reflections, we find from Eq. (4.8) $a_{\mathrm{Ar}} = 3.82$ Å. This is significantly smaller than the distance $\sqrt{3}a_C = 4.26$ Å for the commensurate phase sketched in Fig. 4.11b, thus the Ar monolayer is clearly incommensurate with that of the underlying graphite basal plane. $a_{\mathrm{Ar}} = 3.82$ Å corresponds to the minimum of the Lennard-Jones potential $\Phi(r) = -Ar^{-6} + Br^{-12}$ known for solid Ar. We conclude that the

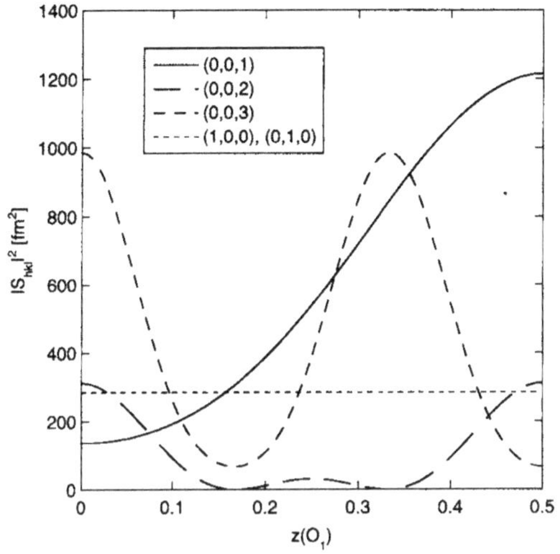

Fig. 4.14 Variation of $|S_{\rm hkl}|^2$ calculated for $YBa_2Cu_3O_6$ vs the parameter z of the O(1) position.

structure of the Ar monolayer is primarily determined by the couplings between the Ar atoms, and interactions with the substrate appear to play a minor role.

Chapter 5

Lattice Dynamics

5.1 Cross section for one-phonon scattering

We start from Eq. (2.17) and describe the position operator $\hat{\boldsymbol{R}}_j(t)$ by

$$\hat{\boldsymbol{R}}_j(t) = \boldsymbol{l}_j + \hat{\boldsymbol{u}}_j(t), \tag{5.1}$$

where $\boldsymbol{l}_j$ denotes the equilibrium position and $\boldsymbol{u}_j(t)$ a small time-dependent displacement from $\boldsymbol{l}_j$. Inserting Eq. (5.1) into Eq. (2.17) yields

$$\frac{\mathrm{d}^2\sigma}{\mathrm{d}\Omega\mathrm{d}\omega} = \frac{k'}{k}\frac{1}{2\pi\hbar}\sum_{j,j'} b_j b_{j'} e^{\imath\boldsymbol{Q}\cdot(\boldsymbol{l}_j-\boldsymbol{l}_{j'})}\int_{-\infty}^{\infty}\langle e^{-\imath\boldsymbol{Q}\cdot\hat{\boldsymbol{u}}_{j'}(0)}e^{\imath\boldsymbol{Q}\cdot\hat{\boldsymbol{u}}_j(t)}\rangle e^{-\imath\omega t}\mathrm{d}t. \tag{5.2}$$

In the harmonic approximation the displacement operators $\hat{\boldsymbol{u}}_j(t)$ can be expressed in terms of the normal modes:

$$\hat{\boldsymbol{u}}_j(t) = \sqrt{\frac{\hbar}{2MN}}\sum_{s,\boldsymbol{q}}\frac{\boldsymbol{e}_s(\boldsymbol{q})}{\sqrt{\omega_s(\boldsymbol{q})}}\left(\hat{a}_s(\boldsymbol{q})e^{\imath[\boldsymbol{q}\cdot\boldsymbol{l}_j-\omega_s(\boldsymbol{q})t]} + \hat{a}_s^+(\boldsymbol{q})e^{-\imath[\boldsymbol{q}\cdot\boldsymbol{l}_j-\omega_s(\boldsymbol{q})t]}\right), \tag{5.3}$$

where $\boldsymbol{q}$ is the wavevector of the normal mode, s its polarization index ($s = 1, 2, 3$ for a Bravais lattice), $\omega_s(\boldsymbol{q})$ and $\boldsymbol{e}_s(\boldsymbol{q})$ its eigenvalue and eigenvector, respectively, and M the atomic mass. The sum extends over the N values of $\boldsymbol{q}$ in the first Brillouin zone. $\hat{a}_s(\boldsymbol{q})$ and $\hat{a}_s^+(\boldsymbol{q})$ are, respectively, the phonon annihilation and creation operators, which obey the commutation relation

$$\left[\hat{a}_s(\boldsymbol{q}), \hat{a}_{s'}^+(\boldsymbol{q}')\right] = \delta_{ss'}\delta_{\boldsymbol{q}\boldsymbol{q}'}. \tag{5.4}$$

We now abbreviate the exponents in Eq. (5.2) by the operators $\hat{A}$ and $\hat{B}$:

$$\hat{A} = -\imath\boldsymbol{Q}\cdot\hat{\boldsymbol{u}}_{j'}(0) = -\imath\sum_{s,\boldsymbol{q}}\left(\alpha_s(\boldsymbol{q})\hat{a}_s(\boldsymbol{q}) + \alpha_s^*(\boldsymbol{q})\hat{a}_s^+(\boldsymbol{q})\right),$$

$$\hat{B} = \imath\boldsymbol{Q}\cdot\hat{\boldsymbol{u}}_j(t) = \imath\sum_{s,\boldsymbol{q}}\left(\beta_s(\boldsymbol{q})\hat{a}_s(\boldsymbol{q}) + \beta_s^*(\boldsymbol{q})\hat{a}_s^+(\boldsymbol{q})\right), \tag{5.5}$$

with

$$\alpha_s(\boldsymbol{q}) = \sqrt{\frac{\hbar}{2MN}} \sum_{s,\boldsymbol{q}} \frac{\boldsymbol{Q}\cdot\boldsymbol{e}_s(\boldsymbol{q})}{\sqrt{\omega_s(\boldsymbol{q})}} e^{\imath \boldsymbol{q}\cdot\boldsymbol{l}_{j'}},$$

$$\beta_s(\boldsymbol{q}) = \sqrt{\frac{\hbar}{2MN}} \sum_{s,\boldsymbol{q}} \frac{\boldsymbol{Q}\cdot\boldsymbol{e}_s(\boldsymbol{q})}{\sqrt{\omega_s(\boldsymbol{q})}} e^{\imath[\boldsymbol{q}\cdot\boldsymbol{l}_j - \omega_s(\boldsymbol{q})t]}. \tag{5.6}$$

The operator term in Eq. (5.2) takes thus the form $\langle e^{\hat{A}} e^{\hat{B}}\rangle$. Since $\hat{A}$ and $\hat{B}$ are Hermite operators, the commutator $[\hat{A},\hat{B}]$ is a complex number. Then, by using the expressions Eqs (5.4) - (5.6), we find:

$$\langle e^{\hat{A}} e^{\hat{B}}\rangle = \langle e^{\hat{A}+\hat{B}}\rangle e^{\frac{1}{2}[\hat{A},\hat{B}]}. \tag{5.7}$$

Now we proceed to evaluate the expectation value $\langle e^{\hat{A}+\hat{B}}\rangle$ in Eq. (5.7). In the harmonic approximation the probability for an atom to be displaced by a small distance u from its equilibrium position is given by a Gaussian distribution:

$$p(u) = \frac{1}{\sqrt{2\pi}\sigma} e^{-\frac{u^2}{2\sigma^2}}, \quad \text{with} \quad \sigma^2 = \frac{\hbar}{2M\omega}\coth\left(\frac{\hbar\omega}{2k_BT}\right). \tag{5.8}$$

From Eq. (5.8) we derive the following expectation values:

$$\langle u^2\rangle = \int u^2 p(u)\mathrm{d}u = \sigma^2, \quad \langle e^u\rangle = \int e^u p(u)\mathrm{d}u = e^{\frac{1}{2}\sigma^2}, \tag{5.9}$$

and thereby

$$\langle e^u\rangle = e^{\frac{1}{2}\langle u\rangle^2}. \tag{5.10}$$

Combining Eqs (5.7) and (5.10) yields:

$$\begin{aligned}\langle e^{\hat{A}} e^{\hat{B}}\rangle &= e^{\frac{1}{2}\langle(\hat{A}+\hat{B})^2\rangle} e^{\frac{1}{2}[\hat{A},\hat{B}]} = e^{\frac{1}{2}\langle \hat{A}^2+\hat{A}\hat{B}+\hat{B}\hat{A}+\hat{B}^2+\hat{A}\hat{B}-\hat{B}\hat{A}\rangle} \\ &= e^{\frac{1}{2}\langle \hat{A}^2+\hat{B}^2+2\hat{A}\hat{B}\rangle}.\end{aligned} \tag{5.11}$$

It can be shown that $e^{\langle \hat{A}^2\rangle}$ corresponds to the Debye-Waller factor:

$$e^{\langle \hat{A}^2\rangle} = e^{-2W(\boldsymbol{Q})}. \tag{5.12}$$

For a Bravais lattice we have $\langle \hat{A}^2\rangle = \langle \hat{B}^2\rangle$, thus Eq. (5.11) transforms to

$$\langle e^{\hat{A}} e^{\hat{B}}\rangle = e^{\langle \hat{A}^2\rangle} e^{\langle \hat{A}\hat{B}\rangle} = e^{-2W(\boldsymbol{Q})} e^{\langle \hat{A}\hat{B}\rangle}. \tag{5.13}$$

Inserting Eq. (5.13) into Eq. (5.2) yields

$$\frac{\mathrm{d}^2\sigma}{\mathrm{d}\Omega\mathrm{d}\omega} = \frac{k'}{k}\frac{1}{2\pi\hbar} e^{-2W(\boldsymbol{Q})} \sum_{j,j'} b_j b_{j'} e^{\imath\boldsymbol{Q}\cdot(\boldsymbol{l}_j - \boldsymbol{l}_{j'})} \int_{-\infty}^{\infty} e^{\langle \hat{A}\hat{B}\rangle} e^{-\imath\omega t}\mathrm{d}t. \tag{5.14}$$

In the above procedure we assumed the moduli of the displacements $\boldsymbol{u}_j(t)$ to be small compared to the lattice parameters, thus $\boldsymbol{Q} \cdot \boldsymbol{u}_j(t) \ll 1$, which means that we can expand the operator term in Eq. (5.14) into a Taylor series:

$$e^{\langle \hat{A}\hat{B}\rangle} = 1 + \langle \hat{A}\hat{B}\rangle + \frac{1}{2}\langle \hat{A}\hat{B}\rangle^2 + \cdots + \frac{1}{n!}\langle \hat{A}\hat{B}\rangle^n + \ldots \tag{5.15}$$

The first term of the Taylor series has to be associated with $\langle \hat{A}\hat{B}\rangle = 0$ and therefore $\langle \boldsymbol{u}_j(t)\rangle = 0$, which corresponds to elastic scattering. We discussed this term already in Chap. 4, however, in the classical procedure the Debye-Waller factor was lost [see Eqs (4.7) and (4.13)]. We now investigate the linear term of Eq. (5.15). We make use of the expressions Eq. (5.5) to calculate the matrix element $\langle \lambda_n | \hat{A}\hat{B} | \lambda_n\rangle$, where $|\lambda_n\rangle$ is the eigenfunction of the system. This procedure yields products of the creation and annihilation operators $\hat{a}_s^+(\boldsymbol{q})$ and $\hat{a}_s(\boldsymbol{q})$, respectively. According to the commutation relation (5.4) only terms involving the operator products $\hat{a}_s(\boldsymbol{q})\hat{a}_s^+(\boldsymbol{q})$ and $\hat{a}_s^+(\boldsymbol{q})\hat{a}_s(\boldsymbol{q})$ yield non-zero contributions to the matrix element, thus

$$\langle \lambda_n | \hat{A}\hat{B} | \lambda_n\rangle = \langle \lambda_n | \sum_{s,\boldsymbol{q}} [\alpha_s(\boldsymbol{q})\beta_s^*(\boldsymbol{q})\hat{a}_s(\boldsymbol{q})\hat{a}_s^+(\boldsymbol{q}) + \alpha_s^*(\boldsymbol{q})\beta_s(\boldsymbol{q})\hat{a}_s^+(\boldsymbol{q})\hat{a}_s(\boldsymbol{q})] | \lambda_n\rangle. \tag{5.16}$$

We explicitly calculate this matrix element for the harmonic oscillator whose eigenvalues are arranged on a ladder as

$$E_n = (n + \frac{1}{2})\hbar\omega, \quad n = 0, 1, 2, ... \tag{5.17}$$

The operators $\hat{a}_s^+(\boldsymbol{q})$ and $\hat{a}_s(\boldsymbol{q})$ are known as ladder operators which convert the eigenfunction $|\lambda_n\rangle$ into the functions $|\lambda_{n+1}\rangle$ and $|\lambda_{n-1}\rangle$ one rung up and down the ladder, respectively:

$$\begin{aligned} \hat{a}_s^+(\boldsymbol{q})|\lambda_n\rangle &= \sqrt{n+1}|\lambda_{n+1}\rangle, \\ \hat{a}_s(\boldsymbol{q})|\lambda_n\rangle &= \sqrt{n}|\lambda_{n-1}\rangle, \end{aligned} \tag{5.18}$$

thus

$$\begin{aligned} \langle \lambda_n | \hat{a}_s(\boldsymbol{q})\hat{a}_s^+(\boldsymbol{q}) | \lambda_n\rangle &= n_s(\boldsymbol{q}) + 1, \\ \langle \lambda_n | \hat{a}_s^+(\boldsymbol{q})\hat{a}_s(\boldsymbol{q}) | \lambda_n\rangle &= n_s(\boldsymbol{q}), \end{aligned} \tag{5.19}$$

where

$$n_s(\boldsymbol{q}) = \left(\exp\left(\frac{\hbar\omega_s(\boldsymbol{q})}{k_B T}\right) - 1\right)^{-1} \tag{5.20}$$

denotes the Bose-Einstein occupation number. Inserting Eqs (5.6), (5.19) and (5.20) into Eq. (5.16), using the substitution $\boldsymbol{l} = \boldsymbol{l}_j - \boldsymbol{l}_{j'}$ (similar to the procedure in Chap. 4, Eq. (4.6)) and considering coherent scattering ($j \neq j'$) only yields

$$\frac{\mathrm{d}^2\sigma}{\mathrm{d}\Omega\mathrm{d}\omega} = \frac{1}{4\pi M} \cdot \frac{k'}{k} \langle b \rangle^2 e^{-2W(\boldsymbol{Q})} \sum_{s,\boldsymbol{q}} \frac{(\boldsymbol{Q} \cdot \boldsymbol{e}_s(\boldsymbol{q}))^2}{\omega_s(\boldsymbol{q})}$$
$$\times \left[(n_s(\boldsymbol{q}) + 1) \sum_{\boldsymbol{l}} e^{\imath(\boldsymbol{Q}-\boldsymbol{q})\cdot \boldsymbol{l}} \int_{-\infty}^{\infty} \mathrm{d}t e^{\imath(\omega_s(\boldsymbol{q})-\omega)t} \right.$$
$$\left. + n_s(\boldsymbol{q}) \sum_{\boldsymbol{l}} e^{\imath(\boldsymbol{Q}+\boldsymbol{q})\cdot \boldsymbol{l}} \int_{-\infty}^{\infty} \mathrm{d}t e^{-\imath(\omega_s(\boldsymbol{q})+\omega)t} \right] . \quad (5.21)$$

By applying Eqs (A.6) and (A.10) we transform the integrals and lattice sums into δ-functions and arrive at the coherent one-phonon neutron cross-section for a Bravais lattice:

$$\frac{\mathrm{d}^2\sigma}{\mathrm{d}\Omega\mathrm{d}\omega} = \frac{4\pi^3}{v_0 M} \cdot \frac{k'}{k} \langle b \rangle^2 e^{-2W(\boldsymbol{Q})} \sum_{s,\boldsymbol{q}} \frac{(\boldsymbol{Q} \cdot \boldsymbol{e}_s(\boldsymbol{q}))^2}{\omega_s(\boldsymbol{q})}$$
$$\times \left[(n_s(\boldsymbol{q}) + 1)\delta\left(\omega - \omega_s(\boldsymbol{q})\right) \sum_{\boldsymbol{\tau}} \delta(\boldsymbol{Q} - \boldsymbol{q} - \boldsymbol{\tau}) \right.$$
$$\left. + n_s(\boldsymbol{q})\delta\left(\omega + \omega_s(\boldsymbol{q})\right) \sum_{\boldsymbol{\tau}} \delta(\boldsymbol{Q} + \boldsymbol{q} - \boldsymbol{\tau}) \right] . \quad (5.22)$$

Equation (5.22) can be generalized for the case of non-Bravais lattices by replacing $\boldsymbol{l}_j$ by $\boldsymbol{l}_j + \boldsymbol{d}_\alpha$ in Eq. (5.14), similar to the procedure adopted in Chap. 4, Eq. (4.9). The general formula for coherent one-phonon scattering reads:

$$\frac{\mathrm{d}^2\sigma}{\mathrm{d}\Omega\mathrm{d}\omega} = \frac{4\pi^3}{v_0} \cdot \frac{k'}{k} \sum_{s,\boldsymbol{q}} \frac{1}{\omega_s(\boldsymbol{q})} \left| \sum_{\boldsymbol{d}} \frac{\langle b_d \rangle}{\sqrt{M_d}} e^{-W_d(\boldsymbol{Q})} e^{\imath \boldsymbol{Q}\cdot\boldsymbol{d}} \left(\boldsymbol{Q} \cdot \boldsymbol{e}_{\boldsymbol{d},s}(\boldsymbol{q})\right) \right|^2$$
$$\times \left[(n_s(\boldsymbol{q}) + 1)\delta\left(\omega - \omega_s(\boldsymbol{q})\right) \sum_{\boldsymbol{\tau}} \delta(\boldsymbol{Q} - \boldsymbol{q} - \boldsymbol{\tau}) \right.$$
$$\left. + n_s(\boldsymbol{q})\delta\left(\omega + \omega_s(\boldsymbol{q})\right) \sum_{\boldsymbol{\tau}} \delta(\boldsymbol{Q} + \boldsymbol{q} - \boldsymbol{\tau}) \right]$$

(5.23)

The cross section contains two terms within $[\,\cdot\,]$ which correspond to phonon emission (first term) and phonon absorption (second term) induced by the

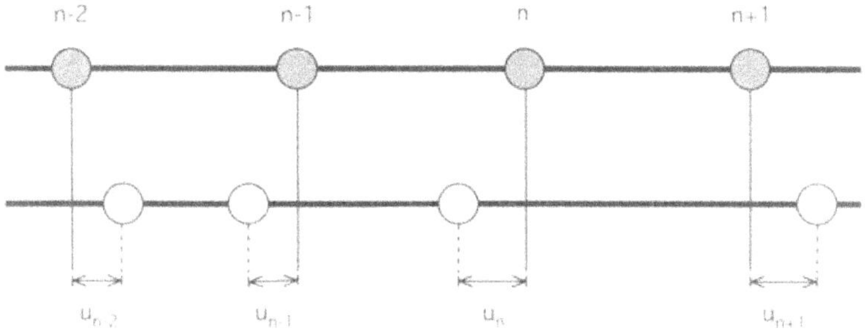

Fig. 5.1 Schematic representation of a linear chain with identical atoms. The equilibrium and displaced atomic positions are indicated by full and open circles, respectively.

neutron in the scattering process which are also known as Stokes and anti-Stokes processes, and from the neutron point of view as energy loss and energy gain processes, respectively (see also Chap. 2.5). From the δ-functions of Eq. (5.23) we can immediately derive the rules for energy and momentum conservation:

$$\pm \hbar\omega_s(\boldsymbol{q}) = \hbar\omega = \frac{\hbar^2}{2m}\left(|\boldsymbol{k}|^2 - |\boldsymbol{k}'|^2\right), \tag{5.24}$$

$$\pm \boldsymbol{q} = \boldsymbol{Q} - \boldsymbol{\tau} = \boldsymbol{k} - \boldsymbol{k}' - \boldsymbol{\tau}. \tag{5.25}$$

5.2 Phonon dispersion relations and phonon polarization vectors

5.2.1 *Linear chain with identical atoms*

Let us consider a linear chain of identical atoms. Each atom exhibits a time-dependent dislocation u_n from its equilibrium position as visualized in Fig. 5.1. In the nearest-neighbor approximation the force acting on the nth atom is given by

$$F_n = \beta(u_{n+1} - u_n) - \beta(u_n - u_{n-1}), \tag{5.26}$$

where β denotes the force constant. We use Eq. (5.26) to write down the equation of motion for the nth atom:

$$M\ddot{u}_n = \beta(u_{n+1} + u_{n-1} - 2u_n) \tag{5.27}$$

with M being the mass of the atom. The solution of the differential equation Eq. (5.27) is

$$u_n = \xi e^{i(\omega t + qna)}. \tag{5.28}$$

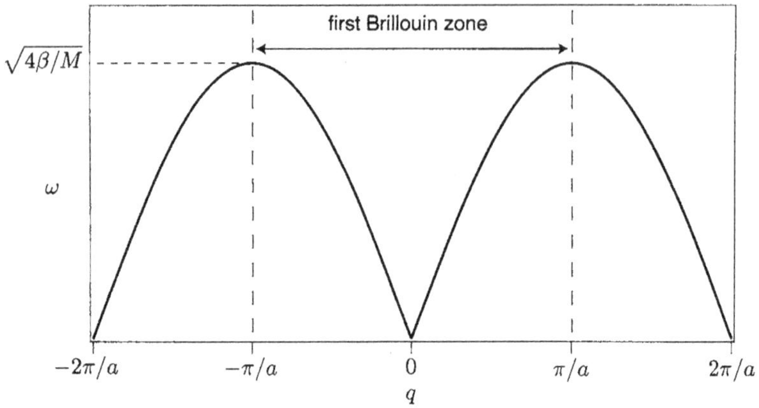

Fig. 5.2 Schematic representation of the phonon dispersion of a linear chain with identical atoms.

Inserting Eq. (5.28) into Eq. (5.27) yields:

$$-\omega^2 M = \beta\left(e^{iqa} + e^{-iqa} - 2\right) = \beta\left(e^{i\frac{qa}{2}} - e^{-i\frac{qa}{2}}\right)^2$$
$$= -4\beta \sin^2 \frac{qa}{2}, \tag{5.29}$$

from which we derive the dispersion relation shown in Fig. 5.2:

$$\omega = \pm\sqrt{\frac{4\beta}{M}} \sin \frac{qa}{2}. \tag{5.30}$$

For small phonon wavevectors q ($q \ll \pi/a$) we recover the Debye model:

$$\omega \approx \sqrt{\frac{\beta}{M}} qa = \sqrt{\frac{c}{\rho}} q = vq, \tag{5.31}$$

where $\rho = M/a$ denotes the density (in one dimension), $c = \beta a$ the elastic constant, and v the sound velocity. The displacement pattern of the atoms results from the ratio $u_n/u_{n+1} = e^{-iqa}$ which is $u_n/u_{n+1} = 1$ for $q = 0$ (center of the Brillouin zone) and $u_n/u_{n+1} = -1$ for $q = \pm\pi/a$ (zone boundary) as visualized in Fig. 5.3. For $q = 0$ the atomic displacement pattern is a mere translational motion of all the atoms by an identical displacement vector which corresponds to a zero-energy motion ($\omega = 0$) as expected from Eq. (5.30). For $q = \pm\pi/a$ neighboring atoms are dynamically displaced in opposite directions.

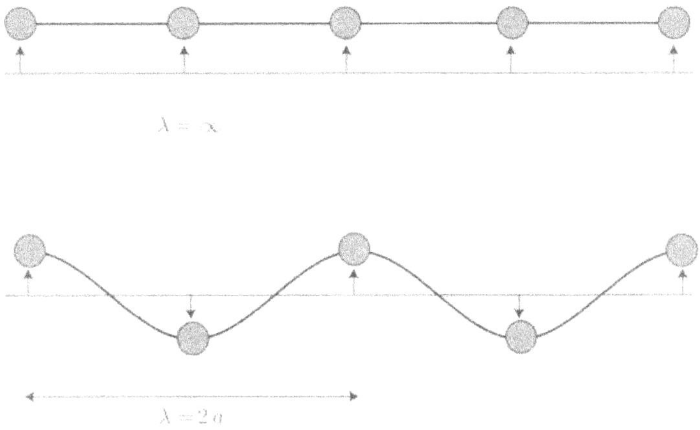

Fig. 5.3 Atomic displacement patterns of a linear chain for phonons with wavevectors $q = 0$ (center of the Brillouin zone) and $q = \pi/a$ (zone boundary).

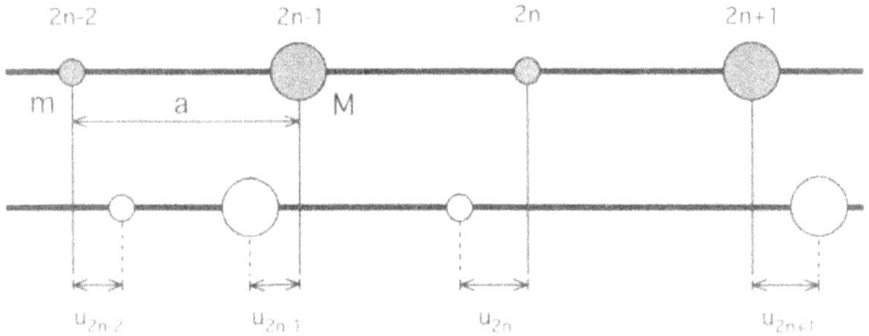

Fig. 5.4 Schematic representation of a linear chain with two different atoms.

5.2.2 *Linear chain with two different atoms*

The dislocation pattern of a linear chain with two different atoms of mass M and m is shown in Fig. 5.4. The equations of motion read:

$$\begin{aligned} m\ddot{u}_{2n} &= \beta(u_{2n+1} + u_{2n-1} - 2u_{2n}) \\ M\ddot{u}_{2n+1} &= \beta(u_{2n+2} + u_{2n} - 2u_{2n+1}). \end{aligned} \tag{5.32}$$

Inserting the solutions

$$u_{2n} = \xi e^{i(\omega t + 2nqa)}; \quad u_{2n+1} = \eta e^{i(\omega t + (2n+1)qa)} \tag{5.33}$$

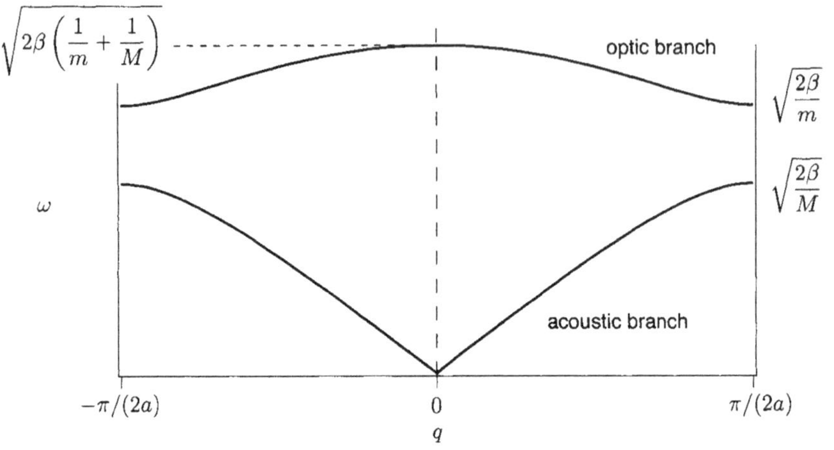

Fig. 5.5 Schematic representation of the acoustic and optic phonon branch of a linear chain with two different atoms ($M = 2m$).

into Eq. (5.32) and solving the corresponding determinant for the coefficients ξ and η yields the dispersion relations

$$\omega_{\pm}^2 = \beta\left(\frac{1}{m}+\frac{1}{M}\right) \pm \beta\sqrt{\left(\frac{1}{m}+\frac{1}{M}\right)^2 - \frac{4\sin^2 qa}{mM}} \tag{5.34}$$

as visualized in Fig. 5.5. Because there are two atoms per unit cell, the phonon dispersion relation is composed of an acoustic and an optic branch. The atomic displacement patterns can be derived from the ratio $u_{2n}/u_{2n+1} = \xi/\eta \cdot e^{-\imath qa}$ as shown for selected phonon wavevectors in Fig. 5.6. For ionic crystals the displacement pattern for optic phonons in the zone center creates an electric polarization which affects the signal in light scattering experiments. This explains historically the term optic phonons.

The atomic displacement patterns give direct information on the phonon polarization vectors $e_{\boldsymbol{d},s}(\boldsymbol{q})$ which are an important quantity in the cross-section formula Eq. (5.23). For the examples displayed in Fig. 5.6 we find:

- $e_{\boldsymbol{d},s}(\boldsymbol{q})$=(1,1) for the acoustic branch in the zone center,
- $e_{\boldsymbol{d},s}(\boldsymbol{q})$=(1,-1) for the optic branch in the zone center,
- $e_{\boldsymbol{d},s}(\boldsymbol{q})$=(0,1) for the acoustic branch at the zone boundary,
- $e_{\boldsymbol{d},s}(\boldsymbol{q})$=(1,0) for the optic branch at the zone boundary.

The situation is more complicated for two- and three-dimensional crys-

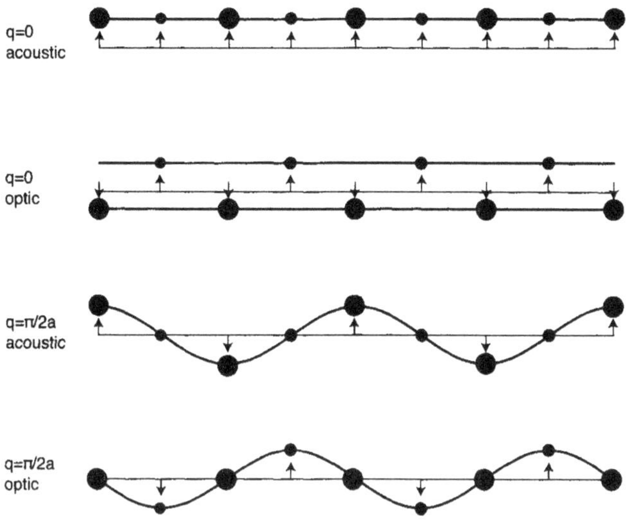

Fig. 5.6 Atomic displacement patterns of a linear chain with two different atoms for $q = 0$ and $q = \pi/2a$.

tals; nevertheless, the simple atomic displacement patterns partially apply when studying phonons with wavevectors along the principal symmetry directions.

5.2.3 *Experimental*

For the determination of the phonon dispersion $\omega_s(\boldsymbol{q})$ one has to be aware that both the energy and momentum conservation laws have to be obeyed. As a consequence, the measurement has be carried out on single crystal samples. These experiments are most easily realized with the help of triple-axis spectrometers introduced in Chap. 3.3.7. With a triple-axis spectrometer, there exists always the possibility during an experiment of keeping one variable $\boldsymbol{Q} = \boldsymbol{q} + \boldsymbol{\tau}$ (or $\boldsymbol{q}$) or ω constant. As a consequence an experiment can be precisely performed at any point of the reciprocal space. Figure 5.7 shows both constant-q and constant-ω scans. For constant-q scans, the momentum transfer $\boldsymbol{Q}$ (or $\boldsymbol{q}$) is kept fixed and the intensity of the scattered neutrons is maximum for $\omega = \omega_s(\boldsymbol{q})$.

By means of the term $(\boldsymbol{Q} \cdot \boldsymbol{e}_s(\boldsymbol{q}))^2$ in Eq. (5.22), the polarization vectors can also be determined. This is necessary since, in principle, there is no

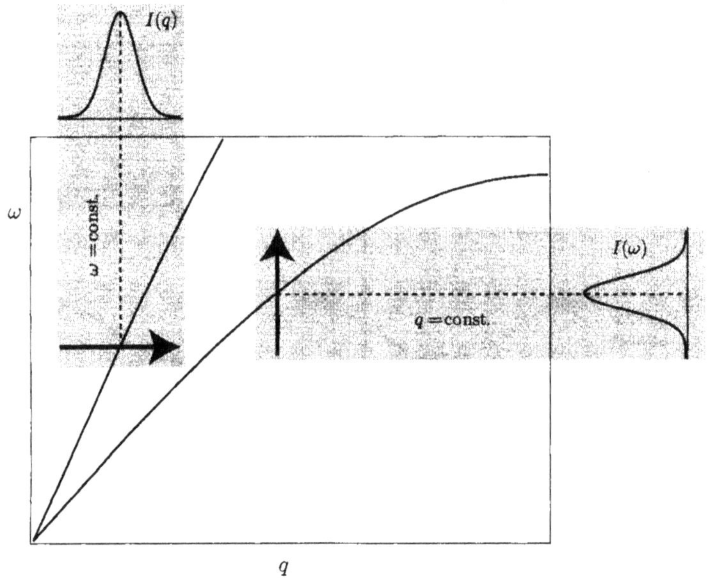

Fig. 5.7 Constant-ω and constant-q scans: depending on the slope of the phonon branch one of the two scan modes is favored.

simple relation between $\boldsymbol{e}_s(\boldsymbol{q})$ and the direction of $\boldsymbol{Q}$.

Figure 5.8 shows low-energy phonon modes of ice I_h (ordinary ice). The individual phonon branches have been measured in different Brillouin zones in order to align $\boldsymbol{Q}$ as close as possible parallel to the respective polarization vector $\boldsymbol{e}_s(\boldsymbol{q})$, thus discriminating between longitudinal and transverse branches and maximizing intensity (Eq. (5.22)).

In a real experiment, the measured linewidth of the excitations is usually non-zero. This is due to both the finite energy resolution of the instrument and the intrinsic broadening caused by anharmonic interactions. If the anharmonicity is small, one can keep the formalism developed in Chap. 5.1 and replace the dispersion relation by the following expression:

$$\omega_s(\boldsymbol{q}) \quad \to \quad \omega_s(\boldsymbol{q}) + \Delta_s(\boldsymbol{q}, w) + \imath\Gamma_s(\boldsymbol{q}, \omega) \tag{5.35}$$

where $\Delta_s(\boldsymbol{q}, w)$ and $\Gamma_s(\boldsymbol{q}, \omega)$ are the energy-shift and the inverse of the life time of the phonon, respectively. The δ-function in Eqs (5.22) and (5.23) has to be replaced by

$$\delta\left(\omega - \omega_s(\boldsymbol{q})\right) \quad \to \quad \frac{\Gamma_s(\boldsymbol{q}, \omega)}{\left(\omega - \omega_s(\boldsymbol{q}) - \Delta_s(\boldsymbol{q}, w)\right)^2 + \Gamma_s^2(\boldsymbol{q}, \omega)} \tag{5.36}$$

that represents a Lorentzian with full-width-half-maximum $\Gamma_s(\boldsymbol{q}, \omega)$.

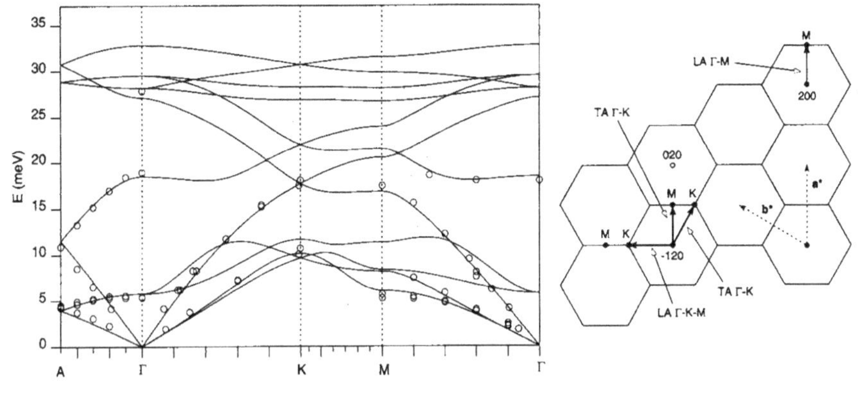

Fig. 5.8 Left: phonon dispersion of D_2O ice I_h (ordinary ice) measured at $T = 140$ K (after [Strässle *et al.* (2004b)]). Circles denote measured points, the curves are a fit to a lattice dynamical model (Born-von-Kármán). Right: scheme of corresponding reciprocal space where the longitudinal acoustic (LA) and transverse acoustic (TA) branches of the basal plane have been measured.

5.3 Incoherent scattering: phonon density of states

In order to calculate the cross section for incoherent scattering, we can make use of the results obtained in Eqs (5.1) - (5.14). With $j = j'$ we obtain:

$$\frac{\mathrm{d}^2\sigma}{\mathrm{d}\Omega\mathrm{d}\omega} = \frac{k'}{k}\frac{1}{2\pi\hbar}e^{-2W(\boldsymbol{Q})}\sum_j b_j^2 \int_{-\infty}^{\infty} e^{\langle\hat{A}\hat{B}\rangle}e^{-\imath\omega t}\mathrm{d}t. \qquad (5.37)$$

Again we only consider the term $\langle\hat{A}\hat{B}\rangle$ in the Taylor expansion of $e^{\langle\hat{A}\hat{B}\rangle}$ (Eq. (5.15)), thus neglecting elastic scattering and multi-phonon processes (see Chap. 5.4 below). For scattering by one phonon, we have $e^{\langle\hat{A}\hat{B}\rangle} = \langle\hat{A}\hat{B}\rangle$; with $j = j'$, and Eq. (5.5), (5.6), (5.16), (5.20) we obtain

$$\langle\hat{A}\hat{B}\rangle = \frac{\hbar}{2MN}\sum_{s,\boldsymbol{q}}\frac{(\boldsymbol{Q}\cdot\boldsymbol{e}_s(\boldsymbol{q}))^2}{\omega_s(\boldsymbol{q})} \times \left[e^{\imath\omega_s(\boldsymbol{q})t}(n_s(\boldsymbol{q})+1) + e^{-\imath\omega_s(\boldsymbol{q})t}n_s(\boldsymbol{q})\right]. \qquad (5.38)$$

By introducing Eq. (5.38) into Eq. (5.37) and using Eq. (A.6) we finally find:

$$\frac{\mathrm{d}^2\sigma}{\mathrm{d}\Omega\mathrm{d}\omega} = \frac{1}{2M}\frac{k'}{k}\left(\langle b^2\rangle - \langle b\rangle^2\right)e^{-2W(\boldsymbol{Q})}\sum_{s,\boldsymbol{q}}\frac{(\boldsymbol{Q}\cdot\boldsymbol{e}_s(\boldsymbol{q}))^2}{\omega_s(\boldsymbol{q})} \times \left[(n_s(\boldsymbol{q})+1)\delta\left(\omega-\omega_s(\boldsymbol{q})\right) + n_s(\boldsymbol{q})\delta\left(\omega+\omega_s(\boldsymbol{q})\right)\right] \quad (5.39)$$

Here again, the cross section contains an emission and an absorption term. As for the coherent cross section, the energy conservation has to be fulfilled, but not the momentum conservation that defines the wavevector $\boldsymbol{q}$ of the created or annihilated phonons. As a consequence, all phonons with energy ω_s will contribute to the scattering independent of their wavevector. The number of phonons per energy interval corresponds to the definition of the density of states $g(\omega)$, which for a 3-dimensional Bravais-lattice is normalized as follows:

$$\int_0^\infty g(\omega)\mathrm{d}\omega = 3N \tag{5.40}$$

From Eq. (5.39) we therefore obtain

$$\boxed{\begin{aligned}\frac{\mathrm{d}^2\sigma}{\mathrm{d}\Omega\mathrm{d}\omega} &= \frac{1}{4M}\frac{k'}{k}\left(\langle b^2\rangle - \langle b\rangle^2\right)e^{-W(\boldsymbol{Q})} \\ &\quad\times \left\langle\left(\boldsymbol{Q}\cdot\boldsymbol{e}_s(\boldsymbol{q})\right)^2\right\rangle\cdot\frac{g(\omega)}{\omega}\cdot\left[\coth\frac{\hbar\omega}{2k_BT}\pm 1\right]\end{aligned}} \tag{5.41}$$

where $\left(\coth(\hbar\omega/2k_BT)\pm 1\right)/2$ stands for the Bose-Einstein occupation factor for phonon emission $n_s(\boldsymbol{q})+1$ and absorption $n_s(\boldsymbol{q})$, respectively.

For cubic Bravais lattices we have $\left\langle\left(\boldsymbol{Q}\cdot\boldsymbol{e}_s(\boldsymbol{q})\right)^2\right\rangle = \frac{1}{3}Q^2$, which gives

$$\begin{aligned}\frac{\mathrm{d}^2\sigma}{\mathrm{d}\Omega\mathrm{d}\omega} &= \frac{1}{12M}\frac{k'}{k}\left(\langle b^2\rangle - \langle b\rangle^2\right)e^{-W(\boldsymbol{Q})}Q^2 \\ &\quad\times\frac{g(\omega)}{\omega}\cdot\left[\coth\frac{\hbar\omega}{2k_BT}\pm 1\right].\end{aligned} \tag{5.42}$$

The method is most useful for samples where the scattering is almost entirely incoherent. The cross section for additional coherent scattering contributions becomes quite laborious to calculate. Moreover, multi-phonon processes have to be considered as well which, however, do not provide useful information, but result in an enhanced background (increasing with Q).

Unfortunately, only for the case of Bravais lattices is $g(\omega)$ directly related to the cross section Eq. (5.42). For all other cases, the experiment yields a so-called generalized phonon density of states which is a weighted sum given by

$$G(\omega) = \sum_i \frac{1}{M_i}\left(\langle b_i^2\rangle - \langle b_i\rangle^2\right)e^{-2W_i}g_i(\omega), \tag{5.43}$$

where the index i refers to the properties of the ith atom in the unit cell. (In principle, a further weighting factor has to be introduced in Eq. (5.43)

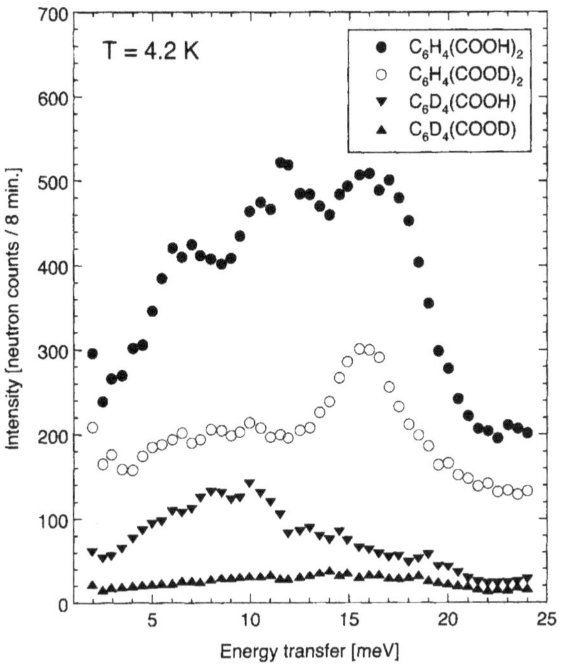

Fig. 5.9 Energy spectra of neutrons scattered from polycrystalline samples of protonated, partly and fully deuterated terephthalic acid HOOC-C_6H_4-COOH. The intensity scale of the two upper spectra is enhanced by 100 counts (after [Zolliker *et al.* (1983)]).

to take account of the average $\langle(\boldsymbol{Q} \cdot e_s(\boldsymbol{q}))^2\rangle$). In case that for a certain atom different isotopes with markedly different incoherent scattering cross sections are available, measurements of $G(\omega)$ taken on samples with different isotope composition can be exploited to extract the partial density of states $g_i(\omega)$. This is visualized in Fig. 5.9 for terephthalic acid HOOC C_6H_4-COOH, for which measurements were taken in the fully protonated and deuterated state as well as with partial deuteration of either the benzol (C_6H_4) or carboxylic (COOH) group. Deuteration results in a drastic suppression of the incoherent scattering cross section, see Appendix B. A comparison of the differently deuterated samples clearly indicates that the line at 16 meV has to be attributed to excitations associated with the benzol group, whereas the lines below 12 meV correspond to modes belonging to the carboxylic group.

5.4 Multi-phonon processes: coherent scattering

We return to Eq. (5.15), which we had linearized, and want now discuss the role of higher order terms. Let us first consider the quadratic term $\frac{1}{2}\langle\hat{A}\hat{B}\rangle^2$. From the definition Eq. (5.16) for $\langle\cdot|\hat{A}\hat{B}|\cdot\rangle$ one readily realizes that now two δ-functions enter in the expression giving the following conservation laws:

$$\hbar\left[\pm\omega_{s_1}(\boldsymbol{q}_1) \pm \omega_{s_2}(\boldsymbol{q}_2)\right] = \hbar\omega = \frac{\hbar^2}{2m}\left(|\boldsymbol{k}|^2 - |\boldsymbol{k}'|^2\right),$$
$$\pm\boldsymbol{q}_1 \pm \boldsymbol{q}_2 = \boldsymbol{Q} - \boldsymbol{\tau} = \boldsymbol{k} = \boldsymbol{k}' - \boldsymbol{\tau}. \qquad (5.44)$$

In the scattering process, two phonons are produced or destroyed simultaneously, or one phonon will be created and another one annihilated. One can easily see that for fixed values of $\boldsymbol{Q}$ and ω, a large number of phonons can produce such two-phonon processes. As a consequence no sharp peak can be observed and the background increases.

The general term $\frac{1}{n!}\langle\hat{A}\hat{B}\rangle^n$ in the Taylor expansion of Eq. (5.15) describes obviously n-phonon scattering processes.

5.5 Further reading

- B. Dorner, *Coherent inelastic neutron scattering in lattice dynamics* (Springer, Berlin, 1982)
- B. Dorner, in *Neutron scattering*, ed. by A. Furrer (Proc. 93-01, ISSN 1019-6447, PSI Villigen, 1999), p. 111: *Structural excitations*
- C. Stassis, in *Methods of experimental physics*, Vol. 23, Part A, ed. by D. L. Price and K. Sköld (Academic Press, London, 1986), p. 369: *Lattice dynamics*

5.6 Exercises

Exercise No. 5.1

The acoustic phonon branches of many "simple" compounds are well explained by the sinusoidal dispersion relation derived in Eq. (5.30). The transverse acoustic phonon branches observed for germanium, however, exhibit an unusual flattening of the dispersion relation upon approaching the zone boundary, see Fig. 5.10. Germanium is a semiconductor with covalent bonds which are usually formed from two electrons, one from each atom participating in the bond. These electrons tend to be partially localized

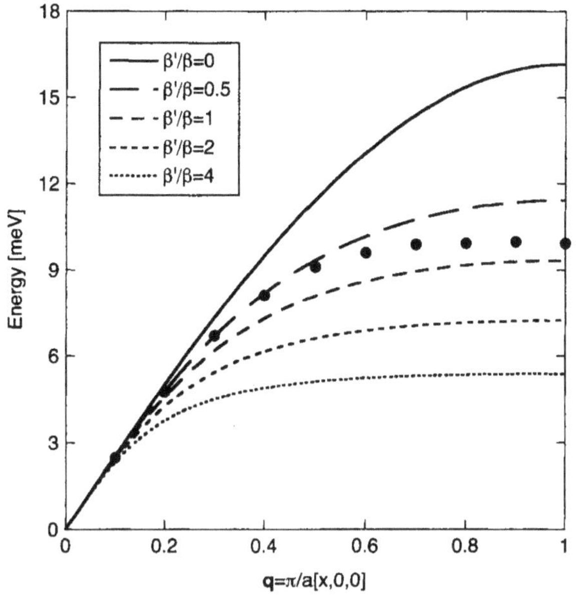

Fig. 5.10 Dispersion relation of the lower transverse acoustic phonon branch measured for Ge at 80 K along the [100] direction (after [Nellin and Nilsson (1972)]). The lines correspond to model calculations based on Eq. (5.49).

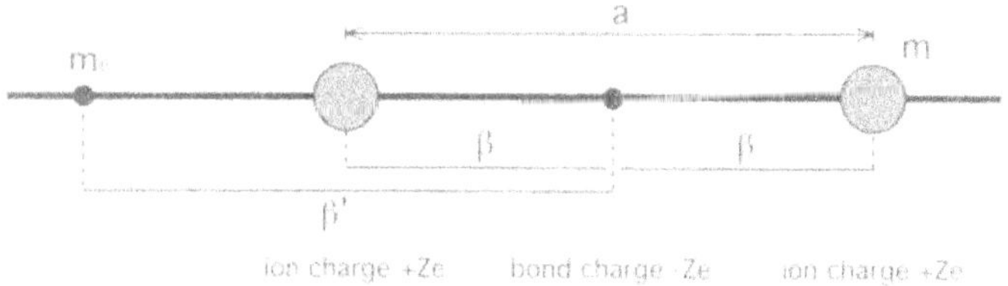

Fig. 5.11 Linear chain formed by alternating ion and bond charges. Bond charges are connected via effective force constants β and β' to neighboring ion and bond charges, respectively.

midway between the two atoms and constitute the so-called bond charge, see Fig. 5.11. Derive the phonon dispersion for the one-dimensional chain illustrated in Fig. 5.11 by following the procedure outlined in Chap. 5.2.2 for a diatomic one-dimensional chain.

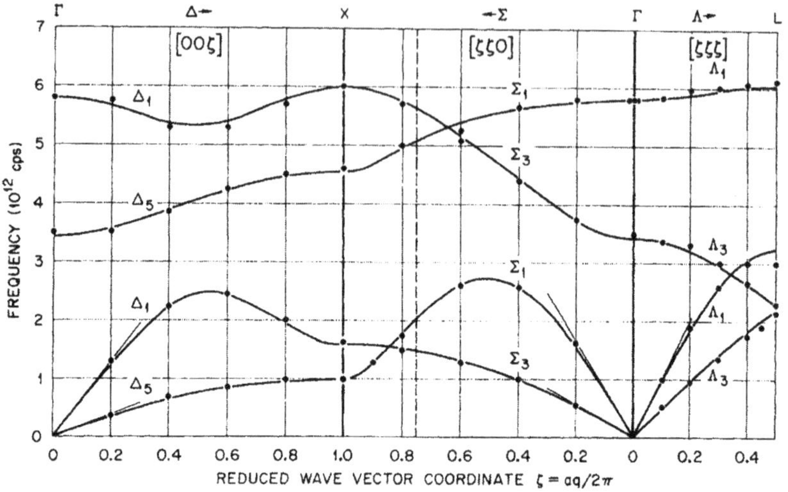

Fig. 5.12 Phonon dispersion relation observed for AgCl at $T = 78$ K. Reprinted with permission from [Vijayaraghavan *et al.* (1970)]. Copyright 1970 by the American Physical Society.

Exercise No. 5.2

Figure 5.12 shows the phonon dispersion relations determined for the ionic compound AgCl by triple-axis neutron spectroscopy [Vijayaraghavan *et al.* (1970)]. AgCl crystallizes in a simple cubic structure with the ions located at $\boldsymbol{d}_{\mathrm{Ag}} = (0,0,0)$ and $\boldsymbol{d}_{\mathrm{Cl}} = a(1/2,1/2,1/2)$. A proper assignment of the observed peaks in the constant-Q scans can be achieved by analyzing the intensities according to the dynamical structure factor

$$g_s^2(\boldsymbol{q},\boldsymbol{\tau}) = \frac{1}{\omega_s(\boldsymbol{q})} \left| \sum_d \frac{\langle b_d \rangle}{\sqrt{M_d}} e^{-W_d(\boldsymbol{Q})} e^{\imath \boldsymbol{Q}\cdot\boldsymbol{d}} \left(\boldsymbol{Q}\cdot\boldsymbol{e}_{d,s}(\boldsymbol{q})\right) \right|^2, \tag{5.45}$$

where $\boldsymbol{Q} = \boldsymbol{\tau} + \boldsymbol{q}$, see Eq. (5.25). Structure factor calculations are particularly important in cases where a crossing of acoustic/optic and transverse/longitudinal phonon branches occurs, as observed in AgCl for phonons propagating along the $[\zeta\zeta 0]$ and $[\zeta\zeta\zeta]$ directions. The structure factor sensitively depends on the polarization vectors $\boldsymbol{e}_{d,s}(\boldsymbol{q})$ which are usually determined from adequate models applied to reproduce the observed phonon energies. Figure 5.13 displays the structure factors calculated for AgCl on the basis of a shell model [Vijayaraghavan *et al.* (1970)].

Verify the essential features of Fig. 5.13 by estimates of the structure factor based on the polarization vectors $\boldsymbol{e}_{d,s}(\boldsymbol{q})$ derived in Chap. 5.2.2.

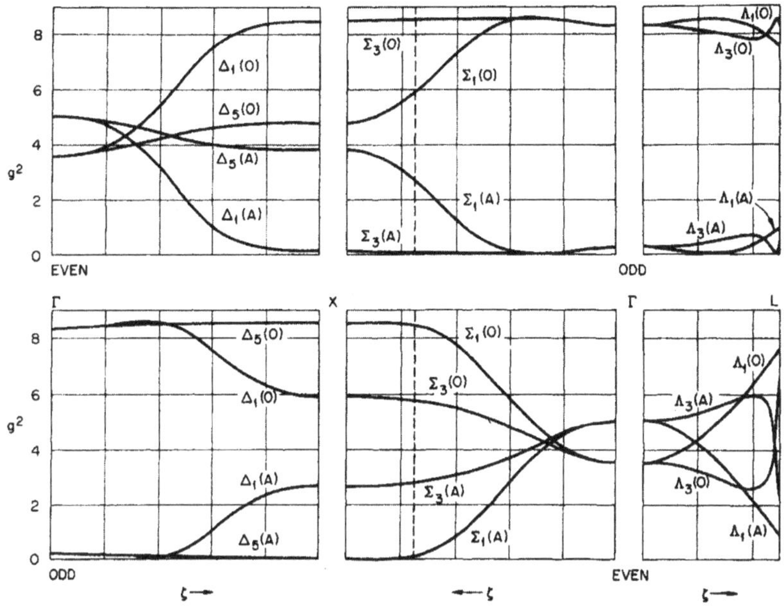

Fig. 5.13 Dynamical structure factor of AgCl calculated for even $(h+k+l=2n)$ and odd $(h+k+l=2n+1)$ triples h,k,l associated with the reciprocal lattice vectors τ_{hkl}. The structure factors are plotted in units of $[(\boldsymbol{e}_s \cdot \boldsymbol{Q})b_{\mathrm{Cl}}]^2/(\omega_s(\boldsymbol{q})M_{\mathrm{Cl}})$. Reprinted with permission from [Vijayaraghavan *et al.* (1970)]. Copyright 1970 by the American Physical Society.

The atomic weights are $M_{\mathrm{Ag}} = 107.88$ and $M_{\mathrm{Cl}} = 35.46$. The coherent scattering lengths are listed in the Appendix B.

5.7 Solutions

Exercise No. 5.1

As sketched in Fig. 5.11, we denote the force constant between an atom of mass m and the bond charge of mass m_e by β. In addition, we introduce the force constant β' to describe the interaction between two bond charges. In analogy to Chap. 5.2.2, the equations of motion read:

$$m\ddot{u}_{2n} = \beta(u_{2n+1} + u_{2n-1} - 2u_{2n}), \tag{5.46}$$

$$m_e\ddot{u}_{2n+1} = \beta(u_{2n+2} + u_{2n} - 2u_{2n+1}) + \beta'(u_{2n+3} + u_{2n-1} - 2u_{2n+1}) = 0,$$

where we set $m_e = 0$ since $m_e \ll m$. Inserting the ansatz for u_{2n+i} defined by Eq. (5.33) into Eq. (5.46) yields

$$m\omega^2\eta = 2\beta\left(\eta - \xi\cos(\frac{qa}{2})\right)$$
$$\beta\left(\eta\cos(\frac{qa}{2}) - \xi\right) + \beta'\xi\left(\cos(qa) - 1\right) = 0 \tag{5.47}$$

from which we obtain a relation between the amplitudes ξ and η:

$$\xi = \frac{\beta\cos(\frac{qa}{2})}{\beta + 2\beta'\sin^2(\frac{qa}{2})}\,\eta. \tag{5.48}$$

Substituting Eq. (5.48) into Eq. (5.47) yields [Brüesch (1982)]

$$\omega(q) = \sqrt{\frac{1}{m}\cdot\frac{2\beta(\beta+2\beta')\sin^2(\frac{qa}{2})}{\beta + 2\beta'\sin^2(\frac{qa}{2})}} \tag{5.49}$$

For $q \ll \pi/a$ Eq. (5.49) transforms to

$$\omega(q) = \sqrt{\frac{\beta+2\beta'}{2m}}\,qa = \sqrt{\frac{c}{\rho}}\,q = v\,q, \tag{5.50}$$

where we use the notation introduced in Eq. (5.31). Figure 5.10 shows dispersion curves calculated from Eq. (5.49) for different ratios β'/β, but the elastic constant $c \sim \beta(1 + 2\beta'/\beta)$ is the same for all curves. We see that the acoustic phonon branch of Ge can be modelled with the ratio $0.5 < \beta'/\beta < 1$.

Exercise No. 5.2

We demonstrate the calculation of the dynamical structure factor for longitudinal phonons propagating along the [0, 0, 1] direction in the Brillouin zone defined by the reciprocal lattice vector $\boldsymbol{\tau}_{hkl} = 2\pi/a \cdot (h,k,l)$. According to Fig. 5.6 the polarization vectors $\boldsymbol{e}_s$=$(\boldsymbol{e}_{\mathrm{Ag}}, \boldsymbol{e}_{\mathrm{Cl}})$ at the zone center (ZC) and at the zone boundary (ZB) are given by:

(a) acoustic branch, ZC, $\boldsymbol{q} = (0,0,0) \rightarrow \boldsymbol{e}_{\mathrm{Ag}} = \xi(0,0,1)$, $\boldsymbol{e}_{\mathrm{Cl}} = \eta(0,0,1)$,
(b) acoustic branch, ZB, $\boldsymbol{q} = \frac{2\pi}{a}(0,0,1) \rightarrow \boldsymbol{e}_{\mathrm{Ag}} = (0,0,1)$, $\boldsymbol{e}_{\mathrm{Cl}} = (0,0,0)$,
(c) optic branch, ZC, $\boldsymbol{q} = (0,0,0) \rightarrow \boldsymbol{e}_{\mathrm{Ag}} = \xi(0,0,1)$, $\boldsymbol{e}_{\mathrm{Cl}} = \eta(0,0,-1)$,
(d) optic branch, ZB, $\boldsymbol{q} = \frac{2\pi}{a}(0,0,1) \rightarrow \boldsymbol{e}_{\mathrm{Ag}} = (0,0,0)$, $\boldsymbol{e}_{\mathrm{Cl}} = (0,0,1)$,

The polarization vectors are unit vectors, thus $\xi^2 + \eta^2 = 1$. Since in AgCl the mean-square displacements $\langle u^2\rangle$ are roughly equal for both the Ag^+ and the Cl^- ions [Vijayaraghavan *et al.* (1970)], we approximate the amplitudes by $\xi = \eta = 1/\sqrt{2}$.

Table 5.1 Dynamical structure factors of AgCl calculated for selected longitudinal phonons propagating along the [0, 0, 1] direction, listed in units of the squared prefactor of $g_s(\boldsymbol{q}, \boldsymbol{\tau})$ introduced in Eq. (5.52).

phonon mode		$g_s^2(\boldsymbol{q}, \boldsymbol{\tau})$	
		$h+k+l=2n$	$h+k+l=2n+1$
acoustic	$\boldsymbol{q}=(0,0,0)$	$\frac{1}{2}\left(l(1+\gamma)\right)^2$	$\frac{1}{2}\left(l(1-\gamma)\right)^2$
acoustic	$\boldsymbol{q}=\frac{2\pi}{a}(0,0,1)$	$(l+1)^2$	$(l+1)^2$
optic	$\boldsymbol{q}=(0,0,0)$	$\frac{1}{2}\left(l(1-\gamma)\right)^2$	$\frac{1}{2}\left(l(1+\gamma)\right)^2$
optic	$\boldsymbol{q}=\frac{2\pi}{a}(0,0,1)$	$(\gamma(l+1))^2$	$(\gamma(l+1))^2$

For the sake of simplicity we set $e^{-W_d(\boldsymbol{Q})} = 1$ and $\omega_s(\boldsymbol{q}) = 1$ in Eq. (5.45). For the case (a) we find:

$$
\begin{aligned}
g_s(\boldsymbol{q}, \boldsymbol{\tau}) &= \frac{\langle b_{\mathrm{Ag}}\rangle}{\sqrt{M_{\mathrm{Ag}}}} e^{\imath(\boldsymbol{\tau}_{hkl}+\boldsymbol{q})\cdot \boldsymbol{d}_{\mathrm{Ag}}} \left((\boldsymbol{\tau}_{hkl}+\boldsymbol{q})\cdot \boldsymbol{e}_{\mathrm{Ag}}\right) \\
&+ \frac{\langle b_{\mathrm{Cl}}\rangle}{\sqrt{M_{\mathrm{Cl}}}} e^{\imath(\boldsymbol{\tau}_{hkl}+\boldsymbol{q})\cdot \boldsymbol{d}_{\mathrm{Cl}}} \left((\boldsymbol{\tau}_{hkl}+\boldsymbol{q})\cdot \boldsymbol{e}_{\mathrm{Cl}}\right) \\
&= \frac{\langle b_{\mathrm{Ag}}\rangle}{\sqrt{M_{\mathrm{Ag}}}} \frac{2\pi}{a} \frac{l}{\sqrt{2}} \\
&+ \frac{\langle b_{\mathrm{Cl}}\rangle}{\sqrt{M_{\mathrm{Cl}}}} \frac{2\pi}{a} \frac{l}{\sqrt{2}} \left(\cos(\pi(h+k+l)) + \imath \sin(\pi(h+k+l))\right),
\end{aligned}
\tag{5.51}
$$

and expressed in dimensionless units:

$$
\begin{aligned}
\frac{a\sqrt{M_{\mathrm{Ag}}}}{2\pi \langle b_{\mathrm{Ag}}\rangle} g_s(\boldsymbol{q}, \boldsymbol{\tau}) = \frac{l}{\sqrt{2}} &\left[1 + \gamma\left(\cos(\pi(h+k+l))\right.\right. \\
&\left.\left. + \imath \sin(\pi(h+k+l))\right)\right],
\end{aligned}
\tag{5.52}
$$

where $\gamma = (\langle b_{\mathrm{Cl}}\rangle \sqrt{M_{\mathrm{Ag}}})/(\langle b_{\mathrm{Ag}}\rangle \sqrt{M_{\mathrm{Cl}}}) = 2.82$. There is a sign change of the cosine function for even $(h+k+l=2n)$ and odd $(h+k+l=2n+1)$ triples (h,k,l), whereas the sine function always vanishes. As a result the dynamical structure factor is much larger in Brillouin zones with even Miller indices (h,k,l) than in Brillouin zones with odd Miller indices, which is very convenient to unambiguously identify the particular phonon mode.

For the cases (b), (c) and (d) we follow the same procedure as indicated above. Table 5.1 lists the corresponding results, which are in reasonable qualitative agreement with the dynamical structure factors displayed in Fig. 5.13.

Chapter 6

Liquids and Amorphous Materials

6.1 Introduction

Matter exists either as a solid or a liquid or a gas. For both the (crystalline) solid and the gas state there is an ideal model which is a good approximation to reality. These are the ideal crystal lattice and the ideal gas models, respectively. In the former, emphasis is on structural order modified slightly by the thermal motion of the atoms, while the latter describes the thermal motion of the atoms on the basis of random atomic positions and motions. Liquids cannot be described by either of the two models. While the transition from the liquid to the solid state always implies a crossing of the phase separation line, this is not the case for the liquid to gas transition due to the coexistence of both states in a certain temperature/pressure region. This suggests some similarities between the liquid and the gas states. On the other hand, the liquid to gas transition is accompanied by a huge volume change, whereas the volume change upon melting is rather small, i.e., both liquids and solids try to pack the atoms as densely as possible. This is nicely confirmed by a comparison of the structure factor $S(Q)$ of liquid zinc with the neutron diffraction pattern taken for polycrystalline zinc as visualized in Fig. 6.1a. The position of the first maximum of the structure factor of liquid zinc corresponds approximately to the positions of the first three Bragg reflections of the polycrystalline material, which means that the nearest-neighbor distances are roughly equal for both solid and liquid zinc.

Figure 6.1b shows the structure factors $S(\boldsymbol{Q})$ observed for the alloy $Pd_{80}Si_{20}$ in the liquid as well as in the amorphous state. The two data sets are very similar, thus the amorphous state can be considered as a frozen liquid state. Therefore the concepts outlined in the present chapter for

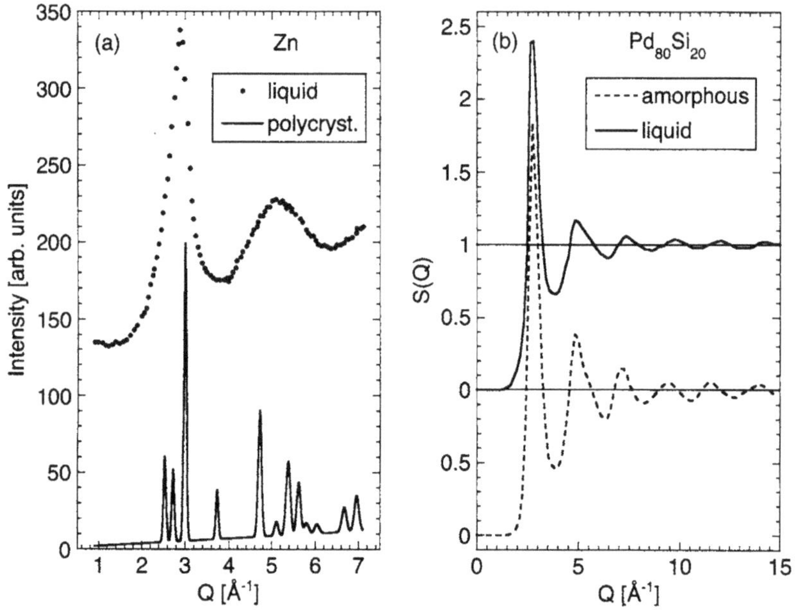

Fig. 6.1 (a) Neutron diffraction patterns observed for liquid Zn at $T = 789$ K and for polycrystalline Zn at $T = 293$ K. The wavelength of the incoming neutrons was 1.3 Å (after [Fischer and Stoll (1968)]). (b) Structure factors for the alloy $Pd_{80}Si_{20}$ observed in the liquid and in the amorphous state at $T = 1253$ K and room temperature, respectively (after [Suzuki (1987)]).

liquids can be applied to amorphous materials and glasses as well.

The following sections concern the static structure factor $S(Q)$ and the dynamical structure factor $S(\boldsymbol{Q},\omega)$ close to $\omega \rightarrow 0$. The latter provides information on the diffusion mechanisms in liquids. We will not discuss $S(\boldsymbol{Q},\omega)$ for $\omega \gg 0$, but we refer to the excitation spectra observed for liquid ^{4}He and ^{3}He which are summarized in Chap. 12.2.2 and Chap. 12.3.2, respectively. Moreover, the detailed understanding of the vibrational dynamics usually has to be accompanied by a *first-principles* pseudopotential treatment of the interatomic potentials, which is beyond the scope of the present chapter.

6.2 Static structure factor

We start with the intermediate scattering function $I(\boldsymbol{Q},t)$, which results from the Fourier inversion of Eq. (2.20):

$$I(\boldsymbol{Q},t) = \hbar \int S(\boldsymbol{Q},\omega) e^{\imath\omega t} \mathrm{d}\omega. \tag{6.1}$$

$I(\boldsymbol{Q},t)$ is dimensionless since $S(\boldsymbol{Q},\omega)$ has the dimension $(\text{energy})^{-1}$. We limit ourselves to elastic scattering, which is related to $I(\boldsymbol{Q},t)$ for $t \to \infty$. As indicated in Fig. 6.2 we split $I(\boldsymbol{Q},t)$ into two parts:

$$I(\boldsymbol{Q},t) = I(\boldsymbol{Q},\infty) + I'(\boldsymbol{Q},t). \tag{6.2}$$

Inserting Eq. (6.2) into Eq. (2.20) and making use of the relations Eqs (A.2) and (A.5) yields:

$$\begin{aligned} S(\boldsymbol{Q},\omega) &= \frac{1}{2\pi\hbar} \int \left(I(\boldsymbol{Q},\infty) + I'(\boldsymbol{Q},t)\right) e^{-\imath\omega t} \mathrm{d}t \\ &= \frac{1}{\hbar}\delta(\omega) I(\boldsymbol{Q},\infty) + \frac{1}{2\pi\hbar} \int I'(\boldsymbol{Q},t) e^{-\imath\omega t} \mathrm{d}t. \end{aligned} \tag{6.3}$$

The first and second terms describe the elastic and inelastic scattering, respectively. The coherent, elastic cross-section is obtained from Eq. (2.28):

$$\left(\frac{\mathrm{d}^2\sigma}{\mathrm{d}\Omega \mathrm{d}\omega}\right)_{\text{coh.el.}} = N\langle b\rangle^2 S(\boldsymbol{Q},\omega) = \frac{N}{\hbar}\langle b\rangle^2 \delta(\omega) I(\boldsymbol{Q},\infty). \tag{6.4}$$

Integration over ω by using Eq. (A.1) yields:

$$\left(\frac{\mathrm{d}\sigma}{\mathrm{d}\Omega}\right)_{\text{coh.}} = N\langle b\rangle^2 I(\boldsymbol{Q},\infty) = N\langle b\rangle^2 \int G(\boldsymbol{r},\infty) e^{\imath \boldsymbol{Q}\cdot\boldsymbol{r}} \mathrm{d}\boldsymbol{r}. \tag{6.5}$$

We recall now the definition of $G(\boldsymbol{r},t)$ described by Eq. (2.23). For $t \to \infty$ correlations between $\boldsymbol{R}_{j'}(0)$ and $\boldsymbol{R}_j(t)$ do no longer persist, thus:

$$G(\boldsymbol{r},\infty) = \frac{1}{N}\sum_{j,j'} \int_{-\infty}^{\infty} \langle\delta(\boldsymbol{r}' - \boldsymbol{R}_{j'})\rangle \langle\delta(\boldsymbol{r}' + \boldsymbol{R}_j - \boldsymbol{r})\rangle \mathrm{d}\boldsymbol{r}'. \tag{6.6}$$

The summations over the δ-functions are nothing else than the particle densities $n(\boldsymbol{r})$ so that we obtain in analogy to the Patterson functions used in x-ray scattering:

$$G(\boldsymbol{r},\infty) = \frac{1}{N} \int_{-\infty}^{\infty} \langle n(\boldsymbol{r}')\rangle \langle n(\boldsymbol{r}' - \boldsymbol{r})\rangle \mathrm{d}\boldsymbol{r}'. \tag{6.7}$$

For a liquid the density $n(\boldsymbol{r})$ does not depend on the position $\boldsymbol{r}$:

$$\langle n(\boldsymbol{r}')\rangle = \langle n(\boldsymbol{r}' - \boldsymbol{r})\rangle = n_0 = \frac{N}{V}, \tag{6.8}$$

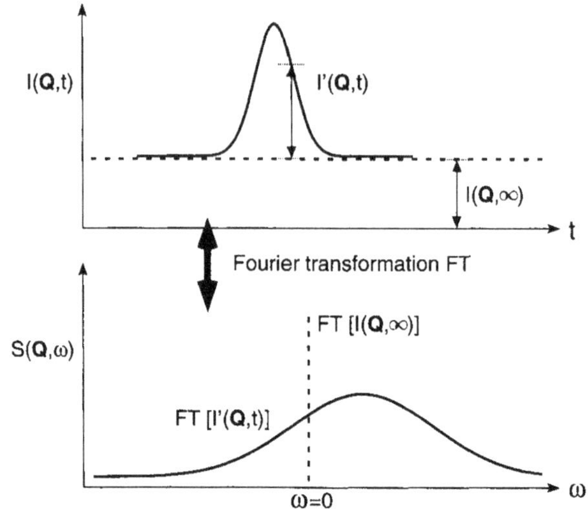

Fig. 6.2 Schematic relation of the intermediate scattering functions $I(\boldsymbol{Q},\infty)$, $I'(\boldsymbol{Q},t)$ and the dynamical structure factors $S(\boldsymbol{Q},\omega=0)$, $S'(\boldsymbol{Q},\infty)$.

where N is the number of atoms in the volume V. We combine Eqs (6.7) and (6.8) to find:

$$G(\boldsymbol{r},\infty)=\frac{1}{N}n_0^2V=n_0. \tag{6.9}$$

$G(\boldsymbol{r},\infty)$ is therefore a constant. Inserting this in Eq. (6.5) yields

$$\left(\frac{\mathrm{d}\sigma}{\mathrm{d}\Omega}\right)_{\mathrm{coh.}}=N\langle b\rangle^2n_0\int e^{i\boldsymbol{Q}\cdot\boldsymbol{r}}\mathrm{d}\boldsymbol{r}=(2\pi)^3N\langle b\rangle^2n_0\delta(\boldsymbol{Q}). \tag{6.10}$$

Elastic scattering is therefore only possible for $Q=0$ which, however, does not constitute a scattering process at all, but just corresponds to the transmission of the incoming neutron beam in the forward direction. Therefore we could conclude that there is *no* coherent elastic scattering in liquids! Of course this is not true, since the pair-correlation function $G(\boldsymbol{r},\infty)$ defined by Eq. (6.6) refers to a distribution of points. However, Eq. (6.9) is only valid for homogeneous systems.

Rather than using the particle density $n(\boldsymbol{r})$, we now want to consider the deviation from the mean value $\langle n(\boldsymbol{r})\rangle$. Therefore we define the function:

$$G'(\boldsymbol{r})=\frac{1}{N}\int_{\infty}^{\infty}\langle\,(n(\boldsymbol{r}'-\boldsymbol{r})-\langle n(\boldsymbol{r}'-\boldsymbol{r})\rangle)\,(n(\boldsymbol{r}')-\langle n(\boldsymbol{r}')\rangle)\,\rangle\,\mathrm{d}\boldsymbol{r}'. \tag{6.11}$$

Combining Eqs (6.8) and (6.11) yields:

$$
\begin{aligned}
G'(\boldsymbol{r}) &= \frac{1}{N}\int_{\infty}^{\infty} \langle\, (n(\boldsymbol{r}'-\boldsymbol{r})-n_0)\,(n(\boldsymbol{r}')-n_0)\,\rangle\,\mathrm{d}\boldsymbol{r}' \qquad (6.12)\\
&= \frac{1}{N}\int_{\infty}^{\infty} \left(\, \langle n(\boldsymbol{r}'-\boldsymbol{r})n(\boldsymbol{r}')\rangle - \langle n(\boldsymbol{r}'-\boldsymbol{r})n_0\rangle - \langle n_0 n(\boldsymbol{r}')\rangle + n_0^2 \,\right) \mathrm{d}\boldsymbol{r}'\\
&= \frac{1}{N}\int_{\infty}^{\infty} \left(\, \langle n(\boldsymbol{r}'-\boldsymbol{r})n(\boldsymbol{r}')\rangle - n_0^2 \,\right) \mathrm{d}\boldsymbol{r}'\\
&= G(\boldsymbol{r},\infty) - \frac{V}{N}n_0^2\\
&= G(\boldsymbol{r},\infty) - n_0,
\end{aligned}
$$

where $G(\boldsymbol{r},\infty)$ is defined by Eq. (6.6). By inserting $G'(\boldsymbol{r})$ in the cross-section Eq. (6.5), the forward scattering is eliminated:

$$
\left(\frac{\mathrm{d}\sigma}{\mathrm{d}\Omega}\right)_{\text{coh.}} = N\langle b\rangle^2 \int \left(G(\boldsymbol{r},\infty)-n_0\right) e^{i\boldsymbol{Q}\cdot\boldsymbol{r}}\mathrm{d}\boldsymbol{r}. \qquad (6.13)
$$

We want now to treat Eq. (6.6) in a correct way. The integration argument is obviously non-zero only when both δ-functions are simultaneously non-vanishing, i.e., when $\boldsymbol{r} = \boldsymbol{R}_{j'} - \boldsymbol{R}_j$. For the integration in Eq. (6.7) one has therefore to sum over these "coincidences", that can be mathematically described by $\delta\left(\boldsymbol{r}-(\boldsymbol{R}_{j'}-\boldsymbol{R}_j)\right)$. We obtain:

$$
\begin{aligned}
G(\boldsymbol{r},\infty) &= \frac{1}{N}\sum_{j,j'}\delta\left(\boldsymbol{r}-(\boldsymbol{R}_{j'}-\boldsymbol{R}_j)\right) \qquad (6.14)\\
&= \frac{1}{N}\left(\sum_{j=j'}\delta\left(\boldsymbol{r}-(\boldsymbol{R}_{j'}-\boldsymbol{R}_j)\right) + \sum_{j\neq j'}\delta\left(\boldsymbol{r}-(\boldsymbol{R}_{j'}-\boldsymbol{R}_j)\right)\right)\\
&= \delta(\boldsymbol{r}) + \frac{1}{N}\sum_{j\neq j'}\delta\left(\boldsymbol{r}-(\boldsymbol{R}_{j'}-\boldsymbol{R}_j)\right),
\end{aligned}
$$

where the second term is a measure of the number of pairs with equal distance $\boldsymbol{r}$ between the two particles forming the pair. It is nothing else than the pair-correlation function $g(\boldsymbol{r})$. By inserting Eq. (6.14) into the cross section Eq. (6.13) one obtains

$$
\begin{aligned}
\left(\frac{\mathrm{d}\sigma}{\mathrm{d}\Omega}\right)_{\text{coh.}} &= N\langle b\rangle^2 \int \left(\delta(\boldsymbol{r})+g(\boldsymbol{r})-n_0\right) e^{i\boldsymbol{Q}\cdot\boldsymbol{r}}\mathrm{d}\boldsymbol{r} \qquad (6.15)\\
&= N\langle b\rangle^2 \left(1+\int \left(g(\boldsymbol{r})-n_0\right) e^{i\boldsymbol{Q}\cdot\boldsymbol{r}}\mathrm{d}\boldsymbol{r}\right)\\
&= N\langle b\rangle^2 S(\boldsymbol{Q}).
\end{aligned}
$$

$S(\boldsymbol{Q})$ is the static structure factor. Liquids are isotropic, thus $g(\boldsymbol{r})$ depends only on $r = |\boldsymbol{r}|$. By averaging the terms with the products $\boldsymbol{Q} \cdot \boldsymbol{r}$ we find:

$$\boxed{\begin{aligned} S(Q) &= 1 + 2\pi \int_0^\infty (g(r) - n_0)\, r^2 \int_{-1}^{1} e^{iQr\cos\theta} \mathrm{d}(\cos\theta)\mathrm{d}r \\ &= 1 + 4\pi \int_0^\infty (g(r) - n_0) \frac{\sin(Qr)}{Qr} r^2 \mathrm{d}r \end{aligned}} \tag{6.16}$$

Equation (6.16) is the fundamental result for the determination of the structure factor $S(Q)$ by means of neutron scattering. The pair-correlation function $g(r)$ is then obtained by Fourier inversion of Eq. (6.16) for which it is of utmost importance to know the limiting values of $S(Q)$ for both $Q \to 0$ and $Q \to \infty$. For the latter, the right-hand side of Eq. (6.16) tends to zero, thus

$$\lim_{Q\to\infty} S(Q) = 1. \tag{6.17}$$

For $Q \to 0$ the following result is obtained [Squires (1996)]:

$$\lim_{Q\to 0} S(Q) = n_0 \kappa_T k_B T, \tag{6.18}$$

where κ_T is the isothermal compressibility.

As an example the structure factor observed for liquid sodium is displayed in Fig. 6.3a. From Eqs (6.16) - (6.18) the pair-correlation function $g(r)$ can be derived as shown in Fig. 6.3b.

For a binary liquid containing N_A atoms of type A and N_B atoms of type B one has to consider the three partial structure factors $S_{AA}(Q)$, $S_{BB}(Q)$, and $S_{AB}(Q)$. In contrast to x-ray scattering, the partial structure factors can be determined in neutron scattering experiments by isotope substitution (A→A′,A″, etc.). In order to achieve this, at least three samples with different isotope composition have to be prepared, e.g., (A′B′, A″B′, A′B″), (A′B, A″B, A‴B), etc. For the latter case, the relevant cross sections

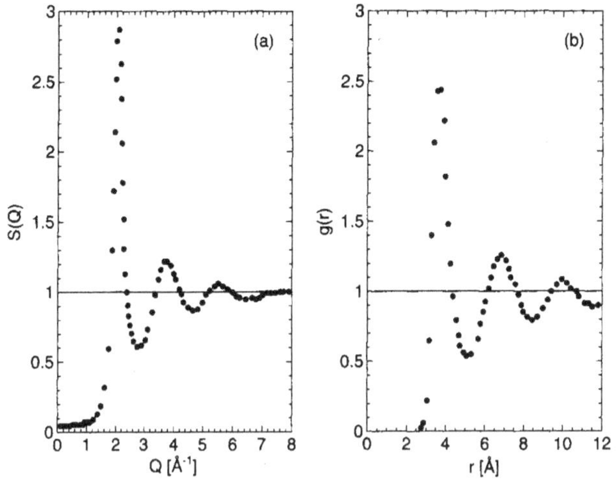

Fig. 6.3 (a) Structure factor $S(Q)$ observed for liquid Na at $T = 373$ K. (b) Pair correlation function $g(r)$ of liquid Na derived from the measured structure factor (after [Hayter *et al.* (1983)]).

can easily be derived from Eq. (6.5):

$$
\begin{aligned}
\left(\frac{\mathrm{d}\sigma}{\mathrm{d}\Omega}\right)_{\mathrm{A'B}} &= \frac{N_A^2}{N_A+N_B}\langle b_{A'}\rangle^2 S_{AA}(Q) + \frac{N_B^2}{N_A+N_B}\langle b_B\rangle^2 S_{BB}(Q) \\
&\quad + \frac{2N_AN_B}{N_A+N_B}\langle b_{A'}\rangle\langle b_B\rangle S_{AB}(Q), \\
\left(\frac{\mathrm{d}\sigma}{\mathrm{d}\Omega}\right)_{\mathrm{A''B}} &= \frac{N_A^2}{N_A+N_B}\langle b_{A''}\rangle^2 S_{AA}(Q) + \frac{N_B^2}{N_A+N_B}\langle b_B\rangle^2 S_{BB}(Q) \\
&\quad + \frac{2N_AN_B}{N_A+N_B}\langle b_{A''}\rangle\langle b_B\rangle S_{AB}(Q), \\
\left(\frac{\mathrm{d}\sigma}{\mathrm{d}\Omega}\right)_{\mathrm{A'''B}} &= \frac{N_A^2}{N_A+N_B}\langle b_{A'''}\rangle^2 S_{AA}(Q) + \frac{N_B^2}{N_A+N_B}\langle b_B\rangle^2 S_{BB}(Q) \\
&\quad + \frac{2N_AN_B}{N_A+N_B}\langle b_{A'''}\rangle\langle b_B\rangle S_{AB}(Q).
\end{aligned}
\tag{6.19}
$$

The application of Eq. (6.19) was demonstrated for the liquid alloy Cu_6Sn_5 [Enderby *et al.* (1966)]. Three samples were investigated consisting of natural tin alloyed with natural copper, with the copper isotope ^{63}Cu (99% enriched), and with the copper isotope ^{65}Cu (99% enriched). The relative scattering strengths $\langle b_{\mathrm{Cu}}\rangle^2$ thus obtained were 1, 0.69, and 1.89 (see Appendix B). The resulting partial structure factors are displayed in Fig. 6.4.

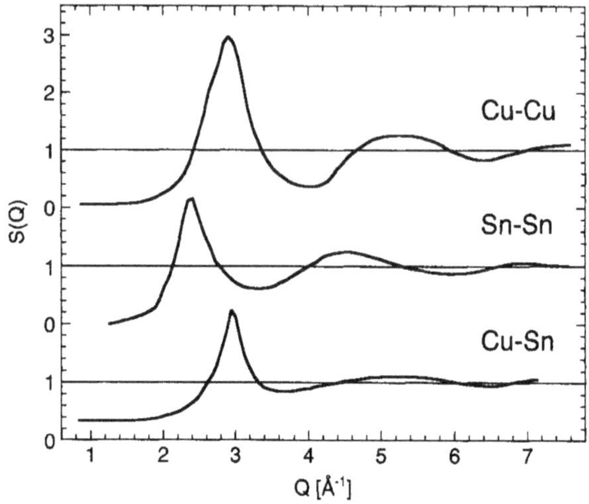

Fig. 6.4 Partial structure factors for the Sn-Sn, Cu-Cu, and Cu-Sn correlations determined from neutron scattering experiments carried out for the liquid alloy Cu_6Sn_5 (after [Enderby *et al.* (1966)]).

It is interesting to note that the extrema for the Cu-Cu and Sn-Sn correlations correspond rather closely with those of pure copper and pure tin, respectively, which is to be expected if the form of the structure factor is dominated by the random packing of hard spheres. However, the first peak of the Cu-Sn correlation does not, as would be expected in a hard sphere model, fall midway between that of the Cu-Cu and Sn-Sn correlations. Another interesting feature is that for the Cu-Sn correlation $S(Q)$ remains substantial when $Q \to 0$, which presumably reflects the importance of long-range fluctuations in the composition.

6.3 Diffusion

Figure 6.5 shows energy spectra of neutrons scattered from liquid argon as a function of the scattering vector Q. The scattering process is incoherent, thus the data correspond to the dynamical structure factor $S_{\text{inc}}(Q, \omega)$ defined by Eq. (2.29). The energy spectra are centered around the elastic position ($\omega = 0$) and have the form of Lorentzians, whose linewidths increase with increasing modulus of Q. This is a typical example for incoherent neutron scattering from liquids, that can be understood in the following way.

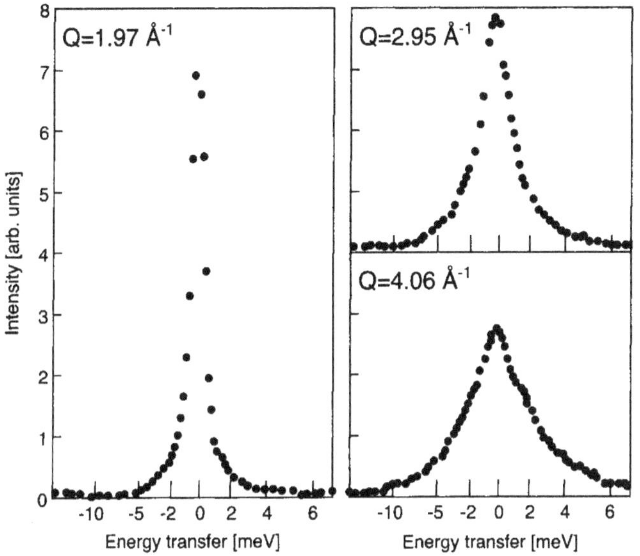

Fig. 6.5 Energy spectra of neutrons scattered from liquid Ar at $T = 85$ K as a function of the scattering vector Q (after [Sköld *et al.* (1972)]).

According to Eqs (2.19) and (2.20), $S_{\text{inc}}(Q, \omega)$ and $G_s(r, t)$ are related by a Fourier transformation. As a consequence, the form of $S_{\text{inc}}(Q, \omega)$ at large Q depends on the form of $G_s(r, t)$ at small values of r. For large Q, $S_{\text{inc}}(Q, \omega)$ extends to large frequencies ω; this means that $G_s(r, t)$ is a highly peaked function for $t \to 0$, thus the short-time behavior of a liquid is therefore similar to the behavior of free atoms (ideal gas behavior). For the other extreme, small values of both Q and ω, $S_{\text{inc}}(Q, \omega)$ depends on the behavior of $G_s(r, t)$ at large values of r and t. The long-time behavior of a liquid is therefore determined by diffusion processes, since in long times the atoms make many collisions with each other. Typical time scales are $t \geq 10^{-12}$ s for diffusion and $t \leq 10^{-13}$ s for the ideal gas behavior.

The diffusive behavior of particles is governed by Fick's law:

$$\frac{\partial n(\boldsymbol{r}, t)}{\partial t} = D \nabla^2 n(\boldsymbol{r}, t), \tag{6.20}$$

where $n(\boldsymbol{r}, t)$ is the number density of atoms and D the diffusion constant. By definition one can replace $n(\boldsymbol{r}, t)$ by $G_s(\boldsymbol{r}, t)$. For isotropic diffusion Eq. (6.20) has the following solution:

$$G_s(\boldsymbol{r}, t) = (4\pi D t)^{-3/2} e^{-\frac{r^2}{4Dt}}, \tag{6.21}$$

which also fulfills the boundary conditions $G_s(\boldsymbol{r},0) = \delta(\boldsymbol{r})$ and $G_s(\boldsymbol{r},\infty) = 0$. Partial Fourier transformation of Eq. (6.21) yields

$$I_s(\boldsymbol{Q},t) = \int G_s(\boldsymbol{r},t)e^{i\boldsymbol{Q}\cdot\boldsymbol{r}} = e^{-Q^2Dt} \tag{6.22}$$

and

$$\boxed{S_{\text{inc}}(Q,\omega) = \frac{1}{2\pi\hbar}\int I_s(Q,t)e^{-i\omega t}\mathrm{d}t = \frac{1}{\pi\hbar}\frac{DQ^2}{\omega^2 + (DQ^2)^2}} \tag{6.23}$$

The right-hand side of Eq. (6.23) corresponds to a Lorentzian centered at $\hbar\omega = 0$ as experimentally observed (see Fig. 6.5). The full-width at half-maximum (FWHM) of the Lorentzian in energy space is

$$\boxed{\Gamma^{\text{fwhm}} = 2\hbar DQ^2} \tag{6.24}$$

thus from the width of the so-called quasi-elastic line the diffusion constant D can be directly determined. However, Eq. (6.24) is only valid for $Q^{-1} \gg a$, where a is the mean distance between neighboring atoms in the liquid.

The macroscopic diffusion theory fails for $Q^{-1} \approx a$, since in that case the neutrons can see the microscopic details of the diffusion process. Microscopically, the mechanism of jump diffusion has proven to be a highly successful model. The idea behind jump diffusion is that the diffusing atom oscillates during a certain time τ_0 around its equilibrium position, then it jumps to another equilibrium position within a jumping time τ_1. At the new equilibrium position the atom oscillates again for a time τ_0 before jumping to the next site, etc. We introduce the mechanism of jump diffusion into Fick's law assuming $\tau_1 \ll \tau_0$:

$$\frac{\partial G_s(\boldsymbol{r},t)}{\partial t} = \frac{\text{Fluctuation}}{\text{Relaxation time}} = \frac{\frac{1}{n}\sum_{\boldsymbol{l}}\left(G_s(\boldsymbol{r}+\boldsymbol{l},t) - G_s(\boldsymbol{r},t)\right)}{\tau_0}, \tag{6.25}$$

where $\boldsymbol{l}$ denotes the jump vector and n the number of possible sites to jump. Fourier transformation of Eq. (6.25) yields:

$$\int \frac{\partial G_s(\boldsymbol{r},t)}{\partial t}e^{i\boldsymbol{Q}\cdot\boldsymbol{r}}\mathrm{d}\boldsymbol{r} = \frac{1}{\tau_0 n}\sum_{\boldsymbol{l}}\int\left(G_s(\boldsymbol{r}+\boldsymbol{l},t) - G_s(\boldsymbol{r},t)\right)e^{i\boldsymbol{Q}\cdot\boldsymbol{r}}\mathrm{d}\boldsymbol{r}. \tag{6.26}$$

By performing the integration we recover the intermediate scattering function:

$$\frac{\partial I_s(\boldsymbol{Q},t)}{\partial t} = -f(\boldsymbol{Q})I_s(\boldsymbol{Q},t) \tag{6.27}$$

with

$$f(\boldsymbol{Q}) = \frac{1}{n\tau_0} \sum_{\boldsymbol{l}} \left(1 - e^{-\imath \boldsymbol{Q}\cdot\boldsymbol{l}}\right) \tag{6.28}$$

The solution of the differential equation Eq. (6.27) is given by

$$I_s(\boldsymbol{Q}, t) = e^{-f(\boldsymbol{Q})t}, \tag{6.29}$$

which satisfies the boundary conditions $G_s(\boldsymbol{r}, 0) = \delta(\boldsymbol{r})$ and $I_s(\boldsymbol{Q}, 0) = \int \delta(\boldsymbol{r}) e^{\imath \boldsymbol{Q}\cdot\boldsymbol{r}} \mathrm{d}\boldsymbol{r} = 1$ (the latter results from application of Eq. (A.8)). The Fourier transform of Eq. (6.29) yields the scattering law:

$$\begin{aligned} S_{\text{inc}}(\boldsymbol{Q}, \omega) &= \frac{1}{2\pi\hbar} \int I_s(\boldsymbol{Q}, t) e^{-\imath\omega t} \mathrm{d}t \\ &= \frac{1}{2\pi\hbar} \int e^{-f(\boldsymbol{Q})t} \cos\omega t = \frac{1}{\pi\hbar} \frac{f(\boldsymbol{Q})}{\omega^2 + f^2(\boldsymbol{Q})} \end{aligned} \tag{6.30}$$

which again corresponds to a Lorentzian with full-width

$$\Gamma^{\text{fwhm}} = 2\hbar f(\boldsymbol{Q}) \tag{6.31}$$

We now evaluate the function $f(\boldsymbol{Q})$ defined by Eq. (6.28) for a liquid. We assume a random orientation of the jump vectors $\boldsymbol{l}$, whose lengths follow a continuous distribution function $\alpha(l)$. Averaging Eq. (6.28) in space yields

$$\begin{aligned} f(Q) &= \frac{1}{\tau_0} \frac{\int\int\int \left(1 - e^{\imath Ql\cos\theta}\right) \alpha(l) l^2 \mathrm{d}l \mathrm{d}(\cos\theta) \mathrm{d}\varphi}{\int\int\int \alpha(l) l^2 \mathrm{d}l \mathrm{d}(\cos\theta) \mathrm{d}\varphi} \\ &= \frac{1}{\tau_0} \frac{\int \left(1 - \frac{\sin(Ql)}{Ql}\right) \alpha(l) l^2 \mathrm{d}l}{\int \alpha(l) l^2 \mathrm{d}l}. \end{aligned} \tag{6.32}$$

We describe the function $\alpha(l)$ by a random distribution:

$$\alpha(l) = e^{-l/l_0}. \tag{6.33}$$

Combining Eqs (6.32) and (6.33) yields:

$$f(Q) = \frac{1}{\tau_0} \left(1 - \frac{1}{\left(1 + (Ql_0)^2\right)^2}\right). \tag{6.34}$$

Of interest are the limiting cases $Q \to \infty$ and $Q \to 0$:

$$\lim_{Q\to\infty} f(Q) = \frac{1}{\tau_0}, \tag{6.35}$$

$$\lim_{Q\to 0} f(Q) = \frac{2}{\tau_0}(Ql_0)^2. \tag{6.36}$$

For $Q \to 0$ we recover the classical behavior. By comparing Eqs (6.23) and (6.30) we find $f(Q) = DQ^2$; using Eq. (6.36) we can express the diffusion constant D by the parameters of the jump diffusion model:

$$D = \frac{2l_0^2}{\tau_0}. \tag{6.37}$$

Finally, we want to establish the relation between the parameter l_0 of the function $\alpha(l)$ defined in Eq. (6.33) and the mean squared jump length:

$$\langle l^2 \rangle = \frac{\int\int\int l^2\alpha(l)l^2 \mathrm{d}l \mathrm{d}(\cos\theta)\mathrm{d}\varphi}{\int\int\int \alpha(l)l^2 \mathrm{d}l \mathrm{d}(\cos\theta)\mathrm{d}\varphi} = \frac{\int l^4 e^{-l/l_0}\mathrm{d}l}{\int l^2 e^{-l/l_0}\mathrm{d}l} = 12l_0^2. \tag{6.38}$$

Combining Eqs (6.37) and (6.38) yields:

$$D = \frac{\langle l^2 \rangle}{6\tau_0}, \tag{6.39}$$

which is in agreement with classical diffusion theory, since with use of Eq. (6.21) we find

$$\begin{aligned}\langle r^2 \rangle &= \int r^2 G_s(\boldsymbol{r},t)\mathrm{d}\boldsymbol{r} = (4\pi Dt)^{-3/2} \int\int\int r^2 e^{-\frac{r^2}{4Dt}} r^2 \mathrm{d}r \mathrm{d}(\cos\theta)\mathrm{d}\varphi \\ &= (4\pi Dt)^{-3/2} 4\pi \int r^4 e^{-\frac{r^2}{4Dt}} \mathrm{d}r = 6Dt. \end{aligned} \tag{6.40}$$

The most abundant liquid on earth – and at the same time the most precious liquid for mankind as well – is water, thus we apply the diffusion mechanisms outlined above to H_2O. Extensive quasielastic neutron scattering data were obtained for water in the temperature range extending from room temperature down to $-20°C$ in the supercooled state [Teixeira *et al.* (1985)]. Some selected results are shown in Fig. 6.6. The linear relationship between the linewidth and the square of the scattering vector predicted for the classical diffusion model (Eq. (6.24)) is only fulfilled for low values of Q. For the data analysis in terms of the jump diffusion model (Eq. (6.34)) we refer to the Exercise No. 6.1.

In some metals hydrogen can be introduced in very high concentrations and will occupy interstitial positions. The hydrogen atoms are usually extremely mobile, so that they perform diffusive jumps between the sites

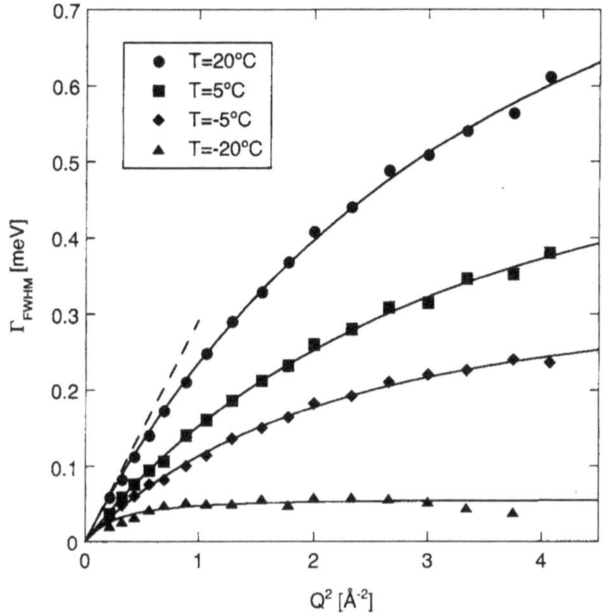

Fig. 6.6 Linewidth as a function of the square of the scattering vector observed in quasielastic neutron scattering experiments performed for water (after [Teixeira *et al.* (1985)]). The full lines are the results of a least-squares fit based on the jump diffusion model, see Exercise No. 6.1. The dashed line corresponds to the classical diffusion model (shown only for the data taken at $T = 20°C$).

they can occupy. The jump diffusion mechanism can therefore be applied to hydrogen storage materials as well. This is exemplified for α-PdH_x which crystallizes in a cubic face-centered structure. The hydrogen atoms can occupy octahedral and tetrahedral sites as shown in Fig. 6.7. Neutron scattering experiments were performed for small hydrogen concentrations x in order to avoid hydrogen correlations [Sköld and Nelin (1967)]. The results are displayed in Fig. 6.8, which exhibit markedly increased diffusion activity upon raising the temperature. The data are in good agreement with a jump diffusion model in which the hydrogen atoms occupy only the octahedral sites. For a detailed data analysis we refer to the exercise No. 6.2.

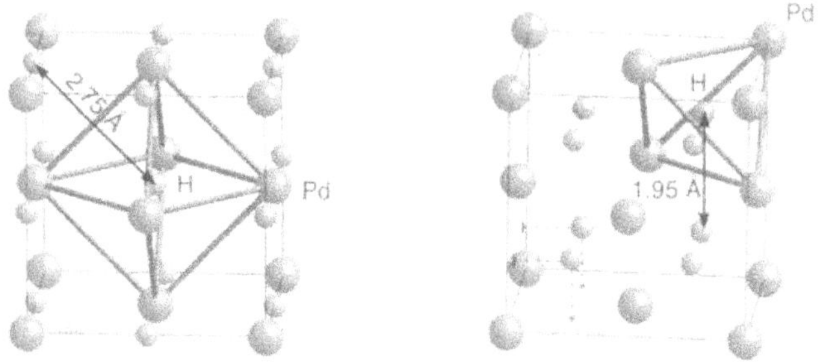

Fig. 6.7 Octahedral (left) and tetrahedral (right) positions of the H atoms in an α-Pd lattice.

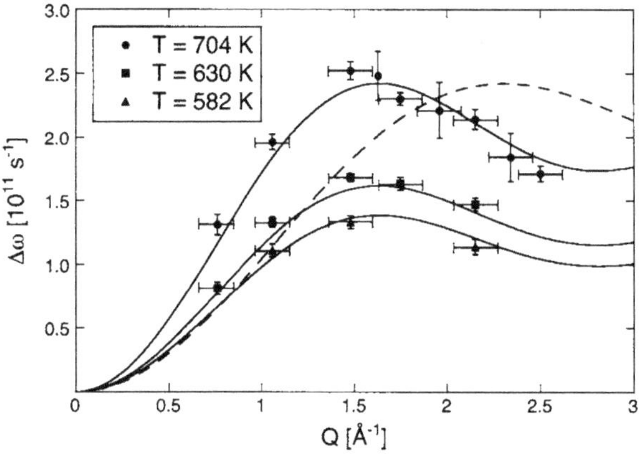

Fig. 6.8 Linewidth $\Gamma \propto \Delta\omega$ vs Q observed for α-PdH$_{0.02-0.04}$ at three temperatures (after [Sköld and Nelin (1967)]). The full and dashed curves correspond to the calculations based on the octahedral and tetrahedral ($T = 704$ K only) model, respectively.

6.4 Further reading

- M. Bée, *Quasielastic neutron scattering* (IOP Publishing Ltd, Bristol, 1988)
- P. A. Egelstaff, *An introduction to the liquid state* (Academic Press, London, 1967)
- P. A. Egelstaff, in *Methods of experimental physics*, Vol. 23, Part B, ed. by D. L. Price and K. Sköld (Academic Press, London, 1987), p. 405: *Classical fluids*
- J. E. Enderby and P. M. N. Gullidge, in *Methods of experimental physics*,

Vol. 23, Part B, ed. by D. L. Price and K. Sköld (Academic Press, London, 1987), p. 471: *Ionic solutions*

- R. Hempelmann, *Quasielastic neutron scattering and solid state diffusion* (Oxford University Press, Oxford, 2000)
- R. Hempelmann, in *Neutron scattering from hydrogen in materials*, ed. by A. Furrer (World Scientific, Singapore, 1994), p. 201: *Jump diffusion of H in metals: neutron scattering*
- P. Lamparter, in *Neutron scattering*, ed. by A. Furrer (Proc. 93-01, ISSN 1019-6447, PSI Villigen, 1993), p. 295: *Neutron and x-ray scattering from amorphous systems*
- S. W. Lovesey, in *Theory of neutron scattering from condensed matter*, Vol. 1 (Clarendon Press, Oxford, 1984), p. 172: *Dense fluids*
- K. Suzuki, in *Methods of experimental physics*, Vol. 23, Part B, ed. by D. L. Price and K. Sköld (Academic Press, London, 1987), p. 243: *Glasses*
- N. Wakabayashi, in *Methods of experimental physics*, Vol. 21, ed. by J. N. Mundy, S. J. Rothman, M. J. Fluss and L. C. Smedskjaer (Academic Press, London, 1983), p. 194: *Diffusion studies*

6.5 Exercises

Exercise No 6.1

(a) Determine the diffusion parameters (τ_0, l_0, $\langle l \rangle$, D) of H_2O from the data displayed in Fig. 6.6.

(b) The diffusion constant D of a liquid exhibits a temperature dependence which can be described by an Arrhenius law

$$D(T) = D_0 \exp\left(-\frac{E_a}{k_B T}\right), \tag{6.41}$$

where E_a is the activation energy (see Chap. 15). Analyze the temperature dependence of D obtained in (a) according to Eq. (6.41).

Exercise No 6.2

(a) Determine the function $f(\boldsymbol{Q})$ defined by Eq. (6.28) for H atoms occupying either octahedral or tetrahedral sites in α-PdH_x, see Fig. 6.7. Consider only nearest-neighbor jumps. The lattice parameter of fcc Pd is $a = 3.8898$ Å, the positions of the H atoms are at $(1/2, 1/2, 1/2)$ and

Table 6.1 Diffusion parameters of H_2O.

T [°C]	τ_0 [ps]	l_0 [Å]	$\langle l \rangle$ [Å]	D [10^{-4} cm^{2}s^{-1}]
20	2.6(1)	0.38(2)	1.3(1)	0.111(8)
5	4.4(3)	0.40(3)	1.4(1)	0.073(8)
-5	7.9(6)	0.48(5)	1.7(2)	0.058(9)
-20	48(14)	1.3(4)	4.5(1.4)	0.070(32)

$(1/4, 1/4, 1/4)$ for the octahedral and tetrahedral case, respectively.

(b) The data displayed in Fig. 6.8 were obtained for polycrystalline α-PdH_x, so that the results of (a) have to be averaged in space.

(c) Compare the result of (b) with the data displayed in Fig. 6.8 in order to find out whether the tetrahedral or octahedral jump diffusion model is appropriate for α-PdH_x. Check the applicability of the Arrhenius law Eq. (6.41).

6.6 Solutions

Exercise No 6.1

(a) Table 6.1 lists the diffusion parameters τ_0 and l_0 obtained from a least-squares fit to the linewidth data displayed in Fig. 6.6 on the basis of Eqs (6.31) and (6.34). The calculated linewidths are also shown in Fig. 6.6. The mean jump length $\langle l \rangle$ and the diffusion constant D are calculated from Eqs (6.38) and (6.37), respectively. As expected, the relaxation time τ_0 is rapidly increasing upon cooling, whereas the jump length remains almost constant. The drastic increase of the diffusion parameters derived for $T = -20°C$ indicates that the nature of a strongly undercooled liquid may be different from the normal liquid state.

(b) As shown in Fig. 6.9, the diffusion constant D obeys the Arrhenius law (Eq. (6.41)) perfectly in the normal liquid state. The fit (excluding the data in the strongly undercooled regime at $T = 20°C$) yields an activation energy $E_a \approx 175$ meV.

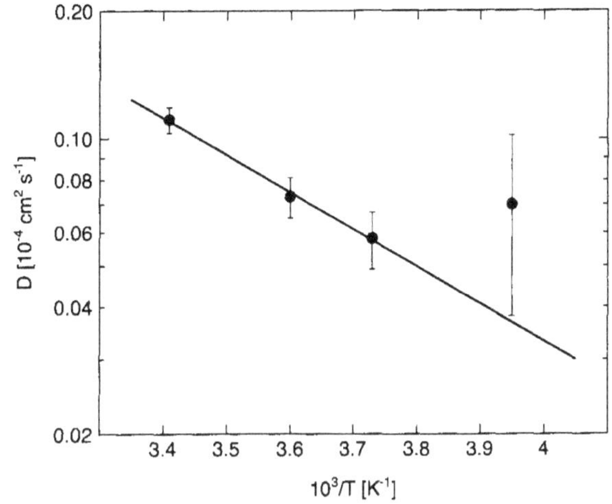

Fig. 6.9 Logarithmic Arrhenius plot of the diffusion constant D of H_2O vs the inverse temperature (see Exercise No 6.1a). The line is the result of a fit to the Arrhenius law (Eq. (6.41)) excluding the data obtained at $T = -20°C$.

Exercise No 6.2

(a) From Fig. 6.7 the following jump vectors $\boldsymbol{l}$ are derived:
tetrahedral sites: $\pm\frac{a}{2}(1,0,0)$, $\pm\frac{a}{2}(0,1,0)$, $\pm\frac{a}{2}(0,0,1)$;
octahedral sites: $\pm\frac{a}{2}(1,1,0)$, $\pm\frac{a}{2}(1,\text{-}1,0)$, $\pm\frac{a}{2}(1,0,1)$, $\pm\frac{a}{2}(1,0,\text{-}1)$, $\pm\frac{a}{2}(0,1,1)$, $\pm\frac{a}{2}(0,1,\text{-}1)$.

Due to the inversion symmetry of the jump vectors, Eq. (6.28) yields terms of the form $(1 - e^{i\boldsymbol{Q}\cdot\boldsymbol{l}})$ and $(1 - e^{-i\boldsymbol{Q}\cdot\boldsymbol{l}})$ which can be expressed pairwise by cosine functions, thus

$$F(\boldsymbol{Q}) = \frac{1}{3\tau_0}\left[3 - \cos\left(\frac{Q_x a}{2}\right) - \cos\left(\frac{Q_y a}{2}\right) - \cos\left(\frac{Q_z a}{2}\right)\right] \tag{6.42}$$

and

$$F(\boldsymbol{Q}) = \frac{1}{3\tau_0}\left[3 - \cos\left(\frac{Q_x a}{2}\right)\cos\left(\frac{Q_y a}{2}\right) - \cos\left(\frac{Q_x a}{2}\right)\cos\left(\frac{Q_z a}{2}\right) - \cos\left(\frac{Q_y a}{2}\right)\cos\left(\frac{Q_z a}{2}\right)\right] \tag{6.43}$$

for the tetrahedral and octahedral sites, respectively.

(b) The averaging procedure is a linear operation with equivalent moduli of the jump vectors $\boldsymbol{l}$, thus

$$\langle F(\boldsymbol{Q})\rangle = \langle \frac{1}{n\tau_0}\sum_{\boldsymbol{l}}\left(1-e^{\imath\boldsymbol{Q}\cdot\boldsymbol{l}}\right)\rangle = \frac{1}{n\tau_0}\langle\sum_{\boldsymbol{l}}\left(1-e^{\imath\boldsymbol{Q}\cdot\boldsymbol{l}}\right)\rangle$$
$$= \frac{1}{\tau_0}\langle\left(1-e^{\imath\boldsymbol{Q}\cdot\boldsymbol{l}}\right)\rangle = \frac{1}{\tau_0}\left(1-\langle e^{\imath\boldsymbol{Q}\cdot\boldsymbol{l}}\rangle\right), \tag{6.44}$$

with

$$\langle e^{\imath\boldsymbol{Q}\cdot\boldsymbol{l}}\rangle = \frac{1}{2}\int_0^{\pi} e^{\imath Ql\cos\theta}\sin\theta \mathrm{d}\theta = \frac{\sin(Ql)}{Ql}, \tag{6.45}$$

thus

$$f(Q) = \frac{1}{\tau_0}\left(1-\frac{\sin(Ql)}{Ql}\right). \tag{6.46}$$

(c) The moduli of the jump vectors are $l = 1.95$ Å and $l = 2.75$ Å for the tetrahedral and octahedral sites, respectively. Using Eq. (6.46), a least-squares fit to the linewidth data displayed in Fig. 6.8 yields the diffusion parameters listed in Table 6.2. The calculated linewidths are also shown in Fig. 6.8 which unambiguously favor the octahedral model. Analyzing the diffusion constant D according to the Arrhenius law (Eq. (6.41)) yields an activation energy $E_a \approx 165$ meV.

Table 6.2 Diffusion parameters determined for α-$PdH_{0.02-0.04}$.

T [K]	l [Å]	τ_0 [ps]	D [10^{-4} cm^{2}s^{-1}]
582	2.75	2.8(2)	0.45(4)
630	2.75	2.4(1)	0.53(3)
704	2.75	1.6(1)	0.79(5)
704	1.95	1.6(3)	0.40(8)

Chapter 7

Magnetic Structures

7.1 General cross section

We start from the master formula Eq. (2.42). For elastic scattering we have $|\lambda\rangle = |\lambda'\rangle$, so that the matrix elements in Eq. (2.43) can be replaced by their expectation values. With $\boldsymbol{l} = \boldsymbol{R}_j - \boldsymbol{R}_{j'}$ and by integration with respect to ω we obtain

$$\frac{\mathrm{d}\sigma}{\mathrm{d}\Omega} = (\gamma r_0)^2 e^{-2W(\boldsymbol{Q})} F^2(\boldsymbol{Q}) \sum_{\alpha,\beta} \left(\delta_{\alpha\beta} - \frac{Q_\alpha Q_\beta}{Q^2} \right) \sum_{l} e^{i\boldsymbol{Q}\cdot\boldsymbol{l}} \langle \hat{S}_0^\alpha \rangle \langle \hat{S}_l^\beta \rangle. \quad (7.1)$$

7.2 Paramagnets

For paramagnetic systems there is no correlation between the spins at sites 0 and l, thus we have for $l \neq 0$

$$\langle \hat{S}_0^\alpha \rangle \langle \hat{S}_l^\beta \rangle = 0.$$

Therefore we have to consider only the case $l = 0$. We find

$$\langle \hat{S}_0^\alpha \rangle \langle \hat{S}_0^\beta \rangle = \delta_{\alpha\beta} \langle \hat{S}_0^\alpha \rangle \langle \hat{S}_0^\beta \rangle = \delta_{\alpha\beta} \langle (\hat{S}_0^\alpha)^2 \rangle = \frac{1}{3} \delta_{\alpha\beta} \langle \hat{\boldsymbol{S}} \rangle^2 = \frac{1}{3} \delta_{\alpha\beta} S(S+1)$$

and

$$\sum_{\alpha,\beta} \left(\delta_{\alpha\beta} - \frac{Q_\alpha Q_\beta}{Q^2} \right) = \sum_{\alpha} \left(1 - \left(\frac{Q_\alpha}{Q} \right)^2 \right) = 2,$$

thus the final cross section reads

$$\boxed{\frac{\mathrm{d}\sigma}{\mathrm{d}\Omega} = \frac{2}{3} N (\gamma r_0)^2 e^{-2W(\boldsymbol{Q})} F^2(\boldsymbol{Q}) S(S+1)} \quad (7.2)$$

i.e., with increasing modulus of $\boldsymbol{Q}$ the paramagnetic scattering is continuously decreasing essentially with the square of the magnetic form factor $F(\boldsymbol{Q})$.

7.3 Ferromagnets

A ferromagnet consists of domains with uniformly arranged spins, but the spin directions in each domain are different. Let us consider a single domain in which the spins are oriented along the z axis. Then

$$\langle \hat{S}_l^x \rangle = \langle \hat{S}_l^y \rangle = 0; \quad \langle \hat{S}_l^z \rangle \neq 0.$$

For a Bravais ferromagnet we can skip the index l and find from Eq. (7.1)

$$\frac{\mathrm{d}\sigma}{\mathrm{d}\Omega} = (\gamma r_0)^2 e^{-2W(\boldsymbol{Q})} F^2(\boldsymbol{Q}) \left(1 - \left(\frac{Q_z}{Q}\right)^2\right) \langle \hat{S}^z \rangle^2 \sum_l e^{\imath \boldsymbol{Q}\cdot \boldsymbol{l}}. \tag{7.3}$$

We introduce $\boldsymbol{e}$ as the unit vector along the magnetization direction z to rewrite

$$\frac{Q_z}{Q} = \frac{\boldsymbol{Q}\cdot\boldsymbol{e}}{Q} = \frac{\boldsymbol{\tau}\cdot\boldsymbol{e}}{\tau}.$$

With use of Eq. (A.10) for the lattice sum we arrive at the final formula for the cross section:

$$\boxed{\frac{\mathrm{d}\sigma}{\mathrm{d}\Omega} = N\frac{(2\pi)^3}{v_0}(\gamma r_0)^2 e^{-2W(\boldsymbol{Q})} F^2(\boldsymbol{Q}) \langle \hat{S}^z \rangle^2 \sum_{\boldsymbol{\tau}} \langle 1 - \left(\frac{\boldsymbol{\tau}\cdot\boldsymbol{e}}{\tau}\right)^2 \rangle\, \delta(\boldsymbol{Q}-\boldsymbol{\tau})} \tag{7.4}$$

The $\langle \cdot \rangle$ brackets denote the average over all the domain orientations which for an arbitrary distribution of domain orientations reduces to

$$\langle 1 - \left(\frac{\boldsymbol{\tau}\cdot\boldsymbol{e}}{\tau}\right)^2 \rangle = \frac{2}{3}. \tag{7.5}$$

Equation (7.5) also holds if for symmetry reasons only a few domain orientations are possible, e.g., (100), (010), and (001) in systems of cubic symmetry.

Equation (7.4) shows that ferromagnetic Bragg scattering occurs at all the reciprocal lattice vectors $\boldsymbol{\tau}$, thus it is often difficult to separate the ferromagnetic from the nuclear Bragg scattering. A discrimination between the two scattering contributions can be achieved by considering the essential factors in Eq. (7.4):

- The magnetic Bragg scattering is proportional to the square of the zero-field magnetization, $\langle \hat{S}^z \rangle^2$, which exhibits a strong temperature dependence, particularly when approaching the Curie temperature.

- The $\boldsymbol{Q}$-dependence of the magnetic scattering follows the square of the magnetic form factor, $F^2(\boldsymbol{Q})$, which decreases rapidly with increasing modulus of $\boldsymbol{Q}$.

- The magnetic scattering depends on the orientation of $\langle \hat{S}^z \rangle$ relative to the reciprocal lattice vector $\boldsymbol{\tau}$.

Furthermore, by applying an external magnetic field along $\boldsymbol{Q}$ the spins will align along that direction (for sufficiently large fields), so that $\frac{\boldsymbol{\tau}\cdot\boldsymbol{e}}{\tau} = 1$, thus the magnetic scattering vanishes. The difference of two measurements (with and without external field) yields therefore directly the magnetic scattering contributions. The most elegant method for discrimination, however, involves the use of polarized neutrons.

7.4 Antiferromagnets

In antiferromagnets the domains consist of two sublattices A und B with antiparallel spin alignment, thus $\langle \hat{S}^z \rangle = 0$ in each domain. One therefore defines S^z as the so-called staggered spin, which corresponds to the zero-field magnetization in each of the two sublattices A and B. For the calculation of the cross section we start from Eq. (7.3) by treating the crystal as non-Bravais with the spins of the sublattice B at site $\boldsymbol{d}$:

$$\frac{d\sigma}{d\Omega} = (\gamma r_0)^2 e^{-2W(\boldsymbol{Q})} F^2(\boldsymbol{Q}) \left(1 - \left(\frac{Q_z}{Q}\right)^2\right) \langle \hat{S}^z \rangle^2 \sum_{\boldsymbol{l}} e^{i\boldsymbol{Q}\cdot\boldsymbol{l}} \sum_{\boldsymbol{d}} \sigma_{\boldsymbol{d}} e^{i\boldsymbol{Q}\cdot\boldsymbol{d}} \tag{7.6}$$

with $\sigma_{\boldsymbol{d}} = +1$ for an ion in sublattice A and $\sigma_{\boldsymbol{d}} = -1$ for an ion in sublattice B. By using Eq. (A.10) we find the final cross-section formula:

$$\boxed{\begin{aligned} \frac{d\sigma}{d\Omega} &= N_m \frac{(2\pi)^3}{v_{0m}} (\gamma r_0)^2 e^{-2W(\boldsymbol{Q})} \\ &\quad \times \sum_{\boldsymbol{\tau}_m} |S_m(\boldsymbol{\tau}_m)|^2 \langle 1 - \left(\frac{\boldsymbol{\tau}_m \cdot \boldsymbol{e}}{\tau_m}\right)^2 \rangle \delta(\boldsymbol{Q} - \boldsymbol{\tau}_m) \end{aligned}} \tag{7.7}$$

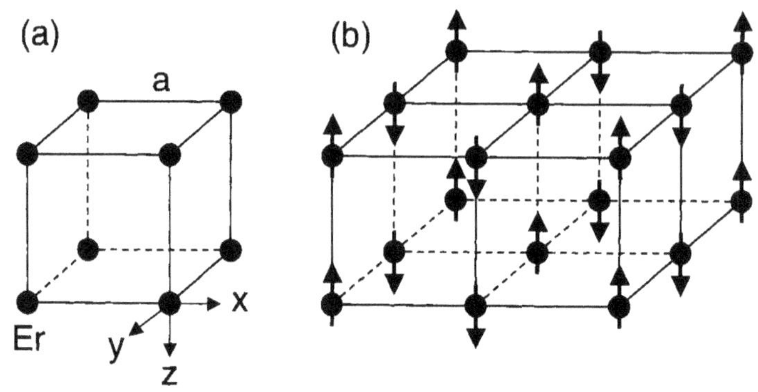

Fig. 7.1 (a) Arrangement of the Er ions in the cubic unit cell of $ErPd_3$ (the positions of the Pd ions are not shown). (b) Antiferromagnetic arrangement of the Er spins in $ErPd_3$ below $T_N = 3$ K.

with the magnetic structure factor

$$\boxed{S_m(\boldsymbol{\tau}_m) = \langle S^z \rangle F(\boldsymbol{\tau}_m) \sum_{\boldsymbol{d}} \sigma_{\boldsymbol{d}} e^{i\boldsymbol{\tau}_m \cdot \boldsymbol{d}}} \tag{7.8}$$

In these equations N_m ($= N/2$) is the number of magnetic unit cells in the crystal, v_{0m} is the volume of the magnetic unit cell, and $\boldsymbol{\tau}_m$ denotes a magnetic reciprocal lattice vector which is defined by the particular spin arrangement.

Let us consider as an example the antiferromagnet $ErPd_3$ ($T_N = 3$ K). The Er ions occupy the positions of a simple cubic lattice as shown in Fig. 7.1(a), thus the reciprocal lattice vectors are defined by $\boldsymbol{\tau} = 2\pi/a \cdot (t_1, t_2, t_3)$ with integer numbers t_i. Below the Néel temperature T_N the antiferromagnetic structure is characterized by the stacking of ferromagnetic (1,1,0) planes in an up-down-up-down sequence, i.e., the magnetic moments of the Er ions alternately point upwards and downwards along the x- and y-directions, whereas the Er spins are parallel to each other along the z-direction as shown in Fig. 7.1(b). This gives rise to a doubling of the magnetic unit cell along the x- and y-directions ($v_{0m} = 4v_0$), thus the magnetic unit cell is composed of four Er ions at positions $\boldsymbol{d}_1 = a(0,0,0)$, $\boldsymbol{d}_2 = a(1,0,0)$, $\boldsymbol{d}_3 = a(0,1,0)$, $\boldsymbol{d}_4 = a(1,1,0)$ with sublattice parameters $\sigma_1 = +1$, $\sigma_2 = -1$, $\sigma_3 = -1$, $\sigma_4 = +1$, respectively. The magnetic recipro-

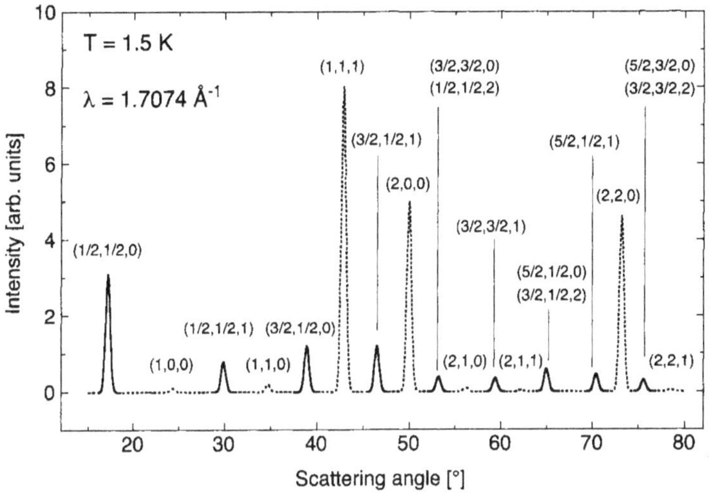

Fig. 7.2 Neutron diffraction pattern of $ErPd_3$ at $T = 1.5$ K (the background is subtracted). The dotted and solid lines refer to nuclear and magnetic Bragg scattering, respectively (after [Elsenhans *et al.* (1991)]).

cal lattice vectors are defined by $\boldsymbol{\tau}_m = 2\pi/a \cdot (t_1 + 1/2, t_2 + 1/2, t_3)$. From Eq. (7.8) we find:

$$\begin{aligned}\sum_{\boldsymbol{d}} \sigma_{\boldsymbol{d}}\, e^{i\boldsymbol{\tau}_m\cdot\boldsymbol{d}} &= 0, \quad \text{for} \quad \boldsymbol{\tau}_m = \frac{2\pi}{a}(t_1, t_2, t_3),\\ &= 4, \quad \text{for} \quad \boldsymbol{\tau}_m = \frac{2\pi}{a}(t_1 + \frac{1}{2}, t_2 + \frac{1}{2}, t_3).\end{aligned}$$

We arrive at the important result that nuclear and magnetic Bragg scattering occurs at different points in the reciprocal lattice. This is visualized in Fig. 7.2 for the neutron diffraction pattern taken for $ErPd_3$ at $T = 1.5$ K. Above T_N only nuclear Bragg scattering (corresponding to the peaks indexed with integer numbers) occurs, whereas below T_N magnetic Bragg scattering (peaks indexed with partly half-integer numbers) shows up.

7.5 Helical spin structures (magnetic spiral structures)

Let us consider the magnetic spiral structure of holmium (magnetic ordering temperature = 133 K) as shown in Fig. 7.3. The Ho ions are arranged in a hexagonal lattice. Within a plane perpendicular to the c axis all the moments are ferromagnetically aligned, but the moment direction turns by

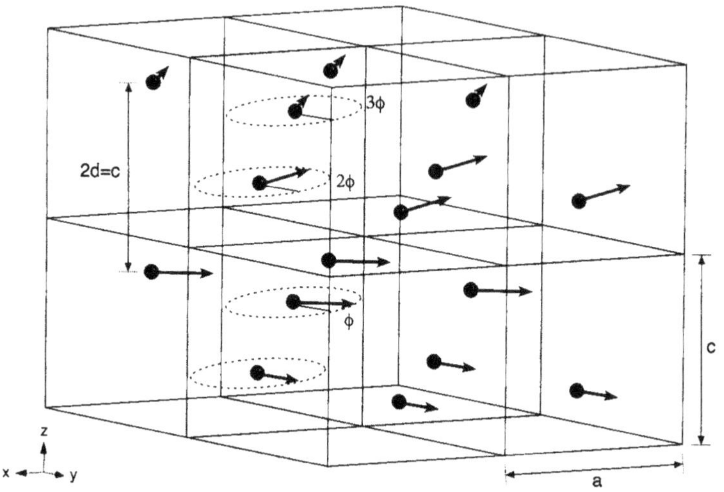

Fig. 7.3 Arrangement of the magnetic moments in Ho below 133 K.

an angle ϕ between adjacent planes. We define a spiral vector $\boldsymbol{P}$ as follows: $\boldsymbol{P}$ is a vector along the z axis; its length equals $(2\pi/\phi)$-times the distance d between adjacent planes. The expectation values of the spin operators are then given by

$$\langle \hat{S}_l^x \rangle = \langle \hat{S} \rangle \cos(\boldsymbol{P}^* \cdot \boldsymbol{l}), \quad \langle \hat{S}_l^y \rangle = \langle \hat{S} \rangle \sin(\boldsymbol{P}^* \cdot \boldsymbol{l}), \quad \langle \hat{S}_l^z \rangle = 0, \tag{7.9}$$

with $\boldsymbol{P}^* = 2\pi/\boldsymbol{P}$. Inserting Eq. (7.9) in Eq. (7.1) yields

$$\begin{aligned} \frac{\mathrm{d}\sigma}{\mathrm{d}\Omega} &= (\gamma r_0)^2 e^{-2W(\boldsymbol{Q})} F^2(\boldsymbol{Q}) \sum_{\boldsymbol{l}} e^{\imath \boldsymbol{Q}\cdot\boldsymbol{l}} \langle \hat{S} \rangle^2 \\ &\times \left[\left(1 - \left(\frac{Q_x}{Q} \right)^2 \right) \cos(\boldsymbol{P}^* \cdot \boldsymbol{l}) - \frac{Q_x Q_y}{Q^2} \sin(\boldsymbol{P}^* \cdot \boldsymbol{l}) \right] . \end{aligned} \tag{7.10}$$

Since $\boldsymbol{P}^*$ is not a reciprocal lattice vector, the summation over $\sin(\boldsymbol{P}^* \cdot \boldsymbol{l})$ vanishes. We express $\cos(\boldsymbol{P}^* \cdot \boldsymbol{l})$ by an exponential:

$$\begin{aligned} \frac{\mathrm{d}\sigma}{\mathrm{d}\Omega} &= (\gamma r_0)^2 e^{-2W(\boldsymbol{Q})} F^2(\boldsymbol{Q}) \langle \boldsymbol{S} \rangle^2 \\ &\times \left(1 - \left(\frac{Q_x}{Q} \right)^2 \right) \frac{1}{2} \sum_{\boldsymbol{l}} \left(e^{\imath(\boldsymbol{Q}+\boldsymbol{P}^*)\cdot\boldsymbol{l}} + e^{\imath(\boldsymbol{Q}-\boldsymbol{P}^*)\cdot\boldsymbol{l}} \right) . \end{aligned} \tag{7.11}$$

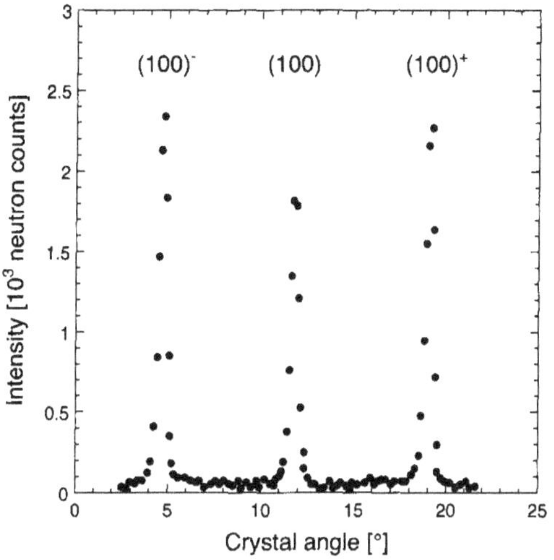

Fig. 7.4 Rocking curve about the nuclear Bragg reflection (100) in Ho at $T = 77$ K. The magnetic satellites are indexed by $(100)^-$ and $(100)^+$ (after [Koehler *et al.* (1966)]).

Application of Eq. (A.10) yields

$$\frac{\mathrm{d}\sigma}{\mathrm{d}\Omega} = \frac{N}{2}\frac{(2\pi)^3}{v_0}(\gamma r_0)^2 e^{-2W(\boldsymbol{Q})} F^2(\boldsymbol{Q})\langle\hat{S}\rangle^2 \left(1 - \left(\frac{Q_x}{Q}\right)^2\right)$$
$$\times \sum_{\boldsymbol{\tau}} \left(\delta(\boldsymbol{Q} + \boldsymbol{P}^* - \boldsymbol{\tau}) + \delta(\boldsymbol{Q} - \boldsymbol{P}^* - \boldsymbol{\tau})\right). \qquad (7.12)$$

The assumptions of Eq. (7.9) were somehow special in the sense that we confined the moment direction to the x axis in the basal plane. We could as well start with the moment direction along the y axis, which would change the polarization factor in Eq. (7.12) from $\left(1 - (Q_x/Q)^2\right)$ to $\left(1 - (Q_y/Q)^2\right)$. Taking the average yields the final cross section

$$\boxed{\frac{\mathrm{d}\sigma}{\mathrm{d}\Omega} = \frac{N}{4}\frac{(2\pi)^3}{v_0}(\gamma r_0)^2 e^{-2W(\boldsymbol{Q})} F^2(\boldsymbol{Q})\langle\hat{S}\rangle^2 \left(1 + \left(\frac{Q_z}{Q}\right)^2\right) \times \sum_{\boldsymbol{\tau}} \left(\delta(\boldsymbol{Q} + \boldsymbol{P}^* - \boldsymbol{\tau}) + \delta(\boldsymbol{Q} - \boldsymbol{P}^* - \boldsymbol{\tau})\right)} \qquad (7.13)$$

We realize that magnetic Bragg scattering occurs for $\boldsymbol{Q} = \boldsymbol{\tau} \pm \boldsymbol{P}^*$, i.e., each nuclear Bragg reflection is flanked by a pair of magnetic satellites. This is visualized in Fig. 7.4 for Ho.

7.6 Magnetic ordering wavevector

Besides the examples treated above there are myriads of other magnetic structure types. It is convenient to characterize them by the magnetic ordering wavevector $\boldsymbol{q}_0$ which is defined by the relation

$$\mu_i = \mu_0 \cos(\boldsymbol{q}_0 \cdot \boldsymbol{R}_i), \tag{7.14}$$

i.e., $\boldsymbol{q}_0$ describes properly the mutual orientation of the magnetic moments at sites $\mathbf{0}$ und $\boldsymbol{R}_i$. For instance, for the antiferromagnet $ErPd_3$ described in Chap. 7.4 we have $\boldsymbol{q}_0 = \frac{2\pi}{a}\left(\frac{1}{2}, \frac{1}{2}, 0\right)$.

It is not always possible to describe the magnetic structure unambiguously by a single ordering wavevector $\boldsymbol{q}_0$, because the magnetic reflections may originate either from the presence of more than one modulation (multiple-$\boldsymbol{q}_0$ structure) or from different domains with a single modulation. This problem can usually not be solved from diffraction data taken for polycrystalline samples, but a discrimination is possible for single crystals by the application of an external magnetic field or by applying a small amount of uniaxial stress (in order to create a single-domain crystal). For a more detailed discussion of this issue we refer to [Chatterji (2006)].

7.7 Zero-field magnetization

An inherent factor in all the cross-section formulae derived above is the square of the expectation value of the spin operator $\hat{\boldsymbol{S}}$, see Eq. (7.4) for the ferromagnet, Eq. (7.7) for the antiferromagnet, and Eq. (7.13) for the helical structure. The expectation value of the spin operator $\hat{\boldsymbol{S}}$ is directly related to the magnetic moment $\boldsymbol{\mu}$:

$$\boldsymbol{\mu} = g\mu_B \langle \hat{\boldsymbol{S}} \rangle. \tag{7.15}$$

This means that both the size and the direction of the magnetic moment can be determined directly by neutron diffraction without the need to apply an external magnetic field (as in conventional magnetization experiments). This property, namely the ability to measure the magnetization without a disturbing magnetic field (therefore the term zero-field magnetization), is unique for neutron scattering. As an example Fig. 7.5 shows the zero-field magnetization of the Dy sublattice in $DyBa_2Cu_3O_7$. Moreover, the zero-field magnetization carries important information about the dimension of the system as well as the dimension of the order parameter (see Chap. 10.4). As demonstrated in Fig. 7.5, the experimental data cannot be explained by

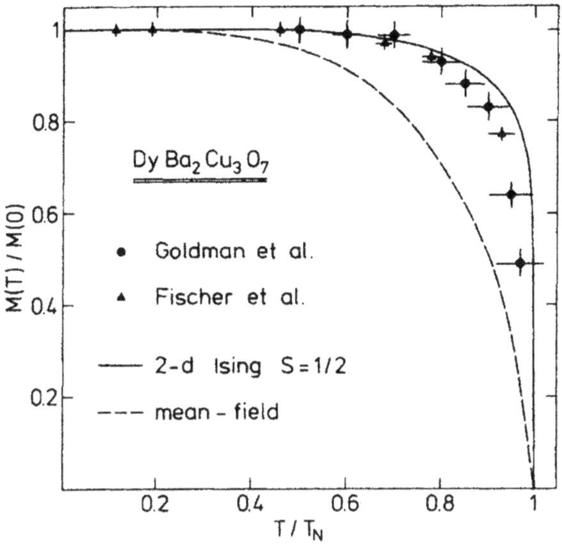

Fig. 7.5 Zero-field magnetization of the Dy^{3+} ions in $DyBa_2Cu_3O_7$. Reprinted with permission from [Allenspach *et al.* (1989)]. Copyright 1989 by the American Physical Society.

a mean-field approach, but are compatible with a two-dimensional Ising-type model.

7.8 Spin densities

As discussed in the Appendix D, the magnetic form factor $F(\boldsymbol{Q})$ is related to the normalized density $s(\boldsymbol{r})$ of the unpaired electrons by

$$F(\boldsymbol{Q}) = \int s(\boldsymbol{r})e^{i\boldsymbol{Q}\cdot\boldsymbol{r}}\mathrm{d}\boldsymbol{r}. \tag{7.16}$$

Thus $s(\boldsymbol{r})$ can be determined by the Fourier inversion of the values of $F(\boldsymbol{Q})$ which is an essential term in the magnetic neutron cross-section Eq. (2.42).

If the measurements are performed with unpolarized neutrons, the cross section is proportional to $b^2 + p^2$, where $p \equiv B\boldsymbol{M} \cdot \hat{\boldsymbol{\sigma}}$ (see Eq. (2.50)) is the magnetic scattering amplitude involving $F(\boldsymbol{Q})$. With $p = \epsilon \cdot b$ and $\epsilon \ll 1$, the magnetic cross section is a small fraction ϵ^2 of the total, thus unpolarized neutrons provide a very insensitive method of measuring $F(\boldsymbol{Q})$.

Now suppose the experiment is done with polarized neutrons, and the cross sections for the processes $|+\rangle \rightarrow |+\rangle$ and $|+\rangle \rightarrow |-\rangle$ defined by

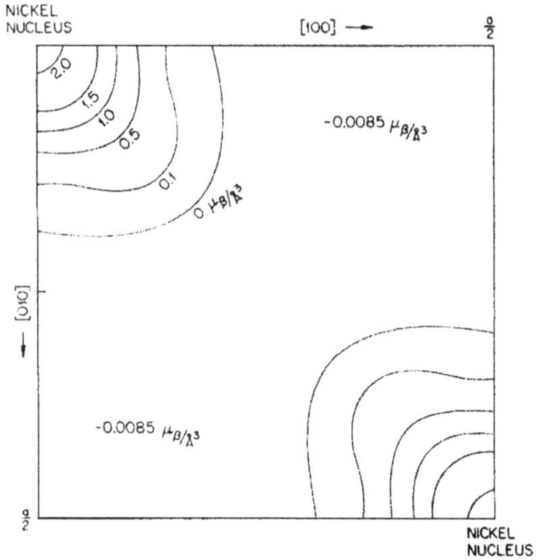

Fig. 7.6 The magnetic moment distribution of nickel in the (100)-plane. Reprinted with permission from [Mook (1966)]. Copyright 1966 by the American Physical Society.

Eq. (2.53) are measured separately. The ratio of the cross sections, known as the flipping ratio, is

$$R = \frac{(b-p)^2}{(b+p)^2} = \frac{(1-\epsilon)^2}{(1+\epsilon)^2} \approx 1 - 4\epsilon, \tag{7.17}$$

thus the fractional change of R is 4ϵ, which is much larger than ϵ^2. Not only is the polarization method more sensitive, but it provides the additional advantage of giving the sign of p relative to b.

As an example the results obtained for nickel are shown in Fig. 7.6. The magnetic form factor $F(\boldsymbol{Q})$ was determined for the first 27 Bragg reflections by the polarization technique. The Fourier inversion of the form factor indicates that the moment density is quite asymmetric about the lattice sites and is negative in the region between the lattice sites.

7.9 Further reading

- P. J. Brown, in *Neutron scattering from magnetic materials*, ed. by T. Chatterji (Elsevier, Amsterdam, 2006), p. 215: *Spherical neutron polarimetry*

- T. Chatterji, in *Neutron scattering from magnetic materials*, ed. by T. Chatterji (Elsevier, Amsterdam, 2006), p. 25: *Magnetic structures*
- P. Fischer, in *Neutron scattering*, ed. by A. Furrer (Proc. 93-01, ISSN 1019-6447, PSI Villigen, 1993), p. 199: *Magnetic structures*
- H. Glättli and M. Goldman, in *Methods of experimental physics*, Vol. 23, Part C, ed. by D. L. Price and K. Sköld (Academic Press, London, 1987), p. 241: *Nuclear magnetism*
- Y.A. Izyumov and R.P. Ozerov, *Magnetic neutron diffraction* (Plenum Press, New York, 1970)
- B. Lebech, in *Magnetic neutron scattering*, ed. by A. Furrer (World Scientific, Singapore, 1995), p. 58: *Magnetic structure determination by neutron diffraction*
- J. M. Rossat-Mignod, in *Methods of experimental physics*, Vol. 23, Part C, ed. by D. L. Price and K. Sköld (Academic Press, London, 1987), p. 69: *Magnetic structures*
- J. Schweizer, in *Neutron scattering from magnetic materials*, ed. by T. Chatterji (Elsevier, Amsterdam, 2006), p. 153: *Polarized neutrons and polarization analysis*

7.10 Exercises

Exercise No 7.1

The quaternary intermetallic compound $HoNi_2B_2C$ exhibits a coexistence of superconductivity and long-range magnetic ordering of the Ho^{3+} sublattice at low temperatures. The compound has a body-centered tetragonal structure as shown in Fig 7.7a. The structural parameters in the paramagnetic state at $T = 10$ K are $a = 3.5087$ Å and $c = 10.5274$ Å, with atomic positions Ho (0,0,0), Ni (1/2,0,1/4), B (0,0,$z = 0.3592$), and C (1/2,1/2,0).

The transition into the superconducting state occurs below $T_c \approx 8$ K. At the same temperature the compound transforms from the paramagnetic state into a magnetically ordered state. The magnetic structure is a superposition of two configurations, one in which the Ho^{3+} spins are coupled antiferromagnetically (Fig. 7.7b), and another which corresponds to a helical spin structure (Fig. 7.7c). The relative contribution of the helical spin structure is strongly reduced with decreasing temperature, and at $T = 2.2$ K only the antiferromagnetic structure is present as shown in Fig. 7.7b. The helical spin structure manifests itself by pairs of magnetic satellites around

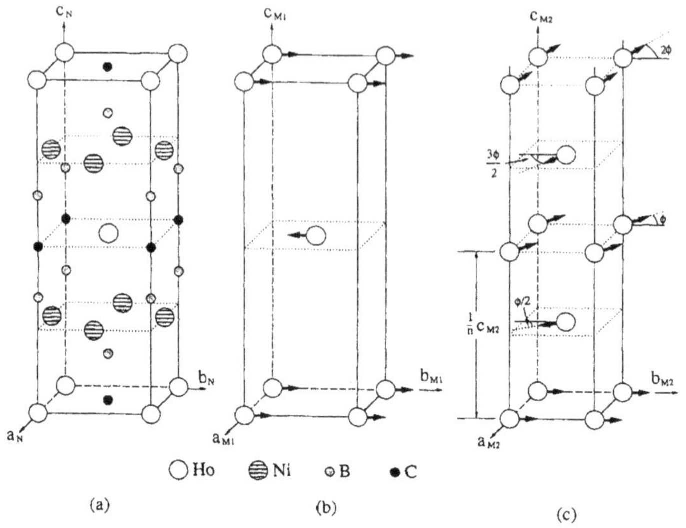

Fig. 7.7 Schematic representation of (a) the unit cell of $HoNi_2B_2C$; (b) the antiferromagnetic ordering of the Ho^{3+} spins; (c) the spin configuration of the magnetic helical structure. In (b) the moments are depicted along the b axis for clarity only (the a and b axes are equivalent). Reprinted from [Huang *et al.* (1995)]. Copyright 1995 by the American Physical Society.

the magnetic Bragg peaks as visualized in Fig. 7.9 for the (0,0,1) reflection observed at $T = 5.1$ K.

(a) Determine the antiferromagnetic structure of $HoNi_2B_2C$ from the neutron diffraction pattern observed at $T = 2.2$ K (Fig. 7.8).
(b) Determine the magnetic ordering wavevector $\boldsymbol{q}_0$ associated with the magnetic spiral structure of $HoNi_2B_2C$ from the neutron diffraction pattern observed at $T = 5.1$ K (Fig. 7.9). What is the turn angle ϕ of the spins between adjacent planes?

7.11 Solutions

Exercise No 7.1

(a) None of the magnetic Bragg peaks are indexed by half integer numbers, thus the magnetic unit cell is the same as the chemical unit cell. This means that the magnetic moments associated with the Ho^{3+} ions at the corners of the unit cell are ferromagnetically coupled.

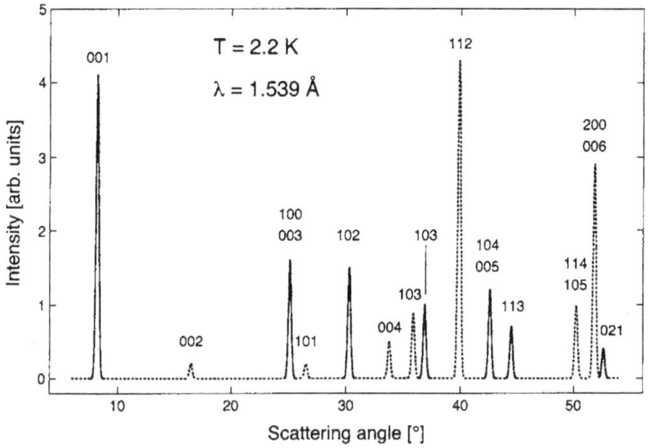

Fig. 7.8 Powder diffraction pattern of $HoNi_2B_2C$ taken at $T = 2.2$ K. The nuclear and magnetic Bragg peaks are indicated by dotted and solid lines, respectively (after [Huang *et al.* (1995)]).

If the magnetic coupling with the Ho^{3+} ion in the body center of the unit cell were also ferromagnetic, then all the magnetic reflections would coincide with the nuclear Bragg peaks. This is clearly not the case, thus the coupling between the Ho^{3+} ion in the body center with those at the corners of the unit cell is antiferromagnetic. This can be proven readily by calculating the magnetic structure factor (Eq. (7.8)) as exemplified below.

There are two Ho^{3+} ions per unit cell located at $\boldsymbol{d}_1$=(0,0,0) and $\boldsymbol{d}_2$=(1/2,1/2,1/2). Magnetic Bragg scattering occurs, e.g., for $\boldsymbol{Q} = \boldsymbol{\tau}_{001}$:

$$\begin{aligned} F_m(\boldsymbol{\tau}_{001}) &\propto \sigma_1 e^{\imath 2\pi (0,0,1)\cdot(0,0,0)} + \sigma_2 e^{\imath 2\pi (0,0,1)\cdot(1/2,1/2,1/2)} \\ &= \sigma_1 e^0 + \sigma_2 e^{\imath\pi} = \sigma_1 - \sigma_2 \neq 0. \end{aligned} \tag{7.18}$$

On the other hand, there is no magnetic Bragg scattering, e.g., for $\boldsymbol{Q} = \boldsymbol{\tau}_{002}$:

$$\begin{aligned} F_m(\boldsymbol{\tau}_{002}) &\propto \sigma_1 e^{\imath 2\pi (0,0,2)\cdot(0,0,0)} + \sigma_2 e^{\imath 2\pi (0,0,2)\cdot(1/2,1/2,1/2)} \\ &= \sigma_1 e^0 + \sigma_2 e^{\imath 2\pi} = \sigma_1 + \sigma_2 = 0. \end{aligned} \tag{7.19}$$

The above conditions require $\sigma_1 = +1$ and $\sigma_2 = -1$, i.e., the antiferromagnetic coupling is confirmed. The magnetic ordering wavevector is $\boldsymbol{q}_0 = (0, 0, 1)$ which follows from Eq. (7.14):

$$\boldsymbol{\mu}_2 = \boldsymbol{\mu}_1 \cos(2\pi(0,0,1)\cdot(1/2,1/2,1/2)) = \boldsymbol{\mu}_1 \cos\pi = -\boldsymbol{\mu}_1. \tag{7.20}$$

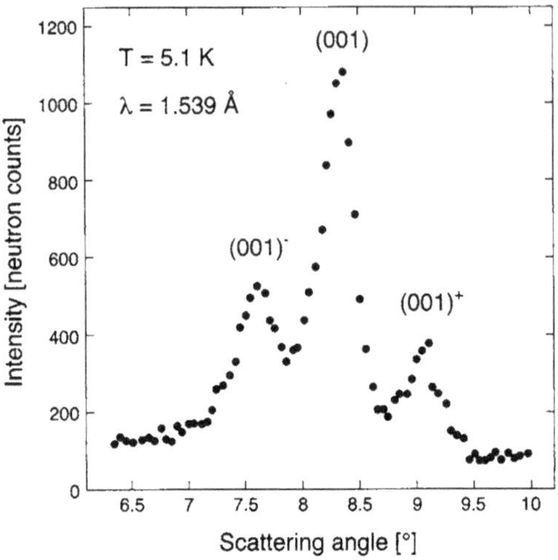

Fig. 7.9 Low-angle part of the diffraction pattern of $HoNi_2B_2C$ taken at $T = 5.1$ K, showing the separation of the helical satellite peaks from the antiferromagnetic (001) reflection (after [Huang *et al.* (1995)]).

Concerning the direction of the magnetic moments $\boldsymbol{\mu}$ we refer to the discussion in Chap. 7.3. For $\boldsymbol{\mu}||\boldsymbol{Q}$ all magnetic reflections $(0,0,l)$ are forbidden. However, the inspection of the diffraction patterns displayed in Fig. 7.8 shows that the antiferromagnetic reflections $(0,0,l)$ are very strong, thus eliminating the solution with moments lying along or in the proximity of the c-axis. In fact, a quantitative analysis shows that the Ho^{3+} spins are oriented perpendicular to the c-axis [Huang *et al.* (1995)].

(b) The reflection $(0,0,1)$ occurs at $2\theta = 8.35°$. With $Q = (4\pi/\lambda)\sin\theta$ (Eq. (4.8)) and $\lambda = 1.539$ Å we derive $Q_{001} = 0.595$ Å^{-1}. The satellites occur at $2\theta^- = 7.65°$ and $2\theta^+ = 9.05°$, thus $Q^-_{001} = 0.545$ Å^{-1} and $Q^+_{001} = 0.644$ Å^{-1}. From Eq. (8.12) we find $P^* = 0.05$ Å^{-1}, thus $P \approx 130$ Å $\approx 12.3\,c$, i.e., the turn angle between adjacent unit cells is $360°/12.3 \approx 30°$ and therefore $\phi \approx 30°/2 \approx 15°$ between adjacent Ho^{3+} planes. The magnetic ordering wavevector is therefore $\boldsymbol{q}_0 = (0,0,1 \pm P^*) = (0,0,1,\pm 0.05)$. The $\pm$ sign indicates that the chirality of the spiral can be either left- or right-handed.

Chapter 8

Magnetic Excitations

8.1 Magnetic cluster excitations

8.1.1 *Dimers*

The simplest magnetic cluster system is the dimer (two coupled spins $\boldsymbol{S}_1$ and $\boldsymbol{S}_2$) for which the Heisenberg Hamiltonian is defined by

$$\hat{\mathcal{H}} = -2J\hat{\boldsymbol{S}}_1 \cdot \hat{\boldsymbol{S}}_2, \tag{8.1}$$

where J is the exchange parameter. $\hat{\mathcal{H}}$ commutes with the total spin $\boldsymbol{S} = \boldsymbol{S}_1 + \boldsymbol{S}_2$, thus S is a good quantum number, and the wave functions of the dimer states can be described by $|S, M\rangle$ with $-S \leq M \leq S$. Assuming identical magnetic ions ($S_1 = S_2$) the eigenvalues of Eq. (8.1) are

$$E(S) = -J\left(S(S+1) - 2S_1(S_1+1)\right), \quad 0 \leq S \leq 2S_1. \tag{8.2}$$

The energy splittings satisfy the Landé interval rule

$$E(S) - E(S-1) = -2JS. \tag{8.3}$$

The energy-level sequence of an antiferromagnetically ($J < 0$) coupled pair of ions is indicated at the right hand side of Fig. 8.1.

Selection rules for transitions $|S\rangle \to |S'\rangle$ can be derived from the differential neutron cross-section for dimer transitions. The corresponding calculation starts from Eqs (2.42) and (2.43). For the evaluation of the matrix elements we replace the spin operators $\hat{S}_j^\alpha$ by irreducible tensor operators $\hat{T}_j^q$ of rank 1:

$$\hat{T}_j^0 = \hat{S}_j^z, \quad \hat{T}_j^{\pm 1} = \mp\frac{1}{\sqrt{2}}\left(\hat{S}_j^x \pm \imath\hat{S}_j^y\right). \tag{8.4}$$

The M dependence of the matrix elements is given by the Wigner-Eckart theorem:

$$\langle S', M'|\hat{T}_j^q|S, M\rangle = (-1)^{S'-M'}\begin{pmatrix} S' & 1 & S \\ -M' & q & M \end{pmatrix}\langle S'||\hat{T}_j||S\rangle. \tag{8.5}$$

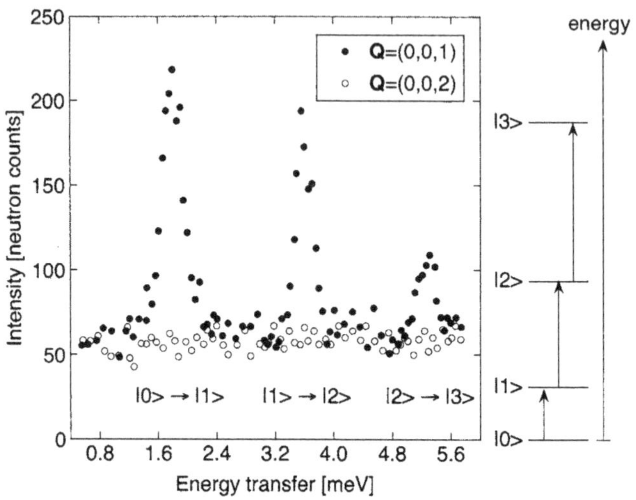

Fig. 8.1 Left: Energy spectra of neutrons scattered from Mn^{2+} pairs in $CsMn_{0.28}Mg_{0.72}Br_3$ at T = 30 K. Full circles: $\boldsymbol{Q}$ =(0,0,1); open circles: $\boldsymbol{Q}$ =(0,0,2) (after [Falk *et al.* (1984)]). Right: Energy-level sequence of an antiferromagnetically coupled spin pair.

Since $\hat{T}_j$ operates only on the jth part of the coupled system, the reduced matrix element of Eq. (8.5) can be further reduced:

$$\begin{aligned}\langle S'||\hat{T}_1||S\rangle &= (-1)^{2S_1+S+1}\sqrt{(2S+1)(2S'+1)}\\ &\quad\times\begin{Bmatrix} S' & S & 1\\ S_1 & S_1 & S_1\end{Bmatrix}\langle S_1|||\hat{T}_1|||S_1\rangle,\\ \langle S'||\hat{T}_2||S\rangle &= (-1)^{S-S'}\langle S'||\hat{T}_1||S\rangle,\\ \langle S_1|||\hat{T}_1|||S_1\rangle &= \sqrt{S_1(S_1+1)(2S_1+1)}.\end{aligned} \tag{8.6}$$

The two-row brackets in Eqs (8.5) and (8.6) are 3-j and 6-j symbols defined in the Appendix F. From the symmetry properties of the 3-j and 6-j symbols the selection rules

$$\Delta S = S - S' = 0, \pm 1; \quad \Delta M = M - M' = 0, \pm 1 \tag{8.7}$$

are derived. Thus inelastic transitions are only possible between adjacent energy levels as indicated by arrows in Fig. 8.1. Since each energy level is $(2S+1)$-fold degenerate in the absence of a magnetic field, we can sum over the quantum numbers M and M':

$$\sum_{M,M'}\langle S,M|\hat{T}_j^q|S',M'\rangle\langle S',M'|\hat{T}_{j'}^q|S,M\rangle = \frac{1}{3}\langle S||\hat{T}_j||S'\rangle\langle S'||\hat{T}_{j'}||S\rangle. \tag{8.8}$$

By making use of the symmetry properties of the matrix elements defined by Eq. (8.6), we find the following cross section for the dimer transition $|S\rangle \rightarrow |S'\rangle$:

$$\frac{d^2\sigma}{d\Omega d\omega} = N(\gamma r_0)^2 \frac{k'}{k} F^2(\boldsymbol{Q}) e^{-2W(\boldsymbol{Q})} p_s \sum_{\alpha} \left(1 - \left(\frac{Q_\alpha}{Q}\right)^2\right) \times \frac{2}{3} \left(1 + (-1)^{\Delta S} \cos(\boldsymbol{Q} \cdot \boldsymbol{R})\right) \cdot |\langle S' || \hat{T}_1 || S \rangle|^2 \times \delta(\hbar\omega + E_S - E_{S'}) \qquad (8.9)$$

where N is the total number of dimers, p_S the Boltzmann population factor, and R the intradimer separation. The structure factor $\left(1 + (-1)^{\Delta S} \cos(\boldsymbol{Q} \cdot \boldsymbol{R})\right)$ is a powerful means to unambiguously identify dimer excitations from other scattering contributions due to its characteristic oscillating behavior. Furthermore, the polarization factor $\left(1 - (Q_\alpha/Q)^2\right)$ permits the discrimination between transverse ($\alpha = x, y$) and longitudinal ($\alpha = z$) transitions by measuring at different scattering vectors $\boldsymbol{Q}$ (as exemplified in Fig. 9.4 for the related topic of crystal-field transitions).

For polycrystalline material Eq. (8.9) has to be averaged in $\boldsymbol{Q}$ space:

$$\frac{d^2\sigma}{d\Omega d\omega} = N(\gamma r_0)^2 \frac{k'}{k} F^2(\boldsymbol{Q}) e^{-2W(\boldsymbol{Q})} p_s \times \frac{4}{3} \left(1 + (-1)^{\Delta S} \frac{\sin(QR)}{QR}\right) \cdot |\langle S' || \hat{T}_1 || S \rangle|^2 \times \delta(\hbar\omega + E_S - E_{S'}). \qquad (8.10)$$

The combination of the polarization factor and the structure factor results in a damped oscillatory Q-dependence of the intensities.

In Fig. 8.1 the theoretical predictions outlined above are exemplified for the magnetic excitations observed for Mn^{2+} pairs introduced into a single crystal of $CsMgBr_3$. Firstly, the Landé interval rule Eq. (8.3) is nicely satisfied by the observed energy splittings. Secondly, the observed intensities are in agreement with the structure factor; with $\boldsymbol{R} = (0, 0, \frac{1}{2})$ the intensity has a maximum for $\boldsymbol{Q} = (0, 0, 1)$ and vanishes for $\boldsymbol{Q} = (0, 0, 2)$.

8.1.2 *Trimers*

The Heisenberg Hamiltonian of a trimer is defined by

$$\hat{\mathcal{H}} = -2\left(J(\hat{\boldsymbol{S}}_1 \cdot \hat{\boldsymbol{S}}_2 + \hat{\boldsymbol{S}}_2 \cdot \hat{\boldsymbol{S}}_3) + J'\hat{\boldsymbol{S}}_1 \cdot \hat{\boldsymbol{S}}_3\right), \qquad (8.11)$$

where J and J' denote the nearest-neighbor and next-nearest-neighbor exchange parameters, respectively. It is convenient to introduce the spin quantum numbers S_{13} and S resulting from the vector sums $\boldsymbol{S}_{13} = \boldsymbol{S}_1 + \boldsymbol{S}_3$ and $\boldsymbol{S} = \boldsymbol{S}_1 + \boldsymbol{S}_2 + \boldsymbol{S}_3$ with $0 \leq S_{13} \leq 2S_i$ and $|S_{13} - S_i| \leq S \leq (S_{13} + S_i)$, respectively. The trimer states are therefore defined by $|S_{13}, S\rangle$, and their degeneracy is $(2S+1)$. With this choice of spin quantum numbers, the Hamiltonian Eq. (8.11) is diagonal; thus, the eigenvalues can readily be derived:

$$\begin{aligned} E(S_{13}, S) = & -J\left[S(S+1) - S_{13}(S_{13}+1) - S_i(S_i+1)\right] \\ & -J'\left[S_{13}(S_{13}+1) - 2S_i(S_i+1)\right]. \end{aligned} \tag{8.12}$$

The evaluation of the differential neutron cross-section follows the procedure outlined in Chap. 8.1.1. For S_1=S_2=S_3 the following reduced matrix elements are obtained:

$$\begin{aligned} \langle S'_{13}S'\|\hat{T}_1\|S_{13}S\rangle = & (-1)^{3S_1+S_{13}+S'_{13}+S} \\ & \times\sqrt{(2S+1)(2S'+1)(2S_{13}+1)(2S'_{13}+1)} \\ & \times\left\{\begin{matrix} S' & S & 1 \\ S_{13} & S'_{13} & S_1 \end{matrix}\right\}\left\{\begin{matrix} S'_{13} & S_{13} & 1 \\ S_1 & S_1 & S_1 \end{matrix}\right\}\langle S'_1\|\|\hat{T}_1\|\|S_1\rangle, \\ \langle S'_{13}S'\|\hat{T}_2\|S_{13}S\rangle = & \ \delta(S_{13}, S'_{13})(-1)^{S_1+S_{13}+S'+1}\sqrt{(2S+1)(2S'+1)} \\ & \times\left\{\begin{matrix} S' & S & 1 \\ S_1 & S_1 & S_{13} \end{matrix}\right\}\langle S'_1\|\|\hat{T}_1\|\|S_1\rangle, \\ \langle S'_{13}S'\|\hat{T}_3\|S_{13}S\rangle = & (-1)^{S_{13}-S'_{13}}\langle S'_{13}S'\|\hat{T}_1\|S_{13}S\rangle. \end{aligned} \tag{8.13}$$

It can be shown that the following relation holds:

$$\langle S'_{13}S'\|\hat{T}_2\|S_{13}S\rangle = -2\langle S'_{13}S'\|\hat{T}_1\|S_{13}S\rangle. \tag{8.14}$$

Inserting Eqs (8.13) and (8.14) into the master formula Eq. (2.42) and applying the Wigner-Eckart theorem Eq. (8.5) yields the following cross section for the trimer transition $|\hat{S}_{13}, S\rangle \rightarrow |S'_{13}, S'\rangle$:

$$\boxed{\begin{aligned}\frac{\mathrm{d}^2\sigma}{\mathrm{d}\Omega\mathrm{d}\omega} &= N(\gamma r_0)^2\frac{k'}{k}F^2(\boldsymbol{Q})e^{-2W(\boldsymbol{Q})}p_s\sum_\alpha\left(1-\left(\frac{Q_\alpha}{Q}\right)^2\right)\\ &\times\frac{2}{3}\big[1+(-1)^{\Delta S_{13}}\cos(\boldsymbol{Q}\cdot\boldsymbol{R}_{13})\\ &+2\delta(S_{13},S'_{13})\left(1-\cos(\boldsymbol{Q}\cdot\boldsymbol{R}_{12})-\cos(\boldsymbol{Q}\cdot\boldsymbol{R}_{23})\right)\big]\\ &\times|\langle S'_{13}S'||\hat{T}_1||S_{13}S\rangle|^2\cdot\delta\left(\hbar\omega+E(S_{13},S)-E(S'_{13},S')\right)\end{aligned}}$$

(8.15)

where N is the total number of trimers, and $\boldsymbol{R}_{jj'}$ denotes the distance between the magnetic ions at sites j and j'. From the symmetry properties of the 3-j and 6-j symbols in Eqs (8.5) and (8.13) we derive the selection rules

$$\begin{aligned}&\Delta S = S - S' = 0, \pm 1; \quad \Delta S_{13} = S_{13} - S'_{13} = 0, \pm 1;\\ &\Delta M = M - M' = 0, \pm 1.\end{aligned} \tag{8.16}$$

For polycrystalline material Eq. (8.15) has to be averaged in $\boldsymbol{Q}$ space. For a trimer with the magnetic ions arranged in a line we find

$$\begin{aligned}\frac{\mathrm{d}^2\sigma}{\mathrm{d}\Omega\mathrm{d}\omega} &= N(\gamma r_0)^2\frac{k'}{k}F^2(\boldsymbol{Q})e^{-2W(\boldsymbol{Q})}p_s\\ &\times\frac{4}{3}\left[\left(1+(-1)^{\Delta\hat{S}_{13}}\frac{\sin(2QR)}{2QR}\right)+2\delta(\hat{S}_{13},S'_{13})\left(1-2\frac{\sin(QR)}{QR}\right)\right]\\ &\times|\langle S'_{13}S'||\hat{T}_1||S_{13}S\rangle|^2\cdot\delta\left(\hbar\omega+E(S_{13},S)-E(S'_{13},S')\right),\end{aligned}$$

(8.17)

where R is the distance between the center spin (at site 2) and the end-standing spins (at sites 1 and 3).

$Ca_3Cu_3(PO_4)_4$ is a magnetic cluster compound in which the copper ions form linear trimers with weak intertrimer interaction, thus the theoretical formalism described above can be applied. Inserting the copper spin $S_i = \frac{1}{2}$ into Eq. (8.12) yields two doublets and a quartet at energies $E(0,\frac{1}{2}) = \frac{3}{2}J'$, $E(1,\frac{1}{2}) = 2J - \frac{1}{2}J'$, and $E(1,\frac{3}{2}) = -J - \frac{1}{2}J'$, respectively. The results of inelastic neutron scattering experiments performed for polycrystalline material at $T = 1.5$ K are shown in Fig. 8.2a. There are two well defined excitations at energies $\epsilon_1 = 9.44(3)$ meV and $\epsilon_2 = 14.22(2)$ meV.

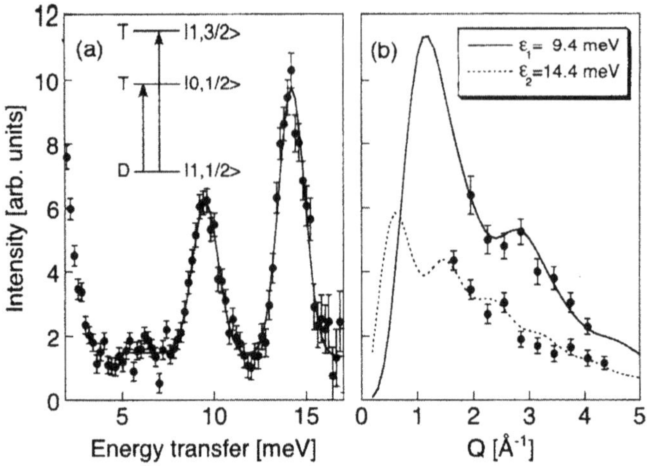

Fig. 8.2 Results of inelastic neutron scattering experiments performed for $Ca_3Cu_3(PO_4)_4$ at $T = 1.5$ K. (a) Energy spectrum observed for $Q = 2.25$ Å^{-1} with inset of the level splitting scheme. (b) Q-dependence of the observed trimer excitations (after [Podlesnyak *et al.* (2007)]).

The ratio $\epsilon_1/\epsilon_2 \approx 2/3$ can only be achieved for antiferromagnetic nearest-neighbor coupling $J < 0$; thus, the doublet $|1, \frac{1}{2}\rangle$ is the ground state, and the peaks at ϵ_1 and ϵ_2 correspond to the excited trimer states $|0, \frac{1}{2}\rangle$ and $|1, \frac{3}{2}\rangle$, respectively, as indicated in the inset of Fig. 8.2a. This identification is nicely confirmed by the Q dependence of the intensities calculated from the cross section Eq. (8.17) as shown in Fig. 8.2b. Analyzing the observed trimer splittings according to Eq. (8.12) yields the coupling parameters $J = -4.74(2)$ meV and $J' = -0.02(3)$ meV.

8.1.3 *Tetramers*

The Heisenberg Hamiltonian of a tetramer with rhombic geometry is given by

$$\hat{\mathcal{H}} = -2J\left(\hat{\boldsymbol{S}}_1 \cdot \hat{\boldsymbol{S}}_3 + \hat{\boldsymbol{S}}_1 \cdot \hat{\boldsymbol{S}}_4 + \hat{\boldsymbol{S}}_2 \cdot \hat{\boldsymbol{S}}_3 + \hat{\boldsymbol{S}}_2 \cdot \hat{\boldsymbol{S}}_4\right)$$
$$-2J'\hat{\boldsymbol{S}}_1 \cdot \hat{\boldsymbol{S}}_2 - 2J''\hat{\boldsymbol{S}}_3 \cdot \hat{\boldsymbol{S}}_4, \qquad (8.18)$$

where J, J', and J'' denote the exchange parameters as indicated in Fig. 8.3a. For a complete characterization of the tetramer states, we need additional spin quantum numbers resulting from the vector sums $\boldsymbol{S}_{12} = \boldsymbol{S}_1 + \boldsymbol{S}_2$, $\boldsymbol{S}_{34} = \boldsymbol{S}_3 + \boldsymbol{S}_4$, and $\boldsymbol{S} = \boldsymbol{S}_{12} + \boldsymbol{S}_{34}$ with $0 \leq S_{12} \leq 2S_i$,

$0 \leq S_{34} \leq 2S_i$, and $|S_{12} - S_{34}| \leq S \leq (S_{12} + S_{34})$, respectively. If the tetramer experiences a magnetic field $\boldsymbol{H}$ along the quantization axis, the Hamiltonian

$$\hat{\mathcal{H}}_{\text{ex}} = -g\mu_B H \sum_{i=0}^{4} \hat{S}_i^z = -g\mu_B H \hat{S}^z, \tag{8.19}$$

has to be added to Eq. (8.18). The joint Hamiltonians Eqs (8.18) and (8.19) are diagonal, thus the eigenvalues can easily be shown to be

$$\begin{aligned} E(S_{12}, S_{34}, S, M) &= -J\left[S(S+1) - S_{12}(S_{12}+1) - S_{34}(S_{34}+1)\right] \\ &-J'\left[S_{12}(S_{12}+1) - 2S_i(S_i+1)\right] - J''\left[S_{34}(S_{34}+1) - 2S_i(S_i+1)\right] \\ &-g\mu_B H M, \end{aligned} \tag{8.20}$$

where $-S \leq M \leq S$. An explicit formula for the neutron cross-section of spin tetramers was given in [Güdel *et al.* (1979)].

As an example, we present in Fig. 8.3b the results of an inelastic neutron scattering investigation of the monoclinic compound α-$MnMoO_4$ which contains tetrameric clusters of Mn^{2+} ions ($S_i = 5/2$) and exhibits long-range antiferromagnetic order below $T_N = 10.7$ K. Within the cluster the Mn^{2+} ions are ferromagnetically aligned, giving rise to a large cluster spin $S = 10$. At $T = 1.5$ K four well defined transitions with varying widths and labeled from I to IV were observed. The energy level sequence of the states $|\Gamma\rangle = |S_{12}, S_{34}, S, M\rangle$ associated with the transitions I-IV is shown as insert in Fig. 8.3b. Applying Eq. (8.20), the energies of the transitions Δ_Γ out of the ground state $|5, 5, 10, 10\rangle$ are given by:

$$\begin{aligned} \Delta_I &= g\mu_B H, \\ \Delta_{II} &= 10J + 10J' + g\mu_B H, \\ \Delta_{III} &= 10J + 10J'' + g\mu_B H, \\ \Delta_{IV} &= 20J + g\mu_B H, \end{aligned} \tag{8.21}$$

where H corresponds to the internal molecular field. Comparing Eq. (8.21) with the experimental data displayed in Fig. 8.3b yields the model parameters $J = 54(3)$ μeV, $J' = -8(7)$ μeV, $J'' = 20(9)$ μeV, and $H = 5.4(3)$ T.

8.1.4 *N-mers*

For clusters with five and more magnetic ions, both the diagonalization of the spin Hamiltonian and the evaluation of the neutron cross-section become cumbersome, but the total spin $\boldsymbol{S}$ remains a good quantum number.

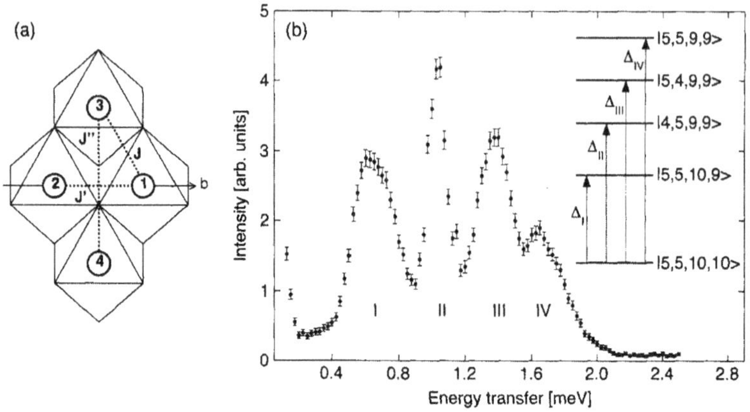

Fig. 8.3 (a) The Mn_4O_{16} cluster in α-$MnMoO_4$ consisting of four edge-sharing MnO_6 octahedra with the Mn^{2+} ions shown as circles and the oxygen ions at the corners. The Mn(1)-Mn(2) bond is along the b-axis. The dotted lines mark the intra-cluster exchange parameters J, J', and J''. (b) Energy spectrum of neutrons scattered from polycrystalline α-$MnMoO_4$ at 1.5 K. The inset shows the resulting energy level sequence in $|S_{12}, S_{34}, S, M\rangle$ tetramer notation (after [Ochsenbein *et al.* (2003)]).

Magnetic clusters are the building blocks of a large class of single-molecule magnets which are presently studied from a fundamental point of view as well as in view of possible applications. A particularly interesting single-molecule magnet is Mn_{12}-acetate, displayed in Fig. 8.4, in which an external ring of eight Mn^{3+} ions with $S = 2$ surrounds a tetrahedron of four Mn^{4+} ions with $S = 3/2$ and opposite spin alignment, giving rise to a ground state with a large total spin $S = 10$. The 21-fold degeneracy of the ground state is partially lifted by the crystal field to produce a sequence of sublevels corresponding to $M = \pm 10, \pm 9, \ldots, 0$ as shown in Fig. 8.5. According to the selection rules Eq. (8.7), this gives rise to ten transitions with $\Delta M = \pm 1$, out of which seven transitions could be observed by inelastic neutron scattering experiments, see Fig. 8.5 (the decreasing energy spacing with decreasing energy transfer prevented the remaining three transitions to be resolved from the elastic line). The observation of transitions in both the energy-gain and energy-loss configuration also serves to illustrate the principle of detailed balance discussed in Chap. 2.5.

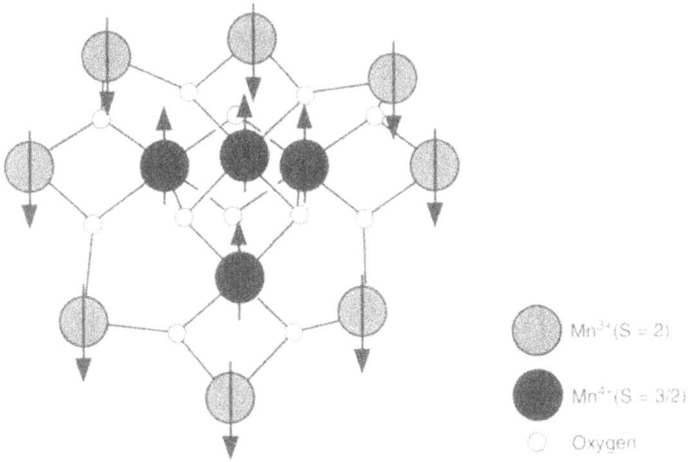

Fig. 8.4 Schematic structure of Mn_{12}-acetate. Only the inner $Mn_{12}O_{12}$ core is shown. Reprinted with permission from [Robinson *et al.* (2000)]). Copyright 2000, Institute of Physics Publishing Ltd.

8.2 Spin waves

8.2.1 *Ferromagnets*

For an extended system comprising N magnetic ions the Heisenberg Hamiltonian reads

$$\hat{\mathcal{H}} = -2\sum_{j,j'}^{N} J_{jj'}\hat{\boldsymbol{S}}_j \cdot \hat{\boldsymbol{S}}_{j'} - g\mu_B H_a \sum_{j}^{N} \hat{S}_j^z, \tag{8.22}$$

where H_a denotes an external magnetic field and/or an anisotropy field along the quantization axis z. The exchange coupling $J_{jj'} > 0$ forces the spins to be perfectly aligned along the z axis at zero temperature. At finite temperatures spin deviations occur which propagate through the lattice giving rise to spin waves. The spin-wave dispersion is given by [Wagner (1972)].

$$\boxed{\hbar\omega(\boldsymbol{q}) = 2S\,(J(0) - J(\boldsymbol{q})) + g\mu_B H_a} \tag{8.23}$$

with the Fourier transformed exchange function

$$\boxed{J(\boldsymbol{q}) = \sum_{j,j'} J_{jj'} e^{i\boldsymbol{q}\cdot(\boldsymbol{R}_j - \boldsymbol{R}_{j'})}} \tag{8.24}$$

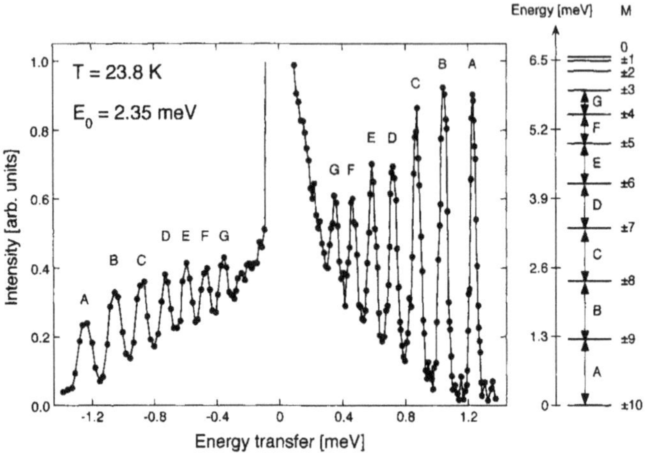

Fig. 8.5 Left: Energy spectrum of neutrons scattered from Mn_{12}-acetate at $T = 23.8$ K (after [Mirebeau *et al.* (1999)]). Right: Crystal-field splitting of the $S = 10$ ground state.

The operators affecting the spin deviation of the jth atom (at site $\boldsymbol{R}_j$) are defined by

$$\hat{S}_j^+(t) = \sqrt{\frac{2S}{N}} \sum_{\boldsymbol{q}} e^{\imath(\boldsymbol{q}\cdot R_j - \omega(\boldsymbol{q})t)} \hat{a}_{\boldsymbol{q}},$$
$$\hat{S}_j^-(t) = \sqrt{\frac{2S}{N}} \sum_{\boldsymbol{q}} e^{-\imath(\boldsymbol{q}\cdot R_j - \omega(\boldsymbol{q})t)} \hat{a}_{\boldsymbol{q}}^+, \tag{8.25}$$
$$\hat{S}_j^z(t) = S - \frac{1}{N} \sum_{\boldsymbol{q},\boldsymbol{q}'} e^{-\imath\left((\boldsymbol{q}-\boldsymbol{q}')\cdot R_j - (\omega(\boldsymbol{q})t - \omega(\boldsymbol{q}')t)\right)} \hat{a}_{\boldsymbol{q}}^+ \hat{a}_{\boldsymbol{q}'},$$

where

$$\hat{a}_{\boldsymbol{q}}|n_{\boldsymbol{q}}\rangle = \sqrt{n_{\boldsymbol{q}}}|n_{\boldsymbol{q}} - 1\rangle; \quad \hat{a}_{\boldsymbol{q}}^+|n_{\boldsymbol{q}}\rangle = \sqrt{n_{\boldsymbol{q}} + 1}|n_{\boldsymbol{q}} + 1\rangle \tag{8.26}$$

are annihilation and creation operators first introduced into spin-wave theory by [Holstein and Primakoff (1940)]. With $\langle \hat{a}_{\boldsymbol{q}}^+ \hat{a}_{\boldsymbol{q}} \rangle = \langle n_{\boldsymbol{q}} \rangle$, $\langle \hat{a}_{\boldsymbol{q}} \hat{a}_{\boldsymbol{q}}^+ \rangle = \langle n_{\boldsymbol{q}} + 1 \rangle$ and the definitions

$$\hat{S}^x = \frac{1}{2}\left(\hat{S}^+ + \hat{S}^-\right); \quad \hat{S}^y = -\frac{\imath}{2}\left(\hat{S}^+ - \hat{S}^-\right) \tag{8.27}$$

we calculate the spin correlation functions $\langle \hat{S}_j^\alpha(0)\hat{S}_{j'}^\beta(t)\rangle$ of Eq. (2.46) to be

$$\begin{aligned}\langle \hat{S}_j^x(0)\hat{S}_{j'}^x(t)\rangle &= \frac{S}{2N}\sum_{\boldsymbol{q}}\left[e^{-\imath(\boldsymbol{q}\cdot(\boldsymbol{R}_j-\boldsymbol{R}_{j'})-\omega(\boldsymbol{q})t)}\langle n_{\boldsymbol{q}}+1\rangle\right.\\ &\left.+e^{\imath(\boldsymbol{q}\cdot(\boldsymbol{R}_j-\boldsymbol{R}_{j'})-\omega(\boldsymbol{q})t)}\langle n_{\boldsymbol{q}}\rangle\right],\\ \langle \hat{S}_j^y(0)\hat{S}_{j'}^y(t)\rangle &= \langle \hat{S}_j^x(0)\hat{S}_{j'}^x(t)\rangle,\\ \langle \hat{S}_j^z(0)\hat{S}_{j'}^z(t)\rangle &= S^2-\frac{2S}{N}\langle n_{\boldsymbol{q}}\rangle,\end{aligned} \tag{8.28}$$

where $\langle n_{\boldsymbol{q}}\rangle$ is the Bose-Einstein occupation number:

$$\langle n_{\boldsymbol{q}}\rangle = \left(\exp\left(\frac{\hbar\omega(\boldsymbol{q})}{k_BT}\right)-1\right)^{-1} \tag{8.29}$$

For the calculation of the neutron cross-section for spin-wave scattering we insert Eq. (8.29) into the scattering function $S^{\alpha\beta}(\boldsymbol{Q},\omega)$ of Eq. (2.46). We can drop the term $\alpha=\beta=z$ which is time-independent, i.e., it describes the elastic magnetic scattering. Moreover, since $\langle \hat{S}_j^x(0)\hat{S}_{j'}^y(t)\rangle = -\langle \hat{S}_j^y(0)\hat{S}_{j'}^x(t)\rangle$, we are left with the terms $\alpha=\beta=x$ and $\alpha=\beta=y$:

$$\begin{aligned}\frac{\mathrm{d}^2\sigma}{\mathrm{d}\Omega\mathrm{d}\omega} &= (\gamma r_0)^2\frac{S}{4\pi\hbar}\frac{k'}{k}F^2(\boldsymbol{Q})e^{-2W(\boldsymbol{Q})}\left[1+\left(\frac{Q_z}{Q}\right)^2\right]\sum_{j,j'}e^{\imath\boldsymbol{Q}\cdot(\boldsymbol{R}_j-\boldsymbol{R}_{j'})}\\ &\times\int_{-\infty}^{\infty}\sum_{\boldsymbol{q}}\left[e^{-\imath(\boldsymbol{q}\cdot(\boldsymbol{R}_j-\boldsymbol{R}_{j'})-\omega(\boldsymbol{q})t)}\langle n_{\boldsymbol{q}}+1\rangle\right.\\ &\left.+e^{\imath(\boldsymbol{q}\cdot(\boldsymbol{R}_j-\boldsymbol{R}_{j'})-\omega(\boldsymbol{q})t)}\langle n_{\boldsymbol{q}}\rangle\right]e^{-\imath\omega t}\mathrm{d}t.\end{aligned} \tag{8.30}$$

Using the relation Eq. (A.10) for the lattice sum and the integral representation Eq. (A.6) for the δ-function we arrive at the final formula:

$$\begin{aligned}\frac{\mathrm{d}^2\sigma}{\mathrm{d}\Omega\mathrm{d}\omega} &= (\gamma r_0)^2\frac{(2\pi)^3S}{2v_0}\frac{k'}{k}F^2(\boldsymbol{Q})e^{-2W(\boldsymbol{Q})}\left[1+\left(\frac{Q_z}{Q}\right)^2\right]\\ &\times\sum_{\boldsymbol{\tau},\boldsymbol{q}}[\langle n_{\boldsymbol{q}}+1\rangle\delta(\boldsymbol{Q}-\boldsymbol{q}-\boldsymbol{\tau})\delta(\hbar\omega(\boldsymbol{q})-\hbar\omega)\\ &+\langle n_{\boldsymbol{q}}\rangle\delta(\boldsymbol{Q}+\boldsymbol{q}-\boldsymbol{\tau})\delta(\hbar\omega(\boldsymbol{q})+\hbar\omega)]\end{aligned} \tag{8.31}$$

The cross section Eq. (8.31) is the sum of two terms, the first corresponding to the creation and the second to the annihilation of a spin wave. We recognize that the two δ-functions in the cross section describe the momentum

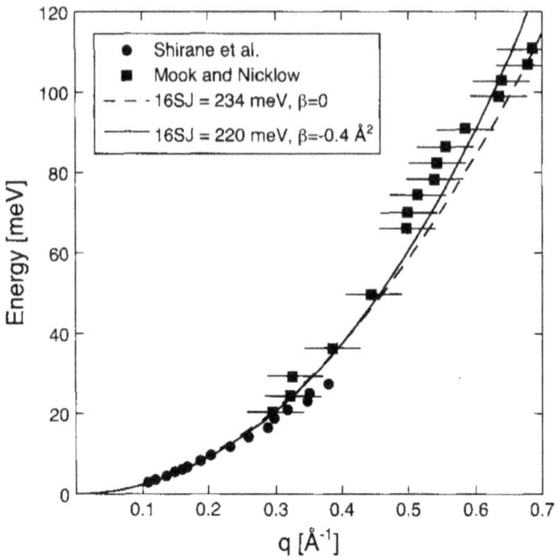

Fig. 8.6 Spin-wave dispersion of bcc iron observed at room temperature (after [Shirane *et al.* (1965)] and [Lynn (1975)]). The line is a least-squares fit through the experimental points as discussed in the text.

and the energy conservation of the neutron scattering process according to Eqs (2.1) and (2.2), respectively.

Spin-wave experiments performed for bcc iron ($T_C = 1041$ K) are shown in Fig. 8.6. For small wavevectors $\boldsymbol{q}$ the spin-wave dispersion is almost isotropic, i.e., independent of the direction of $\boldsymbol{q}$. Well defined spin-wave excitations exist only over about half of the Brillouin zone, since above 100 meV the spin waves are strongly damped by the Stoner excitations (i.e., single-particle excitations). In order to analyze the data in terms of the dispersion relation Eq. (8.23), we assume only the nearest-neighbor exchange interaction J. The positions of the eight nearest neighbors of an atom placed at the origin of a bcc crystal are given by the vectors $\boldsymbol{R} = a\left(\pm\frac{1}{2}, \pm\frac{1}{2}, \pm\frac{1}{2}\right)$, where a is the lattice constant. Insertion of $\boldsymbol{R}$ into Eq. (8.24) yields

$$J(0) - J(\boldsymbol{q}) = 8J\left[1 - \cos\left(\frac{aq_x}{2}\right)\cos\left(\frac{aq_y}{2}\right)\cos\left(\frac{aq_z}{2}\right)\right]. \tag{8.32}$$

The spin-wave energies are smallest for $q = 0$. For small wavevectors we can expand Eq. (8.32) in terms of the wavevector q; the spin-wave dispersion Eq. (8.23) then reads

$$\hbar\omega(q) = 16SJa^2q^2\left(1 - \beta q^2 + \gamma q^4 + \ldots\right) + g\mu_B H_a. \tag{8.33}$$

As can be seen from Fig. 8.6, Eq. (8.33) with $g\mu_B H_a = 0$, $16SJ = 220$ meV and $\beta = -0.4$ Å^2 provides a good description of the spin-wave dispersion up to $q = 0.7$ Å^{-1}.

8.2.2 *Antiferromagnets*

A transfer of the spin-wave concept to antiferromagnetism is not obvious since, in contrast to the case of ferromagnetism, the antiferromagnetic ground state is unknown. The simplest model for an antiferromagnet is that of two sublattices with antiferromagnetic coupling $J < 0$ between nearest neighbors; the corresponding Hamiltonian reads

$$\hat{\mathcal{H}} = -J \sum_{j,j'}^{N/2} \hat{\boldsymbol{S}}_j \cdot \hat{\boldsymbol{S}}_{j'} + g\mu_B H_a \sum_{j}^{N/2} \hat{S}_j^z - g\mu_B H_a \sum_{j'}^{N/2} \hat{S}_{j'}^z, \tag{8.34}$$

where the subscript j labels the spins of the first sublattice, j' those of the second sublattice, and H_a denotes the anisotropy field (along the quantization axis z) which stabilizes the ground state. From the Hamiltonian Eq. (8.34) we obtain the spin-wave dispersion for antiferromagnets [Wagner (1972)]:

$$\hbar\omega(\boldsymbol{q}) = \sqrt{g\mu_B H_a \left(g\mu_B H_a + 4S|J(0)|\right) + 4S^2 \left(J^2(0) - J^2(\boldsymbol{q})\right)}. \tag{8.35}$$

Since the anisotropy field H_a is usually smaller than the internal field $4S|J(0)|/g\mu_B$ by several orders of magnitude, the first term of Eq. (8.35) is often neglected.

The neutron cross-section for scattering by spin waves in antiferromagnets essentially corresponds to Eq. (8.31), but the term $(1 + (Q_z/Q)^2)$ has to be replaced by a function $A(\boldsymbol{Q})$ which depends on the particular coupling model [Lovesey (1984)].

In a first example we apply Eq. (8.35) to a one-dimensional system with antiferromagnetic coupling $J < 0$ between adjacent ions which are separated from each other by the distance a. From Eq. (8.24) we find $J(q) = 2J\cos(qa)$; inserting this result into Eq. (8.35) and setting $H_a = 0$ yields

$$\hbar\omega(q) = 4SJ\sin(qa). \tag{8.36}$$

As shown in Fig. 8.7 for the one-dimensional antiferromagnet $CsMnBr_3$, in which the chains of magnetic Mn^{2+} ions ($S = 5/2$) are oriented parallel to the hexagonal c-direction, Eq. (8.36) provides an excellent description of the measured spin-wave dispersion.

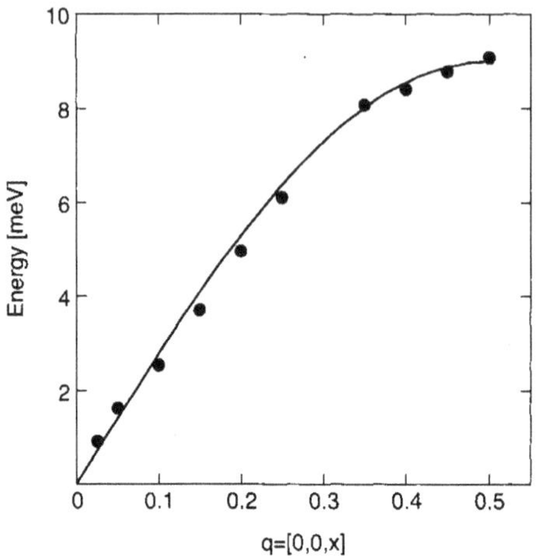

Fig. 8.7 Spin-wave dispersion observed for $CsMnBr_3$ ($S_j = 5/2$) at $T = 5$ K (after [Breitling *et al.* (1977)]). The line is a least-squares fit based on Eq. (8.36) resulting in $J = -0.9$ meV.

Let us consider in the second example the spin waves in the antiferromagnet La_2CuO_4, the parent compound of the high-temperature superconductor $La_{2-x}Sr_xCuO_4$. La_2CuO_4 is a layered structure with an alternate stacking of copper-oxide and lanthanum-oxide planes, i.e., the compound is a typical two-dimensional antiferromagnet. The spin-wave dispersion measured for $\boldsymbol{q} = (0, x, 0)$ is shown in Fig. 8.8. For the data analysis based on Eq. (8.35) we only consider the nearest-neighbor antiferromagnetic exchange interaction $J < 0$ as illustrated in the inset of Fig. 8.8. The positions of the four nearest neighbors of a Cu^{2+} ion ($S_i = 1/2$) placed at the origin of the (x,y)-plane are given by the vectors $\boldsymbol{R} = a\left(\pm\frac{1}{2}, \pm\frac{1}{2}\right)$. Insertion of $\boldsymbol{R}$ into Eq. (8.24) yields

$$J(\boldsymbol{q}) = 4J\cos\left(\frac{aq_x}{2}\right)\cos\left(\frac{aq_y}{2}\right). \tag{8.37}$$

By combining Eqs (8.35) and (8.37) and setting $H_a = 0$ we find

$$\hbar\omega(\boldsymbol{q}) = 4J\sqrt{1-\cos^2\left(\frac{aq_x}{2}\right)\cos^2\left(\frac{aq_y}{2}\right)}. \tag{8.38}$$

We see from Fig. 8.8 that the theoretical form of the dispersion relation Eq. (8.38) with $J = 76(2)$ meV gives an excellent description of the data. Moreover, the theoretical model is also successful to reproduce the observed

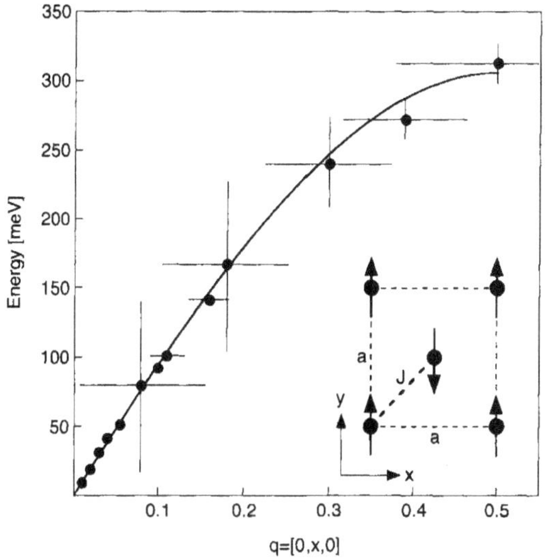

Fig. 8.8 Spin-wave dispersion of La_2CuO_4 observed at $T = 5$ K (after [Hayden *et al.* (1991)]). The line is a least-squares fit based on Eq. (8.38) with $J = -76(2)$ meV. The inset illustrates the antiferromagnetic coupling of the Cu^{2+} ions in the (x,y)-plane.

spin-wave amplitudes with the function

$$A(\boldsymbol{q}) \propto \sqrt{\frac{1 - \cos\left(\frac{aq_x}{2}\right)\cos\left(\frac{aq_y}{2}\right)}{1 + \cos\left(\frac{aq_x}{2}\right)\cos\left(\frac{aq_y}{2}\right)}} \tag{8.39}$$

inserted as structure factor into the cross-section formula Eq. (8.31) (see [Hayden *et al.* (1991)]). Later experiments included other directions in reciprocal space and gave evidence for substantial interactions beyond the nearest-neighbor term [Coldea *et al.* (2001)].

8.2.3 *The random-phase approximation*

The random-phase approximation (RPA) is a convenient means to calculate the collective magnetic excitations of systems whose spin Hamiltonian comprises additional terms $\hat{\mathcal{H}}_0$ beyond the two-ion exchange interaction $\hat{\mathcal{H}}_1$ discussed in the two previous sections:

$$\hat{\mathcal{H}} = \hat{\mathcal{H}}_0 + \hat{\mathcal{H}}_1. \tag{8.40}$$

Typical examples are the $4f$ electron systems for which $\hat{\mathcal{H}}_0$ is the single-ion crystal-field interaction defined by Eq. (9.9) as well as magnetic cluster systems in which $\hat{\mathcal{H}}_0$ corresponds to Eqs (8.1), (8.11) and (8.18) for dimers, trimers and tetramers, respectively, describing the coupling of some ions to form magnetic subunits as basis states. The RPA method for f-electron systems is based on the concept of the generalized magnetic susceptibility [Jensen and Macintosh (1991)]:

$$\chi(\boldsymbol{q},\omega) = \frac{\chi^0(\omega)}{1-\chi^0(\omega)J(\boldsymbol{q})}, \tag{8.41}$$

where $\chi^0(\omega)$ is the single-ion susceptibility, and $J(\boldsymbol{q})$ denotes the Fourier transformed exchange function defined by Eq. (8.24). $\chi(\boldsymbol{q},\omega)$ is related to the spin correlation function $S(\boldsymbol{Q},\omega)$ through Eq. (2.47). For the RPA approach to be valid, $\chi(\boldsymbol{q},\omega)$ has to be dominated by the single-ion susceptibility $\chi^0(\omega)$, which can be expressed as:

$$\chi^0(\omega) = \sum_{\Gamma} \frac{\langle\Gamma|\hat{S}^+|0\rangle\langle 0|\hat{S}^-|\Gamma\rangle}{\Delta_\Gamma - \hbar\omega}, \tag{8.42}$$

where $|0\rangle$ and $|\Gamma\rangle$ denote the ground state and the excited states, respectively, $\langle\Gamma|\hat{S}^+|0\rangle = \langle 0|\hat{S}^-|\Gamma\rangle$ is the transition matrix element, and Δ_Γ corresponds to the mean-field solution of Eq. (8.40). The magnetic excitation spectrum is given by the poles of $\chi(\boldsymbol{q},\omega)$:

$$\hbar\omega(\boldsymbol{q})|_\Gamma = \sqrt{\Delta_\Gamma^2 - 2\Delta_\Gamma|\langle\Gamma|\hat{S}^+|0\rangle|^2 J(\boldsymbol{q})}. \tag{8.43}$$

For magnetic cluster systems Eq. (8.43) is valid by setting $\langle\Gamma|\hat{S}^+|0\rangle = 1$.

We exemplify the application of the RPA model for the tetrameric cluster compound α-$MnMoO_4$ introduced in Chap. 8.1.3. Below $T_N = 10.7$ K the ferromagnetically aligned manganese clusters are antiferromagnetically coupled. The long-range order is defined by the magnetic ordering wavevector $\boldsymbol{q}_0 = (1,0,\frac{1}{2})$, see Eq. (7.14), i.e., the tetramers are located on two sublattices with opposite staggered magnetization. Therefore it is convenient to distinguish the Fourier transform of the intra- and inter-sublattice exchange parameters by $J(\boldsymbol{q})$ and $J'(\boldsymbol{q})$, respectively. Accordingly, in Eqs (8.41) and (8.43) $J(\boldsymbol{q})$ has to be replaced by $J(\boldsymbol{q}) \pm |J'(\boldsymbol{q})|$, which gives rise to a splitting into acoustic and optic branches indicated by the + and - sign, respectively.

The neutron cross-section for scattering by the magnetic excitations in α-$MnMoO_4$ is given by [Jensen and Macintosh (1991)]:

$$\frac{d^2\sigma}{d\Omega d\omega} \propto F^2(\boldsymbol{Q})S(\boldsymbol{Q})\,(1\pm\cos(\varphi))\sum_\alpha\left[1-\left(\frac{Q_\alpha}{Q}\right)^2\right], \tag{8.44}$$

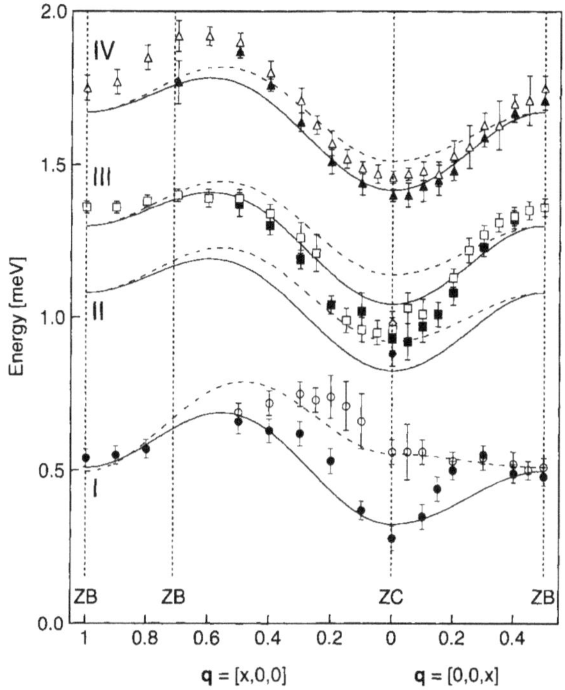

Fig. 8.9 Dispersion of the four excitation branches I-IV observed for α-$MnMoO_4$ for the $[0, x, 0]$ and $[0, x, \frac{\pi}{2}]$ directions at 1.5 K. The circles, diamonds, squares, and triangles refer to the excitation branches I, II, III, and IV, respectively, identified in Fig. 8.3b. The full and open symbols denote the acoustic and optic excitations, respectively. The full and dashed lines correspond to the calculated dispersion of the acoustic and optic branches, respectively. The dotted lines labelled by ZC and ZB mark the zone center and the zone boundary, respectively (after [Häfliger *et al.* (2009)]).

with $S(\boldsymbol{Q})$ being the structure factor of the tetramer

$$S(\boldsymbol{Q}) = \sum_{i,j=1}^{4} e^{\imath \boldsymbol{Q}\cdot(\boldsymbol{R}_i - \boldsymbol{R}_j)}, \tag{8.45}$$

where $\boldsymbol{R}_i$ denotes the position of the Mn^{2+} ions in the tetramer, and the + and − sign refers to the acoustic and optic branch, respectively. The phase φ is defined through

$$J'(\boldsymbol{Q}) = J'(\boldsymbol{q})e^{-\imath\boldsymbol{\tau}\cdot\boldsymbol{\rho}} = |J'(\boldsymbol{q})|e^{-\imath\varphi}, \tag{8.46}$$

with $\boldsymbol{\tau}$ being a reciprocal lattice vector and $\boldsymbol{\rho}$ the vector connecting the two tetramer sublattices.

Fig. 8.9 shows some results obtained from single-crystal experiments. All the observed dispersion curves could be consistently interpreted on the basis of Eq. (8.43). The total magnetic excitation spectrum is composed of eight branches as expected, since there are eight Mn^{2+} ions (or two Mn tetramers) in the magnetic unit cell.

8.3 Solitons

The spin waves discussed in Chap. 8.2 are special solutions of the equation of motion for the spin operators $\hat{S}_j^{\pm}$:

$$-\imath\frac{\partial}{\partial t}\hat{S}_j^{\pm} = \left[\hat{\mathcal{H}}, \hat{S}_j^{\pm}\right]. \tag{8.47}$$

Insertion of a spin Hamiltonian (e.g. Eq. (8.18)) into Eq. (8.47) gives rise to products involving the operators $\hat{S}_j^z$ which are linearized by the replacement $\hat{S}_j^z \rightarrow S$, i.e., the spin waves are small amplitude solutions (see exercise No. 8.3). If the amplitudes are large, the linear spin-wave theory is no longer applicable. A typical example of a non-linear or "solitary" solution is the Bloch wall; the corresponding excitation is called a soliton whose properties are discussed below.

There is a striking similarity between a one-dimensional chain of magnetic atoms and a linear system of coupled pendula. The dynamics of the latter can be described by the Sine-Gordon equation

$$\frac{\partial^2\phi}{\partial z^2} - \frac{1}{c^2}\frac{\partial^2\phi}{\partial t^2} = m^2\sin\phi, \tag{8.48}$$

where ϕ denotes the amplitude of the pendulum, and the constants c and m are related to the specific properties of the coupled pendula [Steiner (1980)]. For small amplitudes ϕ, the Sine-Gordon equation Eq. (8.48) is linearized by setting $\sin\phi = \phi$, which results in the linear solution

$$\phi(z,t) \propto \cos(zt - \omega t). \tag{8.49}$$

For large amplitudes ϕ the solution of Eq. (8.48) has the form

$$\phi(z,t) \propto \arctan(e^{\pm m\xi}) \tag{8.50}$$

with

$$\xi = \frac{z - vt}{\sqrt{1 - \left(\frac{v}{c}\right)^2}}, \tag{8.51}$$

where v denotes the velocity with which the soliton propagates along the z-direction. The $\pm$ sign in Eq. (8.50) refers to a soliton and antisoliton

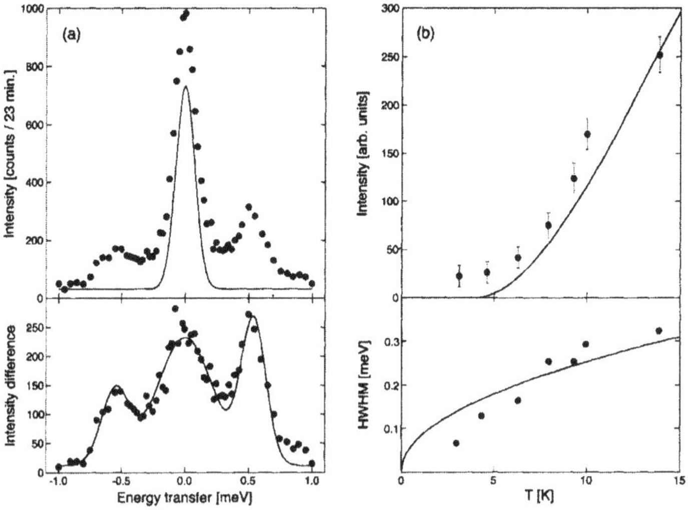

Fig. 8.10 (a) Upper panel: Inelastic neutron scattering spectrum observed for $CsNiF_3$ at $\boldsymbol{Q} = (0, 0, 1.9)$, $T = 9.3$ K and $H = 0.5$ T (circles). The line is the observed profile at $T = 3.1$ K and $H = 3.0$ T which is assumed to be the background. Lower panel: Difference between the two spectra in the upper panel. The line is the result of a Gaussian least-squares fit. (b) Temperature dependence of the integrated intensity (upper panel) and energy half-width at half-maximum (lower panel) of the soliton response at $\boldsymbol{Q} = (0, 0, 1.0)$ and $H = 0.5$ T (circles). The lines are least-squares fits to Eq. (8.54) (after [Kjems and Steiner (1978)]).

propagating in opposite directions. Another solution of the Sine-Gordon equation Eq. (8.48) is the so-called breather which corresponds to a soliton-antisoliton pair [Steiner (1980)].

Magnetic solitons were observed for the first time in the hexagonal compound $CsNiF_3$ which consists of ferromagnetic nickel chains running along c with strong coupling within the chains and without significant coupling between the chains at $T > T_N = 2.7$ K. The magnetic moments are bound to the (ab)-plane by a strong single-ion anisotropy. In this plane they can rotate freely. The appropriate spin Hamiltonian reads

$$\hat{\mathcal{H}} = -2J \sum_i \hat{S}_i \cdot \hat{S}_{i+1} + D \sum_i (\hat{S}_i^z)^2, \tag{8.52}$$

which was used to analyze the spin-wave dispersion along c for $T > T_N$, yielding $J = 1.02$ meV and $D = 0.82$ meV. By applying an external magnetic field $\boldsymbol{H}$ perpendicular to c, the results of inelastic neutron scattering experiments indeed showed the presence of a soliton response. The data were analyzed in terms of the Sine-Gordon equation Eq. (8.48) with the following parameter identification:

$$\phi = \angle(\boldsymbol{H}, \boldsymbol{S}), \quad c = 2S\sqrt{DJ}, \quad m = \sqrt{\frac{g\mu_B H}{JS}}. \tag{8.53}$$

The spin correlation function Eq. (2.43) associated with the solitons was evaluated by Mikeska [Mikeska (1978)]:

$$\boxed{S_{\text{sol}}(\boldsymbol{Q}, \omega) \propto \frac{\beta e^{-8m\beta}}{Rq} \left(\frac{\frac{\pi q}{2m}}{\sinh(\frac{\pi q}{2m})} \right)^2 e^{-\frac{4\beta m\omega^2}{R^2 q^2}}} \tag{8.54}$$

where $\beta = 1/k_B T$ and $R = c/2$ denotes the distance between the nickel ions along c. According to Eq. (8.54) the soliton response is a Gaussian centered at zero energy transfer (quasielastic scattering), which was nicely verified for $CsNiF_3$ as shown in Fig. 8.10a. The soliton response at $\hbar\omega = 0$ is flanked by two inelastic lines at $\hbar\omega = \pm 0.5$ meV corresponding to the excitation and annihilation of spin-wave excitations. Moreover, the observed temperature dependence of both the intensity and the linewidth of the soliton response is in excellent agreement with the theoretical prediction based on Eq. (8.54) as demonstrated in Fig. 8.10b.

8.4 Further reading

- H. P. Andres, S. Decurtins and H. U. Güdel, in *Frontiers of neutron scattering*, ed. by A. Furrer (World Scientific, Singapore, 2000), p. 149: *Neutron scattering of molecular magnets*
- T. Chatterji, in *Neutron scattering from magnetic materials*, ed. by T. Chatterji (Elsevier, Amsterdam, 2006), p. 245: *Magnetic excitations*
- J. Jensen and A. R. Macintosh, *Rare-earth magnetism structures and excitations* (Clarendon Press, Oxford, 1991)
- G. Lander, in *Neutron scattering*, ed. by A. Furrer (Proc. 93-01, ISSN 1019-6447, PSI Villigen, 1993), p. 235: *Magnetic excitations*
- T. G. Perring, in *Frontiers of neutron scattering*, ed. by A. Furrer (World Scientific, Singapore, 2000), p. 190: *Neutron scattering from low-dimensional magnetic systems*
- J. P. Regnault, in *Neutron scattering from magnetic materials*, ed. by T. Chatterji (Elsevier, Amsterdam, 2006), p. 363: *Inelastic neutron polarization analysis*

- W. G. Stirling and K. A. McEwan, in *Methods of experimental physics*, Vol. 23, Part C (Academic Press, London, 1987), p. 159: *Magnetic excitations*

8.5 Exercises

Exercise No 8.1

(a) A detailed analysis of the dimer splitting energies observed for $CsMn_{0.28}Mg_{0.72}Br_3$ (see Fig. 8.1) gave evidence for slight deviations from the Landé interval rule Eq. (8.3) [Falk *et al.* (1984)]:

$$E(1) - E(0) = 1.80(1)\text{meV},$$
$$E(2) - E(1) = 3.60(1)\text{meV},$$
$$E(3) - E(2) = 5.27(2)\text{meV},$$
$$E(4) - E(3) = 6.74(3)\text{meV}.$$

This can be explained by adding a biquadratic exchange term to the Heisenberg Hamiltonian Eq. (8.1):

$$\hat{\mathcal{H}} = -2J\hat{S}_1 \cdot \hat{S}_2 - K(\hat{S}_1 \cdot \hat{S}_2)^2. \qquad (8.55)$$

Determine the eigenvalues of Eq. (8.55) and derive the bilinear and biquadratic exchange parameters J and K, respectively, from the observed dimer splittings.

(b) Kittel [Kittel (1960)] suggested that exchange striction is the likely origin of the biquadratic exchange interaction in magnetic compounds. Exchange striction results from a geometrical degree of freedom which allows the exchange-coupled ions to accommodate their mutual distance r in order to gain magnetic energy at the expense of elastic energy. The elastic energy density for Mn dimers in the hexagonal compound $CsMn_{0.28}Mg_{0.72}Br_3$ reads

$$u_{\text{el}} = \frac{1}{2}c_{11}\left(e_{xx}^2 + e_{yy}^2\right) + \frac{1}{2}c_{33}e_{zz}^2 + c_{12}e_{xx}e_{yy} + c_{13}\left(e_{xx}e_{zz} + e_{yy}e_{zz}\right), \qquad (8.56)$$

where the c_{ik} and $e_{\alpha\beta}$ denote the elastic constants and strain components, respectively. If we set $e_{xx} = e_{yy} = 0$ by considering only a local, longitudinal distortion $e_{zz} = (r - r_0)/r_0$ with $r_0 = c/2$, the elastic energy density is simplified to

$$u_{el} = \frac{1}{2}c_{33}\left(\frac{r - r_0}{r_0}\right)^2. \qquad (8.57)$$

The associated energy per dimer is $W_{\rm el} = v u_{\rm el}$, where $v = \sqrt{3}a^2c/4$ is half the volume of the unit cell occupied by the dimer. The total elastic and magnetic energy is then given by

$$W = \frac{\sqrt{3}}{4}a^2 c \frac{c_{33}}{2}\left(\frac{r - r_0}{r_0}\right)^2 - 2J(r)\hat{S}_1 \cdot \hat{S}_2. \tag{8.58}$$

Determine the distance r corresponding to the minimal total energy from $\mathrm{d}W/\mathrm{d}r = 0$. Then insertion of r back into Eq. (8.58) will give rise to a biquadratic term whose prefactor involves the lattice parameters $a = 7.5304$ Å and $c = 6.4516$ Å, the derivative of the bilinear exchange parameter $\mathrm{d}J/\mathrm{d}r = 3.6$ meV/Å, and the elastic constant c_{33} [Strässle *et al.* (2004a)]. The latter was measured for the isostructural compound $CsNiF_3$ to be $c_{33} = 64$ GPa. The elastic constant c_{33} scales with the acoustic phonon frequencies according to $\omega \propto (c\sqrt{M})^{-1}$ with M being the molar mass, thus we take $c_{33} = 41$ GPa for $CsMn_{0.28}Mg_{0.72}Br_3$. Compare the prefactor with the biquadratic exchange parameter K derived in (a).

Exercise No 8.2

$LaCoO_3$ is a non-magnetic perovskite-type compound, because all the Co^{3+} ions are in a low-spin $S = 0$ state. The substitution of a divalent Sr^{2+} ion for La^{3+} creates a Co^{4+} ion in the lattice which has a nonzero S in any spin-state configuration. The hole introduced by Sr^{2+} doping does not remain localized at the Co^{4+} site, but it is extended over some neighboring Co^{3+} ions, transforming them to a higher magnetic spin state and thereby creating a magnetic cobalt cluster which is called a spin-state polaron. This was evidenced by the observation of a magnetic excitation at an energy transfer of 0.75 meV, which was absent for the undoped parent compound $LaCoO_3$ [Podlesnyak *et al.* (2008)]. The Q dependence of the intensity of the observed excitation exhibits a clear oscillatory behavior as shown in Fig. 8.11, which reflects the size as well as the shape of the polaron through the structure factor. For $\Delta S = 0$ transitions, the neutron cross-section of a cluster comprising n magnetic ions can be approximated for polycrystalline materials by an extension of Eq. (8.10):

$$\frac{\mathrm{d}^2\sigma}{\mathrm{d}\Omega \mathrm{d}\omega} \propto F^2(\boldsymbol{Q}) \sum_{j<j'=1}^{n} \Big(|\langle S||\hat{T}_j||S'\rangle|^2 + 2\frac{\sin(Q|R_j - R_{j'}|)}{Q|R_j - R_{j'}|}\langle S||\hat{T}_j||S'\rangle\langle S'||\hat{T}_{j'}||S\rangle \Big) \tag{8.59}$$

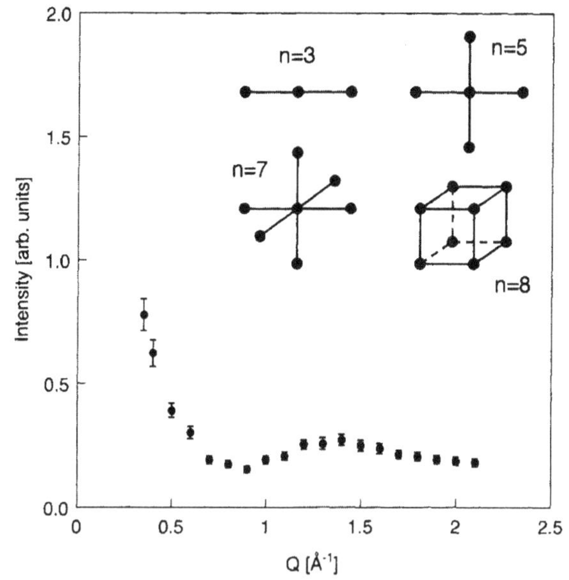

Fig. 8.11 Q-dependence of the intensity of the transition observed at 0.75 meV in $La_{0.998}Sr_{0.002}CoO_3$ (after [Podlesnyak *et al.* (2008)]). The insert sketches different types of Co multimers.

where R_j denotes the position of the j-th ion in the cluster. Calculate the cross section for different types of multimers sketched in the insert of Fig. 8.11 in order to determine the geometry of the spin-state polaron by comparison with the observed intensities (all the reduced matrix elements of Eq. (8.59) can be set to 1). The nearest neighbor Co-Co distance is 3.9 Å. The form factor of cobalt is tabulated in [Freeman and Desclaux (1979)]; it decreases monotonically by less than 25% for $Q \leq 2.5$Å.

Exercise No 8.3

(a) Derive the spin-wave dispersion of a one-dimensional antiferromagnet defined in Eq. (8.36) by solving the equation of motion, e.g., for the spin operator S^+, on the basis of the Heisenberg Hamiltonian

$$\hat{\mathcal{H}} = -2J \sum_i \hat{\boldsymbol{S}}_i \cdot \hat{\boldsymbol{S}}_{i+1} \tag{8.60}$$

(b) How is the spin-wave dispersion modified by extending the Hamiltonian

Eq. (8.60) to include a biquadratic term:

$$\hat{\mathcal{H}} = -2J\sum_i \hat{\boldsymbol{S}}_i \cdot \hat{\boldsymbol{S}}_{i+1} - K\sum_i \left(\hat{\boldsymbol{S}}_i \cdot \hat{\boldsymbol{S}}_{i+1}\right)^2 . \tag{8.61}$$

8.6 Solutions

Exercise No 8.1

(a) It is easy to see that the biquadratic term in Eq. (8.55) has only diagonal matrix elements which determine the eigenvalues:

$$E(S) = -J\eta - \frac{1}{4}K\eta^2, \quad \eta = S(S+1) - 2S_i(S_i+1), \tag{8.62}$$

where $S_i = 5/2$, thus

$$\begin{aligned} E(1) - E(0) &= -2J + \frac{33}{2}K, \\ E(2) - E(1) &= -4J + 27K, \\ E(3) - E(2) &= -6J + \frac{51}{2}K, \\ E(4) - E(3) &= -8J + 6K. \end{aligned}$$

A least-squares fit to the observed dimer splitting energies yields $J = -0.838(5)$ meV and $K = 8.8(8)$ μeV.

(b) We expand the magnetic energy of Eq. (8.58) up to first order:

$$W \approx \frac{\sqrt{3}}{4}a^2 c\frac{c_{33}}{2}\left(\frac{r-r_0}{r_0}\right)^2 - 2\left(J(r_0) + \left(\frac{\mathrm{d}J}{\mathrm{d}r}\right)_{r_0}(r-r_0)\right)\hat{\boldsymbol{S}}_1 \cdot \hat{\boldsymbol{S}}_2.$$

From $\mathrm{d}W/\mathrm{d}r = 0$ we find

$$r - r_0 = \frac{2c}{\sqrt{3}a^2 c_{33}}\left(\frac{\mathrm{d}J}{\mathrm{d}r}\right)_{r_0}\hat{\boldsymbol{S}}_1 \cdot \hat{\boldsymbol{S}}_2. \tag{8.63}$$

Substituting $r - r_0$ back to Eq. (8.58) yields

$$W = -2J(r_0)\hat{\boldsymbol{S}}_1 \cdot \hat{\boldsymbol{S}}_2 - \frac{2c}{\sqrt{3}a^2 c_{33}}\left(\frac{\mathrm{d}J}{\mathrm{d}r}\right)_{r_0}^2\left(\hat{\boldsymbol{S}}_1 \cdot \hat{\boldsymbol{S}}_2\right)^2, \tag{8.64}$$

which we compare with the Hamiltonian Eq. (8.55). For the prefactor of the biquadratic term we find the value 6.6 μeV, which is in reasonable agreement with $K = 8.8(8)$ μeV determined in (a). We therefore conclude that the presence of biquadratic exchange in $CsMn_{0.28}Mg_{0.72}Br_3$ is caused to a major extent by the mechanism of exchange striction.

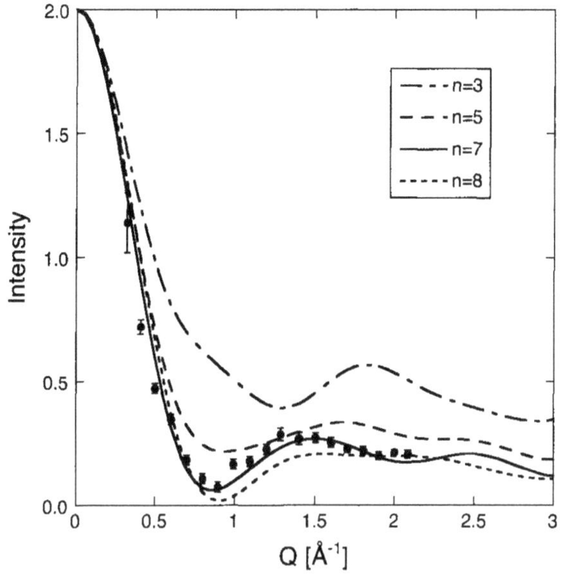

Fig. 8.12 Q-dependence of the intensity of the transition observed at 0.75 meV in $La_{0.998}Sr_{0.002}CoO_3$ (after [Podlesnyak *et al.* (2008)]). The lines are the result of structure-factor calculations based on Eq. (8.59) with different types of Co multimers sketched in Fig. 8.11.

Exercise No 8.2

As shown in Fig. 8.12, the Q-dependence of the intensities observed for the transition at 0.75 meV in $La_{0.998}Sr_{0.002}CoO_3$ is perfectly explained by the scattering from an octahedrally shaped Co heptamer ($n = 7$).

Exercise No 8.3

(a) Following Eq. (C.4) we rewrite the Hamiltonian Eq. (8.60):

$$\hat{\mathcal{H}} = -J\sum_i \left(\hat{S}_i^+\hat{S}_{i+1}^- + \hat{S}_i^-\hat{S}_{i+1}^+ + 2\hat{S}_i^z\hat{S}_{i+1}^z\right)$$
$$= -J\sum_{i,\pm}\left(\frac{1}{2}\left(\hat{S}_i^+\hat{S}_{i\pm1}^- + \hat{S}_i^-\hat{S}_{i\pm1}^+\right) + \hat{S}_i^z\hat{S}_{i\pm1}^z\right). \qquad (8.65)$$

The equation of motion for the operator S^+ reads:

$$-\imath\frac{\partial}{\partial t}\hat{S}_j^+ = [\hat{\mathcal{H}}, \hat{S}_j^+] \tag{8.66}$$

$$= -J\sum_{i,\pm}\left(\frac{1}{2}\left(\hat{S}_i^+\hat{S}_{i\pm 1}^- + \hat{S}_i^-\hat{S}_{i\pm 1}^+\right) + \hat{S}_i^z\hat{S}_{i\pm 1}^z\right), \hat{S}_j^+].$$

Applying the commutation relations

$$\left[\hat{S}_i^+, \hat{S}_j^-\right] = 2\delta_{ij}\hat{S}_i^z, \quad \left[\hat{S}_i^-, \hat{S}_j^+\right] = -2\delta_{ij}\hat{S}_i^z,$$

$$\left[\hat{S}_i^z, \hat{S}_j^+\right] = \delta_{ij}\hat{S}_i^+, \quad \left[\hat{S}_i^z, \hat{S}_j^-\right] = -\delta_{ij}\hat{S}_i^- \tag{8.67}$$

transforms Eq. (8.66) to the form

$$-\imath\frac{\partial}{\partial t}\hat{S}_j^+ = 2J\left(\hat{S}_j^z(\hat{S}_{j-1}^+ + \hat{S}_{j+1}^+) - \hat{S}_j^+(\hat{S}_{j-1}^z + \hat{S}_{j+1}^z)\right). \tag{8.68}$$

We linearize the equation of motion by the following substitutions:

$$\hat{S}_j^z \to S, \quad \hat{S}_{j\pm 1}^z \to -S, \quad \text{i.e.,} \quad \hat{S}_j^z = (-1)^j S. \tag{8.69}$$

Combining Eqs (8.68) and (8.69) yields

$$-\imath\frac{\partial}{\partial t}\hat{S}_j^+ = 2JS(\hat{S}_{j-1}^+ + 2\hat{S}_j^+ + \hat{S}_{j+1}^+). \tag{8.70}$$

By introducing $\hat{S}_j^+$ according to Eq. (8.26) into the equation of motion Eq. (8.70) we obtain:

$$\hbar\omega(q) = 4S|J|\sin(qa). \tag{8.71}$$

(b) The equation of motion for the biquadratic term reads [Falk *et al.* (1984)]

$$-\imath\frac{\partial}{\partial t}\hat{S}_j^+ = K\sum_{\pm}\left((\hat{S}_j^z)^2(\hat{S}_{j\pm 1}^z\hat{S}_{j\pm 1}^+ + \hat{S}_{j\pm 1}^+\hat{S}_{j\pm 1}^z)\right.$$

$$\left. - (\hat{S}_{j\pm 1}^z)^2(\hat{S}_j^z\hat{S}_j^+ + \hat{S}_j^+\hat{S}_j^z)\right). \tag{8.72}$$

By applying the same procedure as in (a), the following result is obtained:

$$\hbar\omega(q) = 4S|J - KS^2|\sin(qa). \tag{8.73}$$

This means that from an analysis of the spin-wave dispersion the relative sizes of the exchange parameters J and K cannot be determined, in contrast to experiments on small clusters of magnetic ions such as dimers (see exercise No. 8.1).

Chapter 9

Crystal-Field Transitions

9.1 Elementary concept of crystal fields

The crystal-field interaction is an essential ingredient in a discussion of the physical properties of magnetic materials. A magnetic ion in a crystal experiences the interaction with the charged surrounding ligand ions, producing an electrostatic field (called crystal field or ligand field). The energy splittings associated with this interaction vary roughly from 1 to 1000 meV, thus inelastic neutron scattering allows a direct determination of the crystal-field states. In the following we will restrict our considerations mainly to the case of a weak crystal field (weak compared, e.g., with the spin-orbit interaction) realized for the $4f$ electron systems.

In order to illustrate the crystal-field concept, we consider the simple case of a p electron defined by the wave function $|\psi_{nlm}\rangle$, where n is the principal quantum number, l the orbital quantum number with $l = 1$, and $-1 \leq m \leq 1$. One differentiates between:

$$
\begin{aligned}
|\psi_+\rangle &= \frac{1}{\sqrt{2}}(x + \imath y) f(r) \\
|\psi_0\rangle &= z f(r) \\
|\psi_-\rangle &= \frac{1}{\sqrt{2}}(x - \imath y) f(r)
\end{aligned}
\tag{9.1}
$$

These eigenstates are orthogonal to each other, and the corresponding energy eigenvalues are threefold degenerate. We now introduce a crystal field by placing point charges $+Q$ at the positions $x = \pm a$, $y = \pm b$, $z = \pm c$ of a Cartesian coordinate system as illustrated in Fig. 9.1a. For this particular problem, the adequate wave functions are the following linear combinations

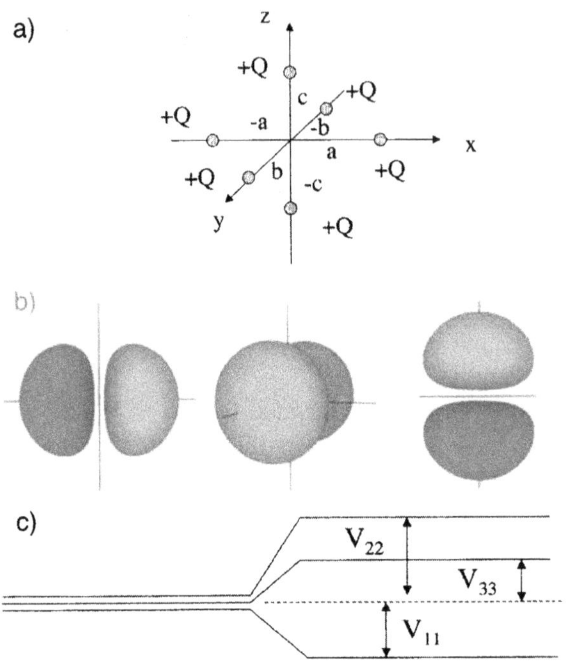

Fig. 9.1 (a) Point charges Q arranged in orthorhombic symmetry. (b) Contours of the wave functions of a p-electron defined by Eq. (9.2). (c) Energy levels of a p-electron in an orthorhombic crystal field.

of Eq. (9.1):

$$\begin{aligned} \frac{1}{\sqrt{2}}(|\psi_+\rangle + |\psi_-\rangle) &= xf(r) \\ \frac{1}{i\sqrt{2}}(|\psi_+\rangle - |\psi_-\rangle) &= yf(r) \\ |\psi_0\rangle &= zf(r) \end{aligned} \tag{9.2}$$

The contours of these wave functions are shown in Fig. 9.1b. It is easy to see that the threefold degeneracy can be lifted depending on the symmetry of the crystal field:

- cubic symmetry ($a = b = c$): no splitting
- tetragonal symmetry ($a = b \neq c$): one singlet + one doublet
- orthorhombic symmetry ($a \neq b \neq c$): three singlets

The crystal-field potential created by the charge $+Q$ at position $x = -a$ is given by:

$$V(x,y,z) = \frac{Q}{\sqrt{(x+a)^2+y^2+z^2}} = \frac{Q}{a\sqrt{1+\frac{2x}{a}+\frac{r^2}{a^2}}}$$
$$\approx \frac{Q}{a}\left(1-\frac{x}{a}-\frac{r^2}{2a^2}+\frac{3x^2}{2a^2}+\dots\right), \quad (9.3)$$

where $r^2 = x^2 + y^2 + z^2$. Similar expressions are found for the charges at the other positions. By summing up over the six charges at $x = \pm a$, $y = \pm b$, $z = \pm c$, all the terms linear in x,y,z disappear, and the resulting crystal-field potential has the form

$$V(x,y,z) = Ax^2 + By^2 + Cz^2 + D. \quad (9.4)$$

The energy eigenvalues E result from the secular determinant:

$$\det\begin{pmatrix} V_{11}-E & V_{12} & V_{13} \\ V_{21} & V_{22}-E & V_{23} \\ V_{31} & V_{32} & V_{33}-E \end{pmatrix} = 0, \quad (9.5)$$

where the V_{ik} are matrix elements based on Eqs (9.2) and (9.4). Examples of V_{ik} are:

$$V_{11} = \int x^2 f^2(r)\left(Ax^2+By^2+Cz^2+D\right)\mathrm{d}\mathbf{r},$$
$$V_{22} = \int y^2 f^2(r)\left(Ax^2+By^2+Cz^2+D\right)\mathrm{d}\mathbf{r}, \quad (9.6)$$
$$V_{12} = \int xy f^2(r)\left(Ax^2+By^2+Cz^2+D\right)\mathrm{d}\mathbf{r}.$$

It follows immediately that only the diagonal elements are non-zero, giving rise to the crystal-field splitting shown in Fig. 9.1c. The case presented here corresponds to the well-known quantum mechanical Stark effect.

We can intuitively see how the crystal field influences the magnetic properties by considering, e.g., the undisturbed wave function $\varphi_0 = xf(r)$ of Eq. (9.2):

$$\varphi_0 = xf(r) = \underbrace{\frac{1}{\sqrt{2}}\underbrace{(x+iy)f(r)}_{\text{right}} + \frac{1}{\sqrt{2}}\underbrace{(x-iy)f(r)}_{\text{left}}}_{\text{resulting current}\,=\,0}.$$

There is a complete cancellation of the right- and left-circular orbital motions of the p-electron, thus there is no net electronic current. However, the perturbation by the crystal field introduces an imbalance $\pm\epsilon$ between the right- and left-circular motions:

$$\varphi = xf(r) = \underbrace{\frac{1}{\sqrt{2}}\underbrace{(1+\epsilon)(x+iy)f(r)}_{\text{right}} + \frac{1}{\sqrt{2}}\underbrace{(1-\epsilon)(x-iy)f(r)}_{\text{left}}}_{\text{resulting current} \neq 0}$$

resulting in a non-zero electronic current, which produces a magnetic field. The parameter ϵ in the wave function $xf(r)$ can be calculated by perturbation theory:

$$\varphi = \varphi_0 + \sum_m \frac{\langle m|\hat{\mathcal{H}}_{\text{CF}}|0\rangle}{E_0 - E_m}\varphi_m,$$

where $\hat{\mathcal{H}}_{\text{CF}}$ is the Hamiltonian corresponding to the crystal-field potential Eq. (9.4).

9.2 Crystal-field interaction of f-electron systems

While the crystal-field potential of a p-electron system is a second-order polynomial defined by Eq. (9.4), fourth-order and sixth-order polynomials result from analogous calculations for d-electron and f-electron systems, respectively. One can show that the crystal field potential for f-electron systems (i.e., lanthanides and actinides) and for cubic symmetry is given by [Hutchings (1964)]:

$$V(x,y,z) = C_4\left[(x^4+y^4+z^4) - \frac{3}{5}r^4\right] + C_6\left[(x^6+y^6+z^6) + \frac{15}{4}(x^2y^4+x^2z^4+y^2x^4+y^2z^4+z^2x^4+z^2y^4) - \frac{15}{14}r^6\right]. \tag{9.7}$$

This is a rather involved expression which was transformed to an elegant and simple scheme by Stevens [Stevens (1967)]. He showed that the crystal-field potential $V(x,y,z)$ being a sum of polynomials in $\boldsymbol{r} = (x,y,z)$ can be replaced by a sum of polynomials of the total angular momentum operators $\hat{\boldsymbol{J}} = (\hat{J}_x, \hat{J}_y, \hat{J}_z)$, which have the same transformation properties as the original expression. They act on the unfilled $4f$ shell as a whole and therefore are much more convenient than the (x,y,z) polynomials which act on an individual $4f$ electron. The rules for transforming an expression from the $\boldsymbol{r}$ space to the $\boldsymbol{J}$ space are such that any products of (x,y,z) are replaced by the corresponding products of $(\hat{J}_x, \hat{J}_y, \hat{J}_z)$, but written in a symmetrized form (e.g.: $xy \rightarrow \frac{1}{2}(\hat{J}_x\hat{J}_y + \hat{J}_y\hat{J}_x)$). Constants of proportionality χ_n have to be introduced which depend on the order n as well as on the quantum numbers L, S and J. Furthermore, the r^n operators have to be replaced by the averages $\langle r^n \rangle$ of the radial part of the $4f$ wave functions. Some illustrative examples are:

$$
\begin{aligned}
3z^2 - r^2 &\equiv \chi_2\langle r^2\rangle \left[3\hat{J}_z^2 - J(J+1)\right] = \chi_2\langle r^2\rangle \hat{O}_2^0 \\
x^2 - y^2 &\equiv \chi_2\langle r^2\rangle \left[\hat{J}_x^2 - \hat{J}_y^2\right] = \chi_2\langle r^2\rangle \hat{O}_2^2 \\
x^4 - 6x^2y^2 + y^4 &\equiv \frac{1}{2}\left[(x+\imath y)^4 + (x-\imath y)^4\right] \\
&= \chi_4\langle r^4\rangle \frac{1}{2}\left[\hat{J}_+^4 + \hat{J}_-^4\right] = \chi_4\langle r^4\rangle \hat{O}_4^4
\end{aligned}
\tag{9.8}
$$

where $\hat{J}_\pm = \hat{J}_x \pm \imath\hat{J}_y$. A complete list of the operators which are called Stevens operators [Stevens (1967)] as well as of the reduced matrix elements χ_n can be found in [Hutchings (1964)]. The radial integrals $\langle r^n \rangle$ were tabulated by Freeman and Desclaux [Freeman and Desclaux (1979)]. In the Stevens notation the crystal-field Hamiltonian reads

$$
\hat{\mathcal{H}}_{\mathrm{CF}} = \sum_n \chi_n \langle r^n \rangle \sum_m A_n^m \hat{O}_n^m = \sum_{n,m} B_n^m \hat{O}_n^m, \tag{9.9}
$$

where the A_n^m and B_n^m are the crystal-field parameters. The crystal-field Hamiltonian Eq. (9.9) is treated as a perturbation of the $(2J+1)$-fold degenerate ground-state J-multiplet. We immediately recognize that the crystal-field Hamiltonian consists of a large number of parameters which, however, can be drastically reduced due to the point symmetry at the rare-earth site. In particular, a center of inversion cancels all the odd n terms, and a p-fold axis of symmetry when chosen as quantization axis reduces the Hamiltonian to terms containing $\hat{O}_n^p$. E.g., for cubic point symmetry (with the four-fold symmetry axis taken as quantization axis) the crystal-field Hamiltonian reads

$$
\hat{\mathcal{H}}_{\mathrm{CF}} = B_4^0\left(\hat{O}_4^0 + 5\hat{O}_4^4\right) + B_6^0\left(\hat{O}_6^0 - 21\hat{O}_6^4\right), \tag{9.10}
$$

i.e., the number of independent crystal-field parameters is reduced to two. Lea, Leask and Wolf [Lea *et al.* (1962)] rewrote Eq. (9.10) in the form

$$
\hat{\mathcal{H}}_{\mathrm{CF}} = W\left[\frac{x}{F_4}\left(\hat{O}_4^0 + 5\hat{O}_4^4\right) + \frac{1-|x|}{F_6}\left(\hat{O}_6^0 - 21\hat{O}_6^4\right)\right], \tag{9.11}
$$

where $B_4^0 F_4 = Wx$ and $B_6^0 F_6 = W(1-|x|)$ with W being an energy scale factor and $-1 \le x \le 1$, so that the entire range $B_4^0, B_6^0 \in (-\infty, \infty)$ is

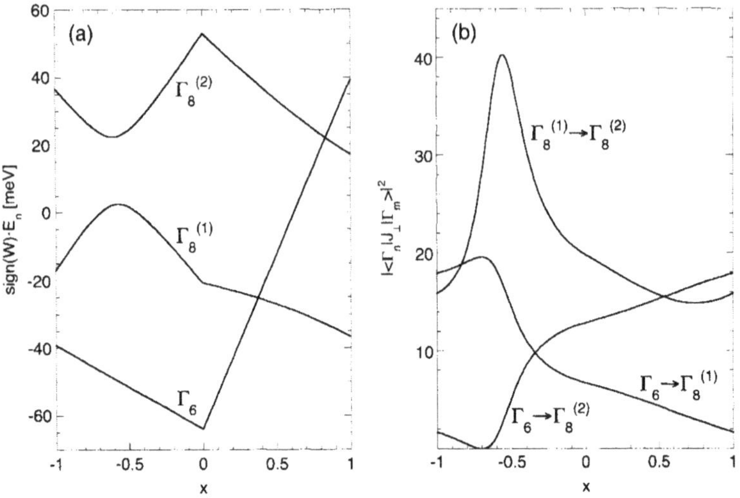

Fig. 9.2 (a) Eigenvalues of Eq. (9.11) for Nd^{3+} ($J = \frac{9}{2}$). The crystal-field interaction splits the tenfold degenerate ground-state multiplet of Nd^{3+} into a doublet Γ_6 and two quartets Γ_8. (b) Matrix elements of the crystal-field transitions $\Gamma_n \to \Gamma_m$ for Nd^{3+}.

covered. Lea at al. [Lea *et al.* (1962)] tabulated the eigenfunctions and eigenvalues as a function of the parameter x for the whole rare-earth series ($2 \leq J \leq 8$). The factors F_4 and F_6 depend only on J and are also listed there. Fig. 9.2a displays the calculations for neodymium ($J = \frac{9}{2}$).

For hexagonal point symmetry and the c-axis taken as quantization axis, the crystal-field Hamiltonian has the form

$$\hat{\mathcal{H}}_{\mathrm{CF}} = B_2^0\hat{O}_2^0 + B_4^0\hat{O}_4^0 + B_6^0\hat{O}_6^0 + B_6^6\hat{O}_6^6 \tag{9.12}$$

with four independent crystal-field parameters. For orthorhombic symmetry, there are as much as nine adjustable crystal-field parameters:

$$\begin{aligned}\hat{\mathcal{H}}_{\mathrm{CF}} = {} & B_2^0\hat{O}_2^0 + B_2^2\hat{O}_2^2 + B_4^0\hat{O}_4^0 + B_4^2\hat{O}_4^2 + B_4^4\hat{O}_4^4 \\ & + B_6^0\hat{O}_6^0 + B_6^2\hat{O}_6^2 + B_6^4\hat{O}_6^4 + B_6^6\hat{O}_6^6 .\end{aligned} \tag{9.13}$$

As an example, Fig. 9.3 shows inelastic neutron scattering data obtained for the compound $NdPd_3$ (isostructural to $ErPd_3$ introduced in Chap. 7.4), in which the Nd^{3+} ions are located at sites of cubic point-symmetry, thus the Hamiltonians of Eqs (9.10) and (9.11) apply. The energy spectrum taken at $T = 4.2$ K exhibits two well resolved inelastic lines, a strong transition at $\hbar\omega = 6$ meV and a weak transition at $\hbar\omega = 10$ meV. Both transitions

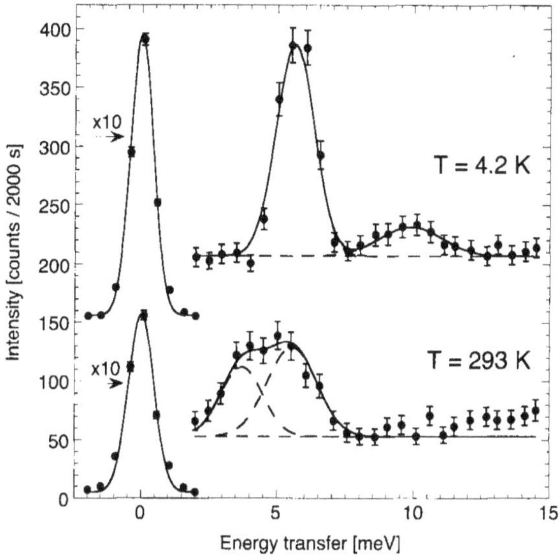

Fig. 9.3 Energy spectra of neutrons scattered from polycrystalline $NdPd_3$ (after [Furrer and Purwins (1976)]).

are excitations out of the ground state. Inspection of the energy splitting scheme shown in Fig. 9.2a indicates that there are six possible solutions to describe the experimental findings, namely the parameter pairs ($x \approx -0.75$, $W > 0$), ($x \approx -0.35$, $W > 0$), ($x \approx 0.65$, $W > 0$), ($x \approx -0.90$, $W < 0$), ($x \approx -0.05$, $W < 0$) and ($x \approx 0.60$, $W < 0$). Obviously a consideration of the energies of the crystal-field transitions alone is not sufficient to determine a unique parameter set, but the additional information provided by the peak intensities has to be considered as well, as discussed in the following section.

9.3 Neutron cross-section

In evaluating the cross-section for the crystal-field transition $\Gamma_n \to \Gamma_m$ we start from the magnetic scattering law $S^{\alpha\beta}(\boldsymbol{Q},\omega)$ defined by Eq. (2.43). We replace the states $|\lambda\rangle$ of the system by the crystal-field states

$$|\Gamma_n\rangle = \sum_{M=-J}^{J} a_M|M\rangle, \quad \text{with} \quad a_M \in \mathbb{C} \tag{9.14}$$

obtained by diagonalization of Eq. (9.9). Since we are dealing with single-ion excitations, we have $j = j'$. For N identical magnetic ions we can even drop the index j. $S^{\alpha\beta}(\boldsymbol{Q},\omega)$ then reduces to

$$S^{\alpha\beta}(\boldsymbol{Q},\omega) = Np_n\langle\Gamma_n|\hat{J}_\alpha|\Gamma_m\rangle\langle\Gamma_m|\hat{J}_\beta|\Gamma_n\rangle\delta(\hbar\omega + E_n - E_m), \tag{9.15}$$

where N is the total number of magnetic ions and p_n the Boltzmann population factor of the initial state $|\Gamma_n\rangle$. Inserting $S^{\alpha\beta}(\boldsymbol{Q},\omega)$ into Eq. (2.42) and making use of the symmetry relations associated with the matrix elements we find the cross section

$$\boxed{\begin{aligned}\frac{\mathrm{d}^2\sigma}{\mathrm{d}\Omega\mathrm{d}\omega} &= N(\gamma r_0)^2\frac{k'}{k}F^2(\boldsymbol{Q})e^{-2W(\boldsymbol{Q})}p_n \\ &\times\sum_\alpha\left[1-\left(\frac{Q_\alpha}{Q}\right)^2\right]\cdot|\langle\Gamma_m|\hat{J}_\alpha|\Gamma_n\rangle|^2\cdot\delta(\hbar\omega+E_n-E_m)\end{aligned}} \tag{9.16}$$

For experiments on polycrystalline material Eq. (9.16) has to be averaged in $\boldsymbol{Q}$ space:

$$\begin{aligned}\frac{\mathrm{d}^2\sigma}{\mathrm{d}\Omega\mathrm{d}\omega} &= N(\gamma r_0)^2\frac{k'}{k}F^2(\boldsymbol{Q})e^{-2W(\boldsymbol{Q})}p_n\cdot|\langle\Gamma_m|\hat{\boldsymbol{J}}_\perp|\Gamma_n\rangle|^2 \\ &\qquad\times\delta(\hbar\omega+E_n-E_m),\end{aligned} \tag{9.17}$$

where $\hat{\boldsymbol{J}}_\perp = \hat{\boldsymbol{J}} - \frac{(\hat{\boldsymbol{J}}\cdot\boldsymbol{Q})}{Q^2}\boldsymbol{Q}$ is the component of the total angular momentum perpendicular to the scattering vector $\boldsymbol{Q}$, and

$$|\langle\Gamma_m|\hat{\boldsymbol{J}}_\perp|\Gamma_n\rangle|^2 = \frac{2}{3}\sum_\alpha|\langle\Gamma_m|\hat{J}_\alpha|\Gamma_n\rangle|^2. \tag{9.18}$$

These matrix elements have been tabulated by Birgeneau [Birgeneau (1972)] for cubic crystal fields. As an example the results for Nd^{3+} are displayed in Fig. 9.2b, which we now use to analyze the energy spectra obtained for $NdPd_3$ at $T = 4.2$ K (see Fig. 9.3). We recognize that only the parameter set ($x \approx -0.9$, $W < 0$) is compatible with the observed intensities, i.e., the strong excitation at $\hbar\omega = 6$ meV corresponds to the $\Gamma_6 \rightarrow \Gamma_8^{(1)}$ transition and the weak excitation at $\hbar\omega = 10$ meV to the $\Gamma_6 \rightarrow \Gamma_8^{(2)}$ transition. This interpretation is confirmed by the data taken at $T = 293$ K (see Fig. 9.3), where an additional line shows up at $\hbar\omega = 4$ meV corresponding to the excited-state transition $\Gamma_8^{(1)} \rightarrow \Gamma_8^{(2)}$, whose strength is comparable to the $\Gamma_6 \rightarrow \Gamma_8^{(2)}$ transition, in agreement with the matrix elements displayed in Fig. 9.2. In conclusion, by considering both the energies and the intensities, the crystal-field splitting of any compound can

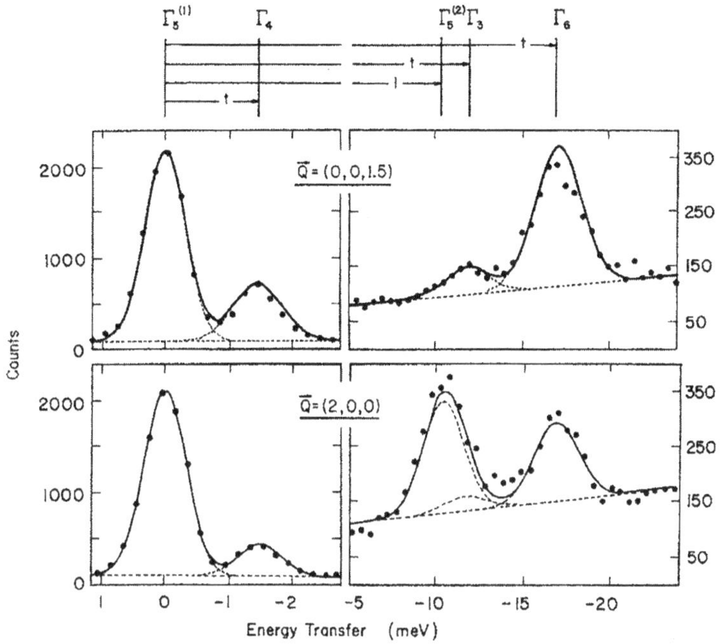

Fig. 9.4 Energy spectra of neutrons scattered from single-crystalline $PrBr_3$ at $T = 1.5$ K for $\boldsymbol{Q}$ parallel and perpendicular to the c-axis. The resulting crystal-field level scheme and the observed transverse (t) and longitudinal (l) ground-state transitions are indicated at the top. Reprinted with permission from [Schmid *et al.* (1987)]. Copyright 1987 by the American Institute of Physics.

unambiguously be assessed. For the case of $NdPd_3$ we find Γ_6 to be the ground state, followed by the excited states $\Gamma_8^{(1)}$ and $\Gamma_8^{(2)}$ at 6 meV and 10 meV, respectively. The corresponding crystal-field parameters turn out to be $B_4^0 = 1.98 \cdot 10^{-3}$ meV and $B_6^0 = -0.49 \cdot 10^{-5}$ meV, which can easily be calculated from Eqs (9.10) and (9.11) and adjusting the scale factor W to the observed energy splittings.

In single-crystal experiments, an additional identification means is provided by the polarization factor of the cross section (see Eq. (9.16)), which permits the discrimination between transverse ($\alpha = x, y$) and longitudinal ($\alpha = z$) crystal-field transitions by measuring at different scattering vectors $\boldsymbol{Q}$. This is nicely demonstrated in Fig. 9.4 for the compound $PrBr_3$ where the Pr^{3+} ions experience a crystal field of hexagonal symmetry defined by Eq. (9.12). For $\boldsymbol{Q} \| c$ only transverse transitions are observed, whereas for $\boldsymbol{Q} \perp c$ the transverse transitions lose half their intensities, and in addition longitudinal transitions appear. The lines in Fig. 9.4 were calculated without any disposable parameters, i.e., the intensities of the crystal-field transitions are excellently described by Eq. (9.16).

9.4 Interactions of the crystal-field split ions

In the preceding Sections we considered crystal-field effects for non-interacting rare-earth ions. However, the crystal-field states are usually subject to interactions with phonons, spin fluctuations, conduction electrons (or more generally charge carriers), etc., which limit the lifetime of the crystal-field states, thus the observed crystal-field transitions exhibit line broadening. Therefore, the δ-function in the neutron cross-section formulae Eqs (9.16) and (9.17) describing the energy conservation has to be replaced by a Lorentzian with a finite linewidth Γ^{fwhm} and a temperature dependence characteristic of the particular type of interaction. Moreover, if the rare-earth ions are coupled by the exchange interaction, the crystal-field excitations can propagate through the lattice giving rise to dispersion (see Chap. 8.2.3). In experiments on polycrystalline material, the crystal-field transitions then exhibit line broadening proportional to the overall bandwidth of the dispersion.

Orbach investigated the line broadening of crystal-field transitions induced by lattice vibrational modes [Orbach (1961)]. For the direct process in which the excited crystal-field state Γ_m decays into the ground state Γ_n through the emission of a phonon with energy $\hbar\omega_{nm} = E_n - E_m$, where E_n and E_m are the energies of the corresponding crystal-field states, the linewidth turns out to be

$$\Gamma_{nm}^{\text{fwhm}}(T) = \frac{3(\hbar\omega_{nm})^2 k_B T}{\pi \hbar^4 \rho v^5 |\xi_{nm}|^2}, \tag{9.19}$$

where ρ is the density and v the sound velocity in the material, and $\xi_{nm} = \langle \Gamma_m | \hat{J}_\alpha | \Gamma_n \rangle$.

The interaction with the charge carriers is by far the dominating relaxation mechanism in metallic rare-earth compounds. Similar to the direct Orbach process (Eq. (9.19)), the linewidth $\Gamma^{\text{fwhm}}(T)$ of crystal-field transitions increases linearly with temperature according to the well-known Korringa law [Korringa (1950)]:

$$\Gamma^{\text{fwhm}}(T) = 4\pi(g-1)^2 J(J+1) \left(N(E_F) \cdot j_{\text{ex}}\right)^2 \cdot T, \tag{9.20}$$

where $N(E_F)$ denotes the density of states of the charge carriers at the Fermi energy E_F, and j_{ex} the exchange integral between the charge carriers

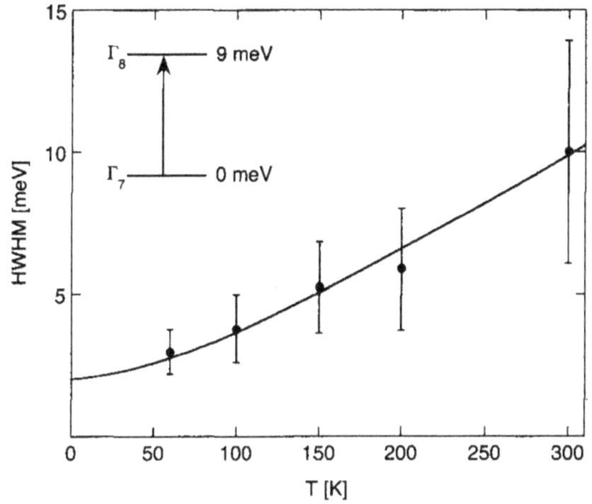

Fig. 9.5 Temperature dependence of the linewidth of the crystal-field transition $\Gamma_7 \rightarrow \Gamma_8$ observed for $Ce_{0.4}La_{0.6}Al_2$ (HWHM: half-width at half-maximum, after [Loewenhaupt and Steglich (1977)]). The line is the result of a least-squares fit based on Eq. (9.21).

and the $4f$ electrons of the rare-earth ions. The inclusion of crystal-field effects slightly modifies the low-temperature limit of Eq. (9.20). According to the theory of Becker, Fulde and Keller [Becker *et al.* (1979)], the linewidth of the crystal-field transition $\Gamma_n \rightarrow \Gamma_m$ is given by

$$\Gamma_{nm}^{\text{fwhm}}(T) = 2j_{\text{ex}}^2 \left[|\xi_{nm}|^2 \coth\left(\frac{\hbar\omega_{nm}}{2k_BT}\right) \chi''(\hbar\omega_{nm}) + \sum_{n \neq k} |\xi_{nk}|^2 \frac{\chi''(\hbar\omega_{nk})}{e^{\frac{\hbar\omega_{nk}}{k_BT}} - 1} \right.$$

$$\left. + \sum_{k \neq m} |\xi_{km}|^2 \frac{\chi''(\hbar\omega_{km})}{e^{\frac{\hbar\omega_{km}}{k_BT}} - 1} \right], \tag{9.21}$$

where $\chi'' = \text{Im}\chi$ is the imaginary part of the susceptibility summed over the Brillouin zone. For a non-interacting Fermi liquid we have

$$\chi''(\hbar\omega_{nm}) = \pi N^2(E_F)\hbar\omega_{nm}. \tag{9.22}$$

As an example we consider inelastic neutron scattering experiments performed for $Ce_{0.4}La_{0.6}Al_2$ in which the crystal-field of cubic symmetry (see Eqs (9.10) and (9.11)) experienced by the Ce^{3+} ions ($J = \frac{5}{2}$) results in a doublet ground-state Γ_7 separated by 9 meV from the excited Γ_8 quartet state. The temperature dependence of the linewidth of the $\Gamma_7 \rightarrow \Gamma_8$ transition is shown in Fig. 9.5. Structural inhomogeneities produce a finite linewidth at $T = 0$, which has to be subtracted before the analysis based on Eqs (9.21) and (9.22) is performed. For the detailed analysis of the data we refer to Exercise No. 9.2. Another example will be discussed in Chap. 11.

9.5 Intermultiplet crystal-field transitions

Often the number of crystal-field splittings observed in the ground-state J-multiplet are not sufficient to unambiguously determine all the parameters of Eq. (9.9). In principle, the number of observables can be increased by measurements of intermultiplet transitions, which have become possible due to the copious flux of epithermal neutrons produced at spallation neutron sources. The Stevens formalism described in Chap. 9.2 is then no longer applicable, since the crystal-field interaction leads to a mixing of different J-multiplets (called J-mixing). Furthermore, the J-multiplets are contaminated by states of different quantum numbers L and S (called intermediate coupling). All these effects are taken into account by diagonalizing the electrostatic, spin-orbit and crystal-field interactions simultaneously as discussed in detail, e.g., by Wybourne [Wybourne (1965)]. This procedure has to be applied as well, if the crystal-field interaction is strong, i.e., comparable to the spin-orbit interaction.

A typical example is $SmBa_2Cu_3O_7$ with orthorhombic symmetry, thus the appropriate Hamiltonian is Eq. (9.13) with nine independent crystal-field parameters. The splitting of the ground-state J-multiplet $^6H_{5/2}$ into three doublets is clearly not sufficient to arrive at a reliable parametrization, thus the observation of crystal-field states in the first-excited J-multiplet $^6H_{7/2}$ was essential to understand the magnetic properties of this compound, see Fig. 9.6.

9.6 Calculation of thermodynamic magnetic properties

In order to check the reliability of the crystal-field parameters determined from inelastic neutron scattering measurements, it is important to calculate various thermodynamic magnetic properties and to compare the results with experimental data. These properties depend explicitly upon both the crystal-field energies E_n and the crystal-field wavefunctions $|\Gamma_n\rangle$. Based on general expressions of statistical mechanics for the free energy $F = -k_BT \ln Z$ and the internal energy $U = F - T\frac{\partial F}{\partial T}|_V$, where Z is the partition

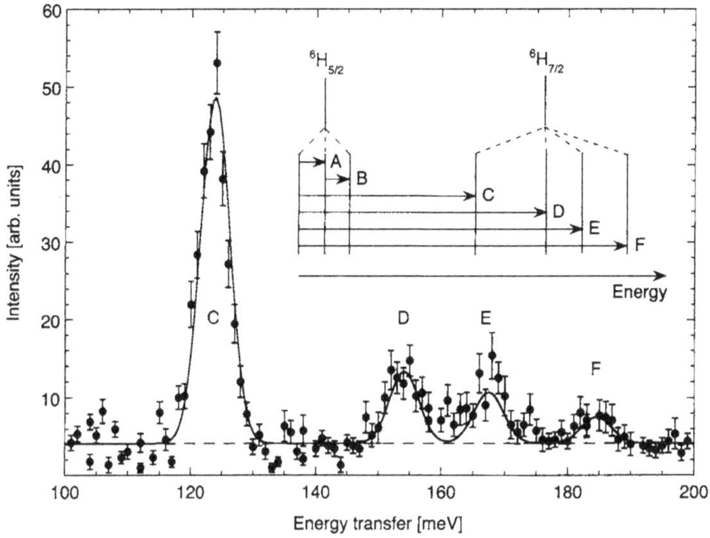

Fig. 9.6 Intermultiplet crystal-field transitions observed for $SmBa_2Cu_3O_7$ (after [Guillaume *et al.* (1993)]). The inset shows the corresponding energy level scheme for the two lowest J-multiplets.

function, we obtain for the magnetization:

$$M_\alpha = \frac{1}{k_B T}\frac{\partial \ln Z}{\partial H_\alpha} = g\mu_B \sum_n p_n \langle \Gamma_n | \hat{J}_\alpha | \Gamma_n \rangle, \tag{9.23}$$

for the single-ion susceptibility:

$$\chi_{\alpha\alpha} = \frac{\partial M_\alpha}{\partial H_\alpha} = (g\mu_B)^2 \left[\sum_n \frac{|\langle \Gamma_n | \hat{J}_\alpha | \Gamma_n \rangle|^2}{k_B T} p_n + \sum_{n \neq m} \frac{|\langle \Gamma_m | \hat{J}_\alpha | \Gamma_n \rangle|^2}{E_n - E_m}(p_m - p_n) \right] \tag{9.24}$$

and for the Schottky heat capacity:

$$c_V = \left(\frac{\partial U}{\partial T}\right)_V = k_B \left[\sum_n \left(\frac{E_n}{k_B T}\right)^2 \cdot p_n - \sum_n \left(\frac{E_n}{k_B T} \cdot p_n\right)^2 \right] \tag{9.25}$$

with the Boltzmann population factor $p_n = \frac{1}{Z} e^{\frac{-E_n}{k_B T}}$.

9.7 Further reading

- P. Fulde and I. Peschel, Adv. Phys. 21, 1 (1972): *Some crystalline field effects in metals*
- P. Fulde, in *Handbook on the physics and chemistry of rare earths*, ed. by K. A. Gschneidner and L. Eyring (North-Holland Publishing Company, Amsterdam, 1978), p. 295: *Crystal fields*
- A. Furrer and A. Podlesnyak, in *Handbook of applied solid state spectroscopy*, ed. by D. R. Vij (Springer, New York, 2006), p. 257: *Crystal-field spectroscopy*
- M. T. Hutchings, in Solid State Physics, Vol. 16, ed. by F. Seitz and D. Turnbull (Academic Press, New York, 1964), p. 227: *Point-charge calculations of energy levels of magnetic ions in crystalline electric fields*
- O. Moze, in *Handbook of magnetic materials*, ed. by K. H. J. Buschow (Elsevier, Amsterdam, 1998), p. 493: *Crystal field effects in intermetallic compounds studied by inelastic neutron scattering*

9.8 Exercises

Exercise No 9.1

In amorphous and liquid rare-earth systems the crystal-field Hamiltonian can be well approximated by

$$\hat{\mathcal{H}}_{\mathrm{CF}} = B_2^0 \hat{O}_2^0, \tag{9.26}$$

where the Stevens operator $\hat{O}_2^0$ is defined in Eq. (9.8).

(a) Calculate the eigenvalues of the Hamiltonian Eq. (9.26) for cerium ($J = 5/2$) and establish the crystal-field energy level diagram.

(b) Calculate the transition matrix elements between all the crystal-field levels according to Eq. (9.18) and thereby define the selection rules for the observation of crystal-field transitions by neutron scattering.

Exercise No 9.2

(a) Calculate j_{ex} from the linewidth data displayed for $\mathrm{Ce_{0.4}La_{0.6}Al_2}$ in Fig. 9.5 with the help of Eq. (9.20). The slope of the Korringa line is $\Gamma^{\mathrm{fwhm}}/T = 0.0577$ meV/K. For cerium the Landé splitting factor is

$g = 6/7$. For a free electron gas the following relations hold for the Fermi wavevector k_F and the Fermi energy E_F:

$$\frac{n}{v} = \frac{k_F^3}{3\pi^2}, \quad E_F = \frac{\hbar^2 k_F^2}{2m_e}, \quad N(E_F) = \frac{v}{4\pi^2}\frac{2m_e}{\hbar^2}k_F, \tag{9.27}$$

where v is the volume of the unit cell, n the number of conduction electrons per unit cell, and m_e the mass of the electron. $Ce_{0.4}La_{0.6}Al_2$ crystallizes in a face-centered cubic (fcc) lattice with lattice parameter $a = 8.11$ Å, with the Ce/La positions identical to the C positions in diamond (see Chap. 4.2).

(b) The effective exchange interaction $J(R)$ between two cerium ions separated by the distance R is related to j_{ex} through the Ruderman-Kittel-Kasuya-Yosida (RKKY) formula [Kasuya (1956); Ruderman and Kittel (1954); Yosida (1957)]

$$J(R) = \frac{3(g-1)^2 j_{ex}^2 k_F^3}{8\pi E_F \cdot (n/v)} \cdot \frac{\sin(2k_F R) - 2k_F R\cos(2k_F R)}{(2k_F R)^4}. \tag{9.28}$$

Determine $J(R)$ for nearest-neighbor cerium ions.

(c) Alternatively, information on the effective exchange interaction $J(R)$ can be obtained via the molecular-field parameter λ from the relation

$$\chi^{-1}(T = T_C) = \pm\lambda = \frac{2}{g\mu_B}\langle S\rangle \sum_r z_r J_r, \tag{9.29}$$

where the magnetic susceptibility is defined by Eq. (9.24), the + and − sign refers to ferromagnets with Curie temperature T_C and antiferromagnets with Néel temperature T_N, respectively, $g\mu_B\langle S\rangle$ is the saturated magnetic moment at low temperatures, z_r the number of r-th nearest-neighbor cerium ions, and J_r the corresponding exchange parameter. For $CeAl_2$ we have $T_N = 3.9$ K and $\langle S\rangle \approx 1$ [Barbara *et al.* (1977)]. Determine the nearest-neighbor exchange parameter J_1 by restricting the sum in Eq. (9.29) to $r = 1$. Compare and discuss the result with the value of J_1 obtained from Eq. (9.28).

9.9 Solutions

Exercise No 9.1

(a) The eigenvalue problem can best be solved by setting up the secular determinant with elements $\langle M'|\hat{\mathcal{H}}_{CF}|M\rangle$, where $-J < M < J$. The

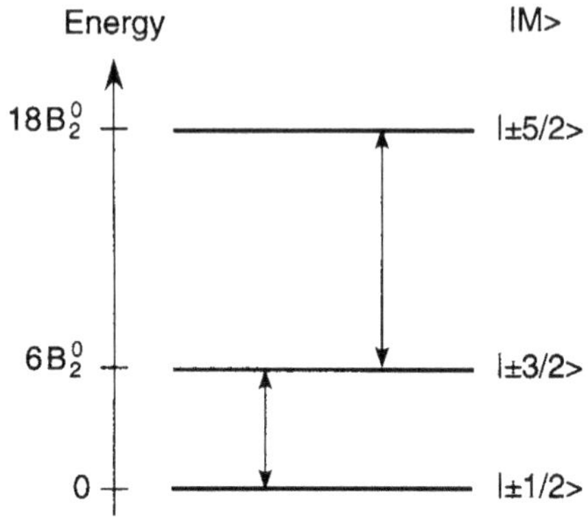

Fig. 9.7 Energy-level splitting of the ground-state J-multiplet of cerium resulting from the crystal-field Hamiltonian Eq. (9.26). The allowed transitions are indicated by arrows.

operator of the Hamiltonian $\hat{\mathcal{H}}_{\text{CF}}$ is defined in Eq. (9.8): $\hat{O}_2^0 = 3\hat{J}_z^2 - J(J+1)$. We can neglect the constant $J(J+1)$ and apply Eq. (C.3):

$$\hat{J}_z|M\rangle = M|M\rangle, \quad \langle M'|\hat{J}_z|M\rangle = \delta_{MM'}M, \quad \langle M'|\hat{J}_z^2|M\rangle = \delta_{MM'}M^2.$$

The secular determinant has the following form:

$$\begin{pmatrix} \frac{75}{4}B_2^0 & 0 & 0 & 0 & 0 & 0 \\ 0 & \frac{27}{4}B_2^0 & 0 & 0 & 0 & 0 \\ 0 & 0 & \frac{3}{4}B_2^0 & 0 & 0 & 0 \\ 0 & 0 & 0 & \frac{3}{4}B_2^0 & 0 & 0 \\ 0 & 0 & 0 & 0 & \frac{27}{4}B_2^0 & 0 \\ 0 & 0 & 0 & 0 & 0 & \frac{75}{4}B_2^0 \end{pmatrix}$$

We recognize that there are no off-diagonal elements, thus the eigenvalues of the crystal-field levels correspond to the diagonal elements. The ground-state J-multiplet of cerium is therefore split into three doublet states as shown in Fig. 9.7.

(b) The operators $\hat{J}_x$ and $\hat{J}_y$ are related to the operators $\hat{J}^+$ and $\hat{J}^-$ through Eq. (C.4). The matrix elements of the operators $\hat{J}_\alpha$ ($\alpha = +, -, z$) are defined by:

$$\begin{aligned} \langle J, M+1|\hat{J}^+|J, M\rangle &= \sqrt{(J-M)(J+M+1)}, \\ \langle J, M-1|\hat{J}^-|J, M\rangle &= \sqrt{(J+M)(J-M+1)}, \\ \langle J, M|\hat{J}_z|J, M\rangle &= M, \end{aligned}$$

i.e., the selection selection rule for crystal-field transitions is $\Delta M = 0, \pm 1$. The following transitions have non-zero matrix elements:

$$|+\frac{1}{2}\rangle \rightarrow |+\frac{3}{2}\rangle \; : \; |\langle \frac{5}{2}, \frac{3}{2}|\hat{J}^+|\frac{5}{2}, \frac{1}{2}\rangle|^2 = \left(\frac{5}{2} - \frac{1}{2}\right)\left(\frac{5}{2} + \frac{1}{2} + 1\right) = 8,$$

$$|-\frac{1}{2}\rangle \rightarrow |-\frac{3}{2}\rangle \; : \; |\langle \frac{5}{2}, -\frac{3}{2}|\hat{J}^-|\frac{5}{2}, -\frac{1}{2}\rangle|^2 = \left(\frac{5}{2} - \frac{1}{2}\right)\left(\frac{5}{2} + \frac{1}{2} + 1\right) = 8,$$

$$|+\frac{3}{2}\rangle \rightarrow |+\frac{5}{2}\rangle \; : \; |\langle \frac{5}{2}, \frac{5}{2}|\hat{J}^+|\frac{5}{2}, \frac{3}{2}\rangle|^2 = \left(\frac{5}{2} - \frac{3}{2}\right)\left(\frac{5}{2} + \frac{3}{2} + 1\right) = 5,$$

$$|-\frac{3}{2}\rangle \rightarrow |-\frac{5}{2}\rangle \; : \; |\langle \frac{5}{2}, -\frac{5}{2}|\hat{J}^+|\frac{5}{2}, -\frac{3}{2}\rangle|^2 = \left(\frac{5}{2} - \frac{3}{2}\right)\left(\frac{5}{2} + \frac{3}{2} + 1\right) = 5.$$

The transition $|\pm\frac{1}{2}\rangle \leftrightarrow |\pm\frac{5}{2}\rangle$ is not allowed. In a neutron scattering experiment we expect to observe two crystal-field transitions at energies $6B_2^0$ and $12B_2^0$ with an intensity ratio 8/5 which was nicely confirmed in experiments on liquid cerium [Millhouse and Furrer (1976)], see Fig. 9.8.

Exercise No 9.2

(a) We assume that each of the eight Ce^{3+}/La^{3+} and sixteen Al^+ ions in the fcc unit cell provide three and one conduction electrons, respectively, thus $N = 3 \times 8 + 16 \times 1 = 40$. From Eq. (9.27) we find $k_F = 1.3$ Å^{-1} and $N(E_F) = 4.6$ eV^{-1}. Inserting the latter into Eq. (9.20) yields $j_{ex} = 0.12$ eV.

(b) The value of the first term of Eq. (9.28) is 1.4 meV. The second term with $2k_F R = 9.13$ ($R_1 = 3.51$ Å) yields $-1.2 \cdot 10^{-3}$. The product of the two terms is therefore $J_1 = -1.7$ μeV.

(c) The energy of the crystal-field transition $\Gamma_7 \rightarrow \Gamma_8$ of 9 meV is obtained from Eq. (9.10) for the parameters $B_4^0 = 0.025$ meV and $B_6^0 = 0$. From Eq. (9.24) we find $\chi(T = T_N = 3.9 \text{ K}) = 0.061$ emu/mole. Inserting into Eq. (9.29) yields $2\sum_r z_r J_r = -0.39$ meV. We restrict the summation to the four nearest-neighbor cerium ions, thus $2 \times 4 \times J_1 = -0.39$ meV and $J_1 = -49$ μeV, which exceeds the result obtained from Eq. (9.28) by more than an order of magnitude. This discrepancy tells us that the application of the RKKY formula, being derived for a free electron gas, has to be considered with caution.

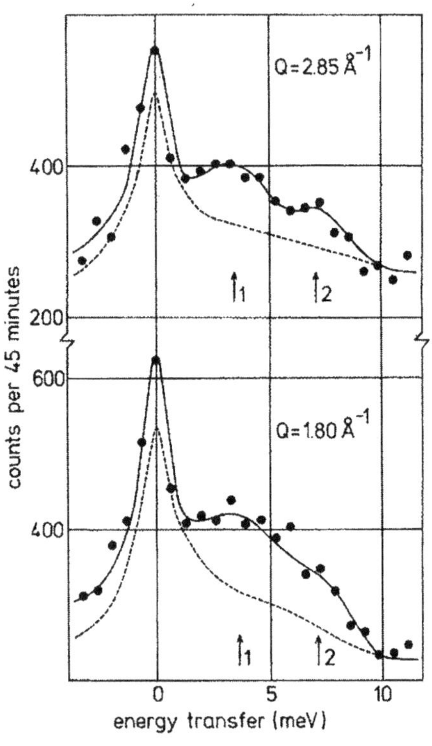

Fig. 9.8 Energy spectrum of neutrons scattered from liquid Ce at $T = 1285$ K. The full curve is the result of a least-squares fit with Gaussian lines for the elastic and inelastic peaks. The dashed line is the "background" corresponding to the energy spectrum observed for liquid lanthanum. Reprinted with permission from [Millhouse and Furrer (1976)]. Copyright 1976 by the American Institute of Physics.

Chapter 10

Phase Transitions

10.1 Introduction

Phase transitions are dramatic events in materials. The phase transition occurs at a critical value of thermodynamic variables x_i like the temperature T, pressure P, magnetic field $\boldsymbol{H}$, etc., where two phases with different degree of ordering are separated from each other. For the description of phase transitions it is convenient to introduce the Gibbs free enthalpy $G(x_1, x_2, x_3, \dots)$ defined by

$$G = -k_B T \ln Z + P \cdot V, \quad \text{or} \quad G = -k_B T \ln Z - \boldsymbol{H} \cdot \boldsymbol{M}, \tag{10.1}$$

where

$$Z = \sum_{i=1}^{N} e^{-\frac{E_i}{k_B T}} \tag{10.2}$$

is the partition function depending on the number N of excited states with energy E_i. The partial derivatives of G

$$\phi_i = \left(\frac{\partial G}{\partial x_i}\right)_{x_1, x_2, \dots, x_{i-1}, x_{i+1}, \dots} \tag{10.3}$$

are called extensive variables and serve to classify the phase transitions. A particular extensive variable ϕ can be identified as the so-called order parameter, which takes a finite value in one phase and vanishes in the other phase.

First-order and second-order phase transitions are characterized by a discontinuous and continuous change of the order parameter ϕ, respectively. The continuous phase transitions, in contrast to the discontinuous ones, exhibit a universal behavior in the sense that the critical phenomena only depend on the dimensionality of both the system and the order parameter, but not on the details of the underlying interactions. In the Landau theory of second-order phase transitions, the Gibbs free enthalpy is described as a Taylor series in terms of the order parameter ϕ:

$$G = \int \mathrm{d}V \left(A\,(\nabla\phi)^2 + \frac{r}{2}\phi^2 + u\phi^4 + \dots \right) - H\phi. \tag{10.4}$$

It is assumed that the constants A and u in Eq. (10.4) are positive and constant. In the following we discuss the response of the system at vanishing external magnetic field, i.e., $H = 0$. In order to realize a phase transition, the parameter r has to change sign at the critical value x_c:

$$r(x) = a(x_c - x), \quad \text{with} \quad a > 0. \tag{10.5}$$

This is visualized in Fig. 10.1, which demonstrates the presence of different minima of G above and below x_c.

From the derivative of Eq. (10.4) we readily obtain the behavior of the extensive variables defined by Eq. (10.3) around the critical value x_c:

$$\frac{\partial G}{\partial \phi} = r(x)\phi + 4u\phi^3 = 0 \quad \rightarrow \quad \phi \propto \left(\frac{x}{x_c} - 1\right)^{1/2}. \tag{10.6}$$

The exponent on the right hand side of Eq. (10.6) is the so-called critical mean-field exponent, which describes the critical behavior of the extensive variable ϕ in a universal manner. Similarly, we find for the second derivative of Eq. (10.4):

$$\frac{\partial^2 G}{\partial \phi^2} = r(x) + 12u\phi^2 = \chi^{-1} \quad \rightarrow \quad \chi \propto \left(1 - \frac{x}{x_c}\right)^{-1}. \tag{10.7}$$

where χ is a generalized susceptibility diverging at x_c with a critical mean-field exponent of size -1. There is a close analogy of Eq. (10.7) in the field of magnetism where the inverse susceptibility obeys the Curie-Weiss law.

Neutron scattering has proven to be an outstanding tool in the elucidation of phase transitions in many fields like structure investigations, magnetism, superconductivity, superfluidity, valence changes, etc. Some of these fields are discussed in separate Chapters of the book. The following Sections will focus on second-order phase transitions, because the neutron provides relevant information on the mechanisms driving the system through the critical value in an unprecedented manner.

10.2 Structural phase transitions

The change of structure at a phase transition in a solid can occur in two distinct ways. In first-order structural phase transitions, the atoms of a

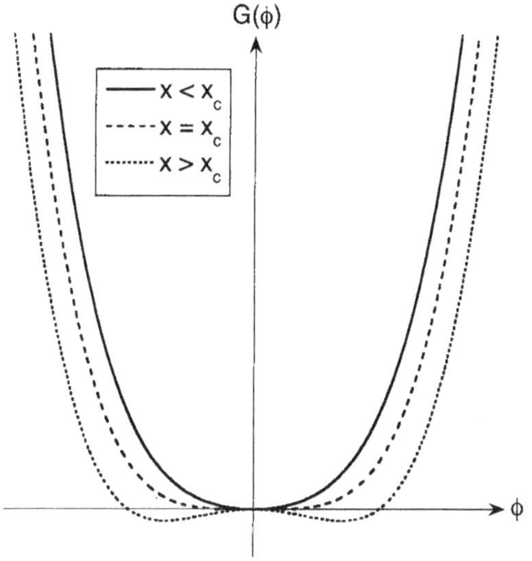

Fig. 10.1 Gibbs free enthalpy calculated from Eq. (10.4) for (a) $x < x_c$, (b) $x = x_c$, (c) $x > x_c$.

solid reconstruct a new lattice, for example, when graphite transforms into diamond or if an amorphous solid changes to a crystalline state. In second-order structural phase transitions, a regular lattice is distorted slightly without in any other way disrupting the linkage of the net. This can occur as a result of small displacements in the lattice position of single atoms or small rotations of molecular units. In this Section we will only consider the latter category.

Fig. 10.2 shows schematically displacive phase transformations from one ordered structure to another ordered structure. The order parameter of the phase transition corresponds to the static displacements $\boldsymbol{w}$, which are zero in the phase of higher symmetry and non-zero in the low-symmetry phase. We associate the displacements $\boldsymbol{w}$ with a particular phonon mode in the high-symmetry phase, which has the same eigenvector as the order parameter. It is plausible that the restoring forces for this mode will decrease upon approaching the critical temperature T_c. Consequently the frequency of this mode will decrease (or *soften*) and approach zero for $T \searrow T_c$ and finally freeze into the static displacement $\boldsymbol{w}$ of the low-temperature phase. Therefore these phonon modes are called soft modes.

The displacement patterns displayed in Fig. 10.2 can be attributed to particular phonon modes introduced in Chap. 5. The displacements $\boldsymbol{w}$ in

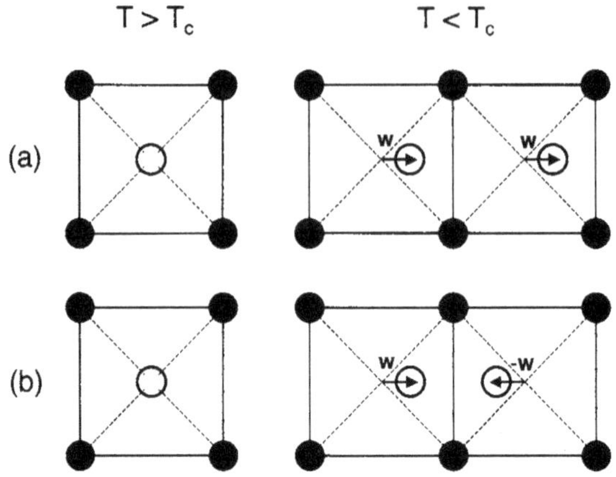

Fig. 10.2 Schematic representation of displacive phase transitions in a two-dimensional lattice with two different atoms.

Fig. 10.2a are characteristic of an optic phonon in the center of the Brillouin zone ($q = 0$), while the displacement pattern of Fig. 10.2b is typical of an optic phonon at the zone boundary ($q = \pm\pi/a$). The latter corresponds to the situation encountered in the ferroelectric compound $SrTiO_3$ which transforms from a cubic to a tetragonal structure at $T_c \approx 110$ K. The structural phase transition is characterized by a rotation of the oxygen octahedra as depicted in Fig. 10.3a. The octahedra on the layer above rotate the opposite way. The order parameter is either the rotation angle φ or the oxygen shift $\boldsymbol{w}$ which is, of course, directly related to φ. In the low-temperature phase the R point of the cubic reciprocal lattice becomes a superlattice point, enlarging the unit cell, thus the soft mode can be identified as an optic phonon mode at $\boldsymbol{q} = (\frac{1}{2}, \frac{1}{2}, \frac{1}{2})$. This was evidenced by inelastic neutron scattering measurements as shown in Fig. 10.3b. The soft mode follows closely the relation

$$\hbar\omega \propto \sqrt{T - T_c}. \tag{10.8}$$

ω^{-2} can be considered as the inverse generalized susceptibility which indeed exhibits above T_c the linear temperature dependence predicted by Eq. (10.7). The soft-mode concept outlined above covers the essential mechanism for all second-order structural phase transitions in solids.

In ferroelectric compounds, the phonon frequency is related to the dielectric constant ϵ through

$$\omega^2 \propto \frac{4\pi}{9v}\frac{\epsilon + 2}{\epsilon^2} \sim \epsilon^{-1}, \tag{10.9}$$

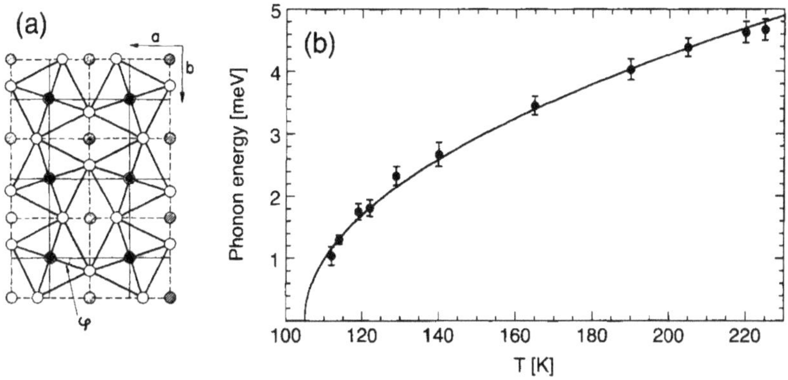

Fig. 10.3 (a) Low-temperature structure of $SrTiO_3$, demonstrating the rotation of the oxygen octahedron surrounding Ti. Reprinted with permission from [Shirane (1974)]. Copyright 1974 by the American Physical Society. (b) Soft-phonon energy approaching the phase transition at $T_c \approx 110$ K (after [Shirane (1974)]. The line is a least-squares fit to Eq. (10.8).

where v is the volume of the unit cell. ϵ diverges for $T \searrow T_c$ as $\epsilon \propto (T - T_c)^{-1}$, hence

$$\omega \propto \epsilon^{-1/2} \propto \sqrt{T - T_c}, \tag{10.10}$$

in agreement with Eq. (10.8).

10.3 Phase transitions in ice

Water exhibits an astonishingly rich phase diagram with fifteen crystalline and three amorphous phases identified so far (Fig. 10.4). The hexagonal phase of ice Ih (ordinary ice) represents all forms of ice familiar to us in daily life and, together with cubic ice Ic identified in the upper atmosphere, the only two phases present on Earth. Some phases of ice are suspected to exist on the satellites of Jupiter and Saturn, whereas amorphous ice has been proposed to constitute the main part of water in the universe.

The ice phases are stereotypical examples of molecular solids and hydrogen-bonded open-network structures. The H_2O molecule forms a rigid entity with the two protons covalently bound to the oxygen at a distance of 0.96 Å and an angle of 104.52°. From the fact that individual molecules are linked by hydrogen bonds follows the so-called 'Bernal-Fowler ice rules':

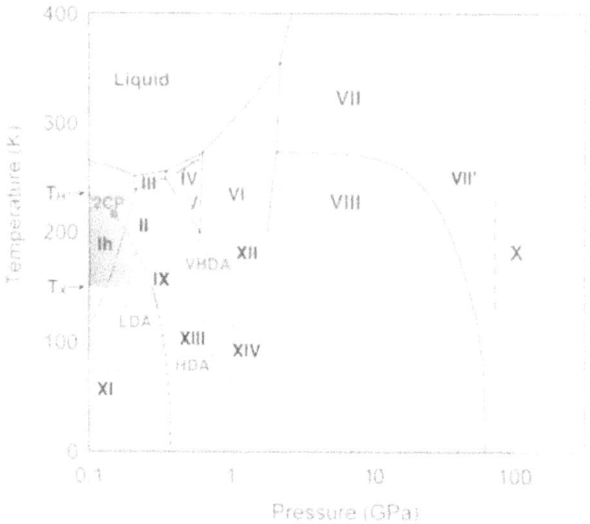

Fig. 10.4 Phase diagram of water with hydrogen disordered phases Ih, III, IV, V, VI, VII, XII and ordered phases II, VIII, IX, XI, XIII, XIV. Cubic ice Ic (not shown) is a metastable variant of ice Ih. Also shown in gray are the amorphous forms of ice: LDA, HDA and VHDA (very-high density amorphous), together with the hypothesized 2*nd* critical point of water. The gray shaded region depicts the experimentally inaccessible region where liquid water and amorphous ice spontaneously crystallize. Note the logarithmic scale of the x-axis.

(i) There are two hydrogens adjacent to each oxygen.
(ii) There is only one hydrogen per bond. (10.11)

The proximity of the H-O-H angle to the tetrahedral angle of 109.47° suggests a preference for tetrahedral coordination of the molecules as found, e.g., in ice Ih. However, we may immediately identify six possible orientations of an H_2O molecule at a given tetrahedral lattice site. In a crystal of N molecules we may thus find 6^N possible arrangements with 50% chance that each of the $2N$ hydrogen bonds is correctly formed (Eq. (10.11)). Therefore we are left with $6^N(1/2)^{2N}$ equivalent configurations and an apparent discrepancy to the third law of thermodynamics that claims an exact configuration at $T = 0$. It follows that ice Ih cannot represent the structure of lowest energy for $T \to 0$ and that in the average structure it cannot be decided to which of the two adjacent oxygens a given hydrogen is covalently bound. Ice Ih represents hence one of the so-called hydrogen-disordered ice phases that all are not ground-state structures for $T \to 0$ due to their finite entropy. The averaged structure of ice Ih has been proposed by Pauling in 1935. The location of the hydrogen atoms could not be determined crystallographically until the availability of neutron beams. Indeed, neutron diffraction on deuterated ice Ih has been one of the very first applications of neutron scattering in condensed matter physics.

Crystalline phase transitions

Upon cooling no phase transition of ice Ih to its ordered analogue is observed due to the lack of thermal motion of defects in the compound and hindrance of molecule rotation. However, small doping with potassium hydrate (KOH) introduces defects that allow the molecules to reorient collectively at 72 K and form the hydrogen-ordered ice XI phase.

Order-disorder phase transitions naturally involve a large change in entropy ΔS and minor change in volume ΔV. According to the equation of Clausius-Clapeyron

$$\frac{\mathrm{d}P}{\mathrm{d}T} = \frac{\Delta S}{\Delta V} \tag{10.12}$$

these transitions are readily identified by horizontal phase boundaries in Fig. 10.4 (small $\Delta V/\Delta S$). Similarly to ice Ih and ice XI, the two dominant high-pressure phases of ice, phases VII and VIII, form a pair of disordered and ordered ice phases. They transform without addition of a dopant at temperatures near room temperature. Ice XIII and ice XIV are the two most recently identified ice phases and represent the ordered analogues of ice V and ice XII upon doping with HCl. Figure 10.5 shows the subtle changes in the neutron diffraction patterns between the disordered and ordered phases. Note that unlike to x-rays, neutrons are dominantly scattered by deuterium and hence allow for a precise discrimination between the ordered and disordered phases.

Phase transitions with vertical phase boundaries in Fig. 10.4 are volume driven with progressively larger densities, e.g., ice VII and VIII exhibit roughly twice as large densities compared to ordinary ice (ice Ih). These two phases are the last phases with intact H_2O molecules before the solid eventually transforms into hydrogen-symmetrized ice X. The ice VIII$\rightarrow$X(VII') transition exhibits a pronounced isotope effect (H$\rightarrow$D) hinting to the importance of zero-point motion of the proton in this transition. All other phase transitions in water show minor isotope effects with the phase boundaries shifted by a few K only.

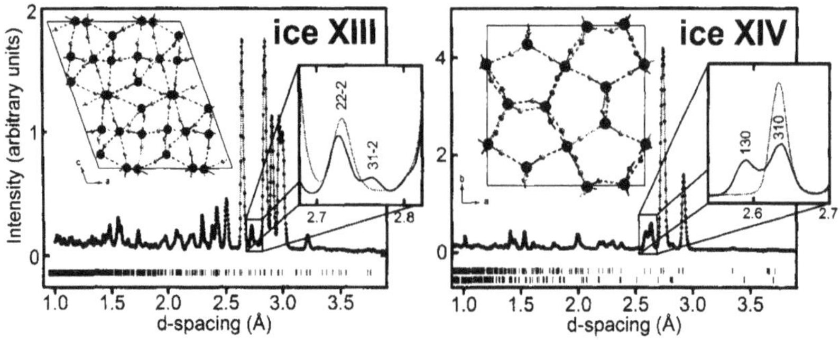

Fig. 10.5 Observed, calculated, and difference profiles of (deuterated) ice XIII and ice XIV. The insets compare the ordered with the disordered phases. Ice XIII (monoclinic) distinguishes from ice V (monoclinic) by the appearance of $(31\bar{2})$ reflections. The (310) reflection of tetragonal ice XII is split in orthorhombic ice XIV into (130) and (310) reflections. From [Salzmann *et al.* (2006)]. Reprinted with permission from the American Association for the Advancement of Science.

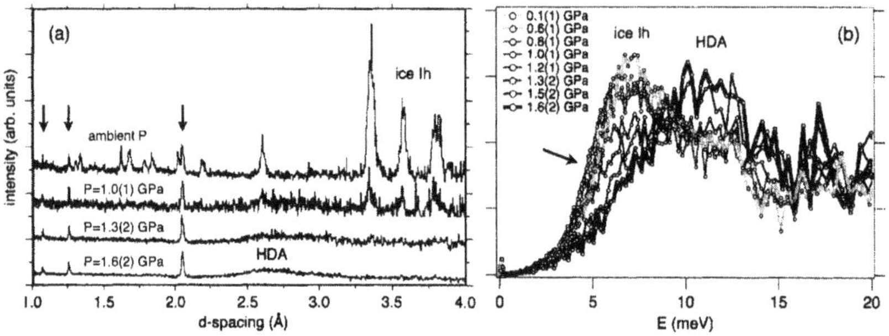

Fig. 10.6 (a) Neutron diffraction patterns throughout the amorphization of (deuterated) ice Ih to 1.6 GPa. The broad bump around 2.7 Å is characteristic of the amorphous product. Arrows denote peaks from the pressure cell. (b) Phonon density of states throughout the amorphization (after [Strässle *et al.* (2007)]). The arrow highlights the softening of phonon modes in the 5 meV region before the amorphization commences.

Amorphization of ice Ih

When compressed at temperatures below $\sim$ 130 K, ordinary ice Ih transforms irreversibly at $P_c \approx 1.2$ GPa into an amorphous phase. The transition is witnessed with neutron diffraction by a complete loss of sharp Bragg peaks and with inelastic neutron scattering by substantial changes of the phonon density of states (Fig. 10.6). As a precursor of amorphization an increased density of low-energy phonon modes is observed. The responsi-

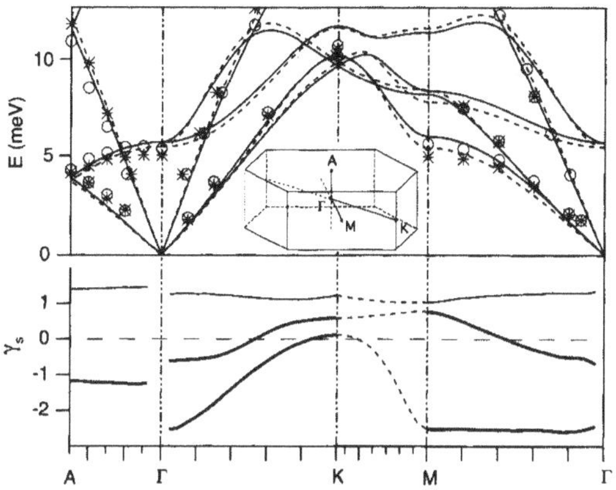

Fig. 10.7 Phonon dispersion of (deuterated) ice Ih measured at P=0.05 GPa (o) and 0.5 GPa (*) together with mode Grüneisen parameters $\gamma_s = -\partial \ln \omega / \partial \ln V$ of the acoustic phonon branches (after [Strässle *et al.* (2004b)]). The latter are a measure for the relative change of the phonon frequency as a function of volume change. The transverse acoustic phonon branches with basal polarization show pronounced uniform softening along the $\Gamma - M$ and $\Gamma - A$.

ble acoustic phonon branches could be determined by measurements on a D_2O single crystal (Fig. 10.7). Unlike the soft-mode scenario described in Chap. 10.2 these phonons exhibit uniform softening over the entire branch and not at a specific wavevector that otherwise would define the resulting new crystalline structure. As a result a manifold of wavevectors and hence structures may be selected simultaneously yielding to topological frustration and eventually amorphization via so-called mechanical melting of the system.

Polyamorphism

When heated below $P = 0.2$ GPa the above described amorphous phase of ice transforms irreversibly at $T \approx 125$ K into another amorphous phase characterized by smaller density. Correspondingly, they are referred to as high-density (HDA, $\rho = 1.17$ g/cm^3) and low-density amorphs (LDA, $\rho = 0.94$ g/cm^3), respectively. Subsequent application of pressure at temperatures around 120 K then allows the reversible transformation between LDA and HDA which constitutes one of the rare examples of polyamorphism in condensed matter physics. The close link between polyamorphism and liquid-liquid phase transitions puts forward the scenario of a second critical point (CP) in water near $T \approx 220$ K and $P \approx 0.1$ GPa. This sce-

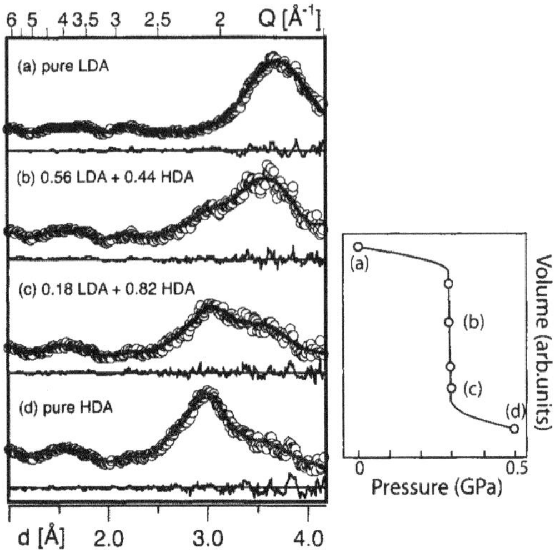

Fig. 10.8 Representative neutron diffraction patterns of the LDA-HDA transition measured at $T = 130$ K (after [Klotz *et al.* (2005a)]. The lines denote linear interpolations of pure LDA and pure HDA. The right figure depicts the measured pressure throughout the transition.

nario suggested by theory may explain part of the many anomalies in the thermodynamic properties of water near ambient conditions, which all are amplified in supercooled liquid water. Experimentally, the $2nd$ CP cannot be accessed since supercooled water and amorphous ice both spontaneously crystallize for $T \lesssim 235$ K and $T \gtrsim 150$ K, respectively (see shaded region in Fig. 10.4). The observed two amorphous phases, however, may be regarded as the amorphous proxies to the two liquid phases into which water would separate if cooled below the $2nd$ CP.

The above scenario requires that the LDA-HDA transition is of first order, a necessity being fulfilled as revealed by neutron diffraction measurements shown in Fig. 10.8. The LDA phase is prepared in a pressure cell that allows the *in-situ* change of pressure. With increasing force load, the volume available to the sample is gradually reduced and portions of LDA are transformed into HDA. Characteristic of a first-order transition, all diffraction patterns shown in Fig. 10.8 can be reproduced as linear combinations of the pure LDA and HDA patterns and the pressure during the transition is found constant until the sample is completely transformed to HDA. In contrast to a temperature-induced phase transition, here the experimentally varied parameter (force, i.e., available sample volume) is directly linked to the order parameter of the phase transition (density) which allows to halt the transition at stages before complete transformation into the final state.

10.4 Magnetic phase transitions

Magnetic compounds offer a rich variety of phase transitions which often cannot be described by the Landau theory introduced in Chap. 10.1. The Landau theory is a mean-field approach which neglects the effect of fluctuations as well as the symmetry of the order parameter. If we allow the exchange Hamiltonians introduced in Chap. 8 to exhibit anisotropic interactions of the form

$$\hat{\mathcal{H}} = -2\sum_{j,j'} J_{jj'}\left[r\hat{S}_j^z\hat{S}_{j'}^z + (1-r)\left(\hat{S}_j^x\hat{S}_{j'}^x + \hat{S}_j^y\hat{S}_{j'}^y\right)\right], \tag{10.13}$$

the dimension of the order parameter is $n = 1$ for $r = 1$ (Ising model), $n = 2$ for $r = 0$ (XY model), and $n = 3$ for $r = 1/2$ (Heisenberg model). All these models can be realized in one-dimensional ($d = 1$), two-dimensional ($d = 2$), and three-dimensional ($d = 3$) magnets. In most cases, the critical exponents associated with particular dimensions d and n cannot be calculated exactly, but result from specific models and/or computer simulations. Table 10.1 summarizes the available critical components which are relevant for the interpretation of neutron scattering experiments. The corresponding thermodynamic properties, expressed in reduced temperature units

$$t = \frac{T - T_c}{T_c}, \tag{10.14}$$

are the magnetization

$$M(t) = -\left(\frac{\partial G}{\partial H}\right)_T = M_0(-t)^\beta, \tag{10.15}$$

the magnetic susceptibility

$$\chi(t) = -\left(\frac{\partial^2 G}{\partial H^2}\right)_T = \chi_0|t|^{-\gamma}, \tag{10.16}$$

and the correlation length

$$\xi(t) = |t|^{-\nu}, \tag{10.17}$$

which is related to the wavevector dependent susceptibility through the Ornstein-Zernike formula

$$\chi(q,t) = \chi(q=0,t)\cdot\frac{\kappa^2(t)}{q^2+\kappa^2(t)}, \tag{10.18}$$

Table 10.1 Critical exponents derived for magnetic systems. The data given in the first two rows result from exactly soluble models, those in the last three rows from approximations. (after [Gebhardt and Krey (1980)]).

			α	β	γ	ν
Landau theory			0 [a]	0.50	1.0	0.50
dimension of system	parameter	order parameter				
$d=2$	$n=1$	S^z	0 [b]	0.125	1.75	1.0
$d=3$	$n=1$	S^z	0.1100(24)	0.325(1)	1.2462(9)	0.6300(8)
$d=3$	$n=2$	S^x,S^y	-0.0079(30)	0.3454(15)	1.3160(12)	0.6693(10)
$d=3$	$n=3$	S^x,S^y,S^z	-0.1162(30)	0.3646(12)	1.3866(12)	0.7054(11)

[a]jump discontinuity, [b]logarithmic singularity

where $\kappa(t) = 2\pi/\xi(t)$ is the inverse correlation length. $\chi(q,t)$ is related to the scattering law $S(\boldsymbol{Q},\omega)$ through Eq. (2.47). As can be seen from Table 10.1, the critical exponents obey the general relations

$$\alpha + 2\beta + \gamma = 2,$$
$$\alpha = 2 - d \cdot \nu,$$

with α being the critical exponent of the heat capacity at constant field:

$$c(t) = -T\left(\frac{\partial^2 G}{\partial T^2}\right)_H \propto |t|^{-\alpha}. \tag{10.19}$$

As an example we present the results of a detailed critical neutron scattering study performed for a single crystal of the rare-earth compound CeBi, which below $T_N = 25.35$ K exhibits antiferromagnetic ordering of type I characterized by a stacking of ferromagnetic (001) planes in the sequence $+ - +-$ along the z-axis. The octahedral crystal field defined by Eq. (9.10) with $B_6^0 = 0$ splits the ground-state J-multiplet of the Ce^{3+} ions into a doublet Γ_7 and a quartet Γ_8, favoring the (111) axis as easy axis of magnetization. However, (001) spin alignment is realized because of anisotropic magnetic interactions according to Eq. (10.13). The magnetic ordering wavevector is therefore $\boldsymbol{q}_0 = (0, 0, \frac{1}{2})$.

The development of the order parameter $\langle S^z \rangle$ was studied by measuring the intensity of the (110) antiferromagnetic Bragg reflection below T_N. In Fig. 10.9a the observed reduced sublattice magnetization M/M_0 is shown as a function of reduced temperature in a doubly logarithmic plot. A least-squares fitting procedure based on Eq. (10.15) yields the results

$$T_N = 25.35(1)\ \text{K}, \quad \beta = 0.317(5).$$

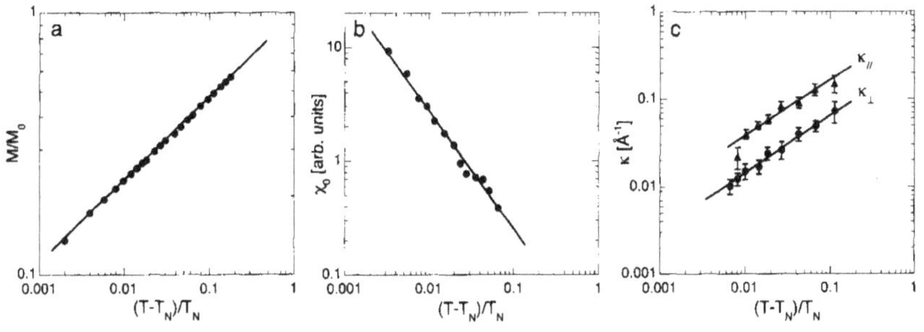

Fig. 10.9 Thermodynamic properties of CeBi obtained by neutron scattering experiments plotted in reduced temperature units: (a) reduced magnetization below T_N; (b) staggered susceptibility above T_N; (c) inverse correlation lengths parallel and perpendicular to the (001)-direction above T_N (after [Hälg *et al.* (1982)]).

The susceptibility $\chi(\boldsymbol{q}, t)$ was measured by scans across (110) above T_N as summarized in Fig. 10.10. The peaks have a Lorentzian shape predicted by Eq. (10.18) and exhibit broader linewidths (i.e., larger inverse correlation lengths κ) for $\boldsymbol{q}_{\parallel} = (00x)$ than for $\boldsymbol{q}_{\perp} = (xx0)$, i.e., the critical scattering is cigar-like with its long axis parallel to (001). The temperature dependence of the intensities and linewidths is plotted in Fig. 10.9b and Fig. 10.9c and analyzed according to Eqs (10.16) and (10.17), respectively, yielding

$$T_N = 25.32(3)\text{ K}, \qquad \gamma = 1.16(12);$$

$$T_N = 25.37(4)\text{ K}, \qquad \nu = 0.63(6).$$

The anisotropy in κ is seen to be independent of temperature with a mean value $\kappa_{\parallel}/\kappa_{\perp} = 2.5(2)$, thus the Ce^{3+} spins are much more highly correlated within the (001) sheets than they are between adjacent sheets.

In summary, the thermodynamic magnetic properties of CeBi were found to follow simple power laws predicted by Eqs (10.15) - (10.17) with the same Néel temperature $T_N = 25.35$ K. A comparison of the observed values of the critical exponents β, γ and ν with those expected from theoretical models listed in Table 10.1 shows the best agreement with the predictions for a three-dimensional Ising system ($d = 3$, $n = 1$), thus the spin Hamiltonian appropriate for CeBi is Eq. (10.13) with $r = 1$.

10.5 Quantum phase transitions

Quantum phase transitions are quantitatively different from the classical phase transitions discussed in the preceding Sections, since they occur only

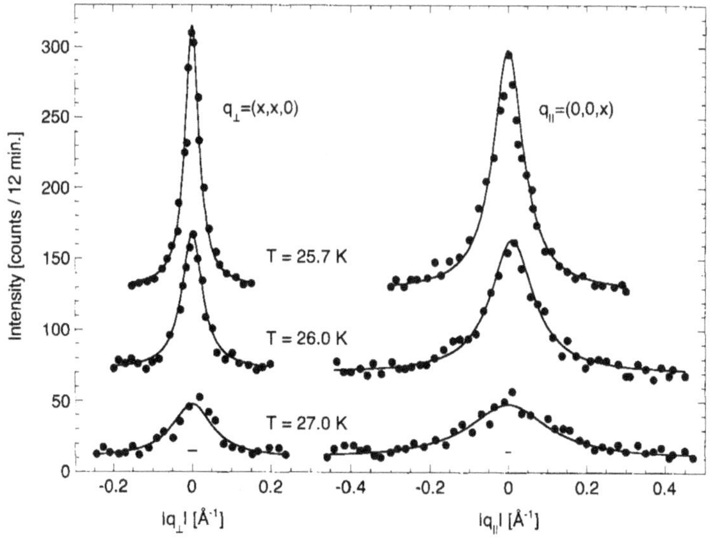

Fig. 10.10 Intensity distributions of the critical scattering in CeBi across (110) for scans parallel ($\boldsymbol{q}_{\parallel}$) and perpendicular ($\boldsymbol{q}_{\perp}$) to the (001)-direction (after [Hälg *et al.* (1982)]). The instrumental resolution is indicated by the horizontal bars. The lines are fits to Eq. (10.18).

at the absolute zero of temperature. The transition is not governed by the competition between order and thermal fluctuations, but it takes place at the quantum critical value of some other parameter such as pressure, composition or magnetic field strength. A quantum phase transition occurs when co-operative ordering of the system disappears, but this loss of order is driven solely by the quantum fluctuations. The physical properties of these quantum fluctuations are quite distinct from those of the thermal fluctuations responsible for traditional, finite-temperature phase transitions. Therefore, the description of a quantum system near the quantum critical point requires novel theories that have no analogue in the traditional framework of phase transitions. Magnetism is an ideal testing ground for quantum phase transitions. Novel collective quantum phenomena were observed in two-dimensional $S = 1/2$ antiferromagnets and most clearly in the monoclinic compound $TlCuCl_3$ as outlined below.

The magnetic properties of $TlCuCl_3$ are determined by the Cu^{2+} ions which are arranged in centrosymmetric, antiferromagnetically coupled pairs. According to the Hamiltonian Eq. (8.1) the dimer ground-state is a

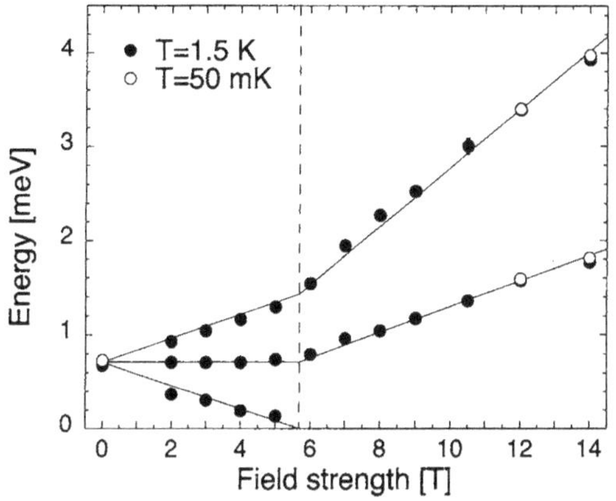

Fig. 10.11 Field dependence of the magnetic excitation energies measured in $TlCuCl_3$ at $\boldsymbol{Q} = (0,4,0)$. The solid lines reflect a linear Zeeman model. The critical field $H_c = 5.7$ T is indicated by a dashed line (after [Rüegg *et al.* (2003)]).

singlet with wavefunction $|S, M\rangle = |0,0\rangle$, separated by an energy gap Δ from the excited triplet states with components $|1,+1\rangle$, $|1,0\rangle$ and $|1,-1\rangle$. At a critical external magnetic field H_c the energy of the Zeeman split triplet component $|1,+1\rangle$ intersects the ground-state singlet as shown in Fig. 10.11, thus H_c is a quantum critical point separating a gapped spin-liquid state ($H < H_c$) from a field-induced magnetically ordered state ($H > H_c$).

The triplet components $|1,+1\rangle$ can be regarded as bosons, thus the question arises whether Bose-Einstein condensation occurs at H_c. Bose-Einstein condensation denotes a collective quantum ground state of identical particles with integer spin. The realization of Bose-Einstein condensation requires the chemical potential $\mu = \Delta - g\mu_B H$ to vanish, which is fulfilled at H_c. Thus, at H_c the gas of triplet bosons undergoes a quantum phase transition into a novel condensate state with macroscopic occupation of the single-particle ground-state which can be described by the coherent superposition

$$|\psi\rangle = a_s|0,0\rangle + a_t e^{i\phi}|1,+1\rangle, \tag{10.20}$$

where $a_s \approx 1$ and $a_t \ll 1$ are the singlet and triplet amplitudes, respectively, and ϕ denotes the phase factor. a_t and ϕ are determined by the spin expectation values of the two Cu^{2+} ions of the dimer which are given by

$$\begin{aligned} \langle S_{x_1}\rangle &= -\langle S_{x_2}\rangle \propto a_t \cos\phi, \\ \langle S_{y_1}\rangle &= -\langle S_{y_2}\rangle \propto a_t \sin\phi, \\ \langle S_{z_1}\rangle &= \langle S_{z_2}\rangle \propto a_t^2. \end{aligned} \tag{10.21}$$

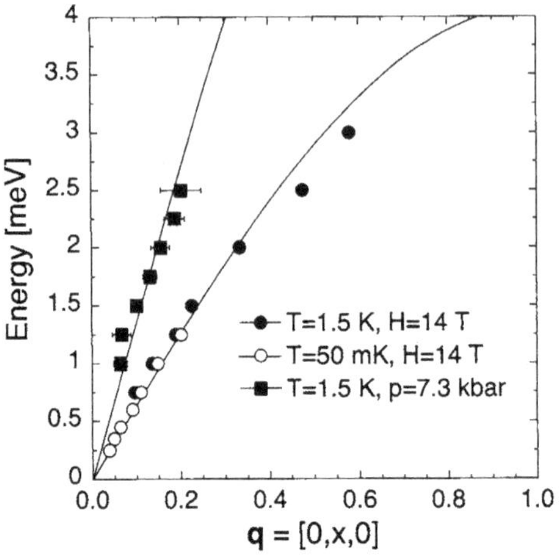

Fig. 10.12 Energy dispersion of the low-lying magnetic excitations measured in $TlCuCl_3$ at $H = 14$ T, $P = 0$ and $T = 1.5$ K (full circles) and $T = 50$ mK (open circles) as well as at $H = 0$, $P = 7.3$ kbar and $T = 1.5$ K (full squares). Reprinted with permission from [Rüegg *et al.* (2003, 2004)]. Copyright 2004 by the American Physical Society. The lines demonstrate the linear dispersion behavior of the Goldstone mode.

What is the experimental proof that $TlCuCl_3$ is Bose-Einstein condensed at H_c? The spin dynamics of the Bose-Einstein condensate state has been theoretically predicted to be governed by a gapless Goldstone mode associated with the breaking of rotational symmetry by the staggered magnetic order, thus the presence of a spin-wave-like mode with a linear dispersion is a convincing signal for the existence of the Bose-Einstein condensate. This theoretical prediction was indeed observed as evidenced in Fig. 10.12.

The energy gap Δ in $TlCuCl_3$ can also be closed by the application of hydrostatic pressure. This offers the unique possibility to realize and explore the quantum phase transition into a magnetically ordered state by another external parameter. Fig. 10.13 shows the pressure dependence of the energy gap which is softening with increasing pressure and vanishes at a critical pressure $P_c \approx 1$ kbar. For $P > P_c$ long-range antiferromagnetic order is observed with Néel temperatures indicated in Fig. 10.13. The spin excitations observed above P_c are consistent with the theoretically predicted two degenerate gapless Goldstone modes, see Fig. 10.12. In comparison to the field-induced ordered phase we recognize an increase of the spinwave stiffness by about a factor of two, which can be understood by the considerable

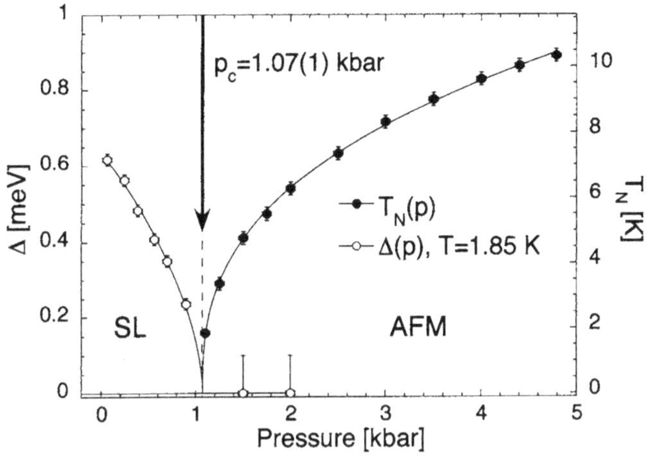

Fig. 10.13 Pressure-induced quantum phase transition in $TlCuCl_3$ between the spin-liquid (SL) and the antiferromagnetically ordered (AFM) phases. The singlet-triplet gap $\Delta(P)$ and the Néel temperature $T_N(P)$ were measured at $\boldsymbol{Q} = (0, 0, 1)$. Reprinted with permission from [Rüegg *et al.* (2004)]. Copyright 2004 by the American Physical Society.

increase of the exchange coupling parameters under compression.

10.6 Further reading

- R. A. Cowley, in *Magnetic neutron scattering*, ed. by A. Furrer (World Scientific, Singapore, 1995), p. 99: *Magnetic phase transitions*
- R. A. Cowley, in *Methods of experimental physics*, Vol. 23, Part C, ed. by D. L. Price and K. Sköld (Academic Press, London, 1987), p. 1: *Phase Transitions*
- V. F. Petrenko and R. W. Whitworth, *Physics of ice* (Oxford University Press, Oxford, 1999)
- H. E. Stanley, *Introduction to phase transitions and critical phenomena* (Oxford University Press, Oxford, 1987)

Chapter 11

Superconductivity

11.1 Introduction

The phenomenon of superconductivity was discovered in 1911 by Kamerlingh-Onnes who observed a sudden drop of the resistance in mercury below a critical temperature $T_c = 4.2$ K. Thanks to this effect, the electron transport occurs without losses and very large currents can be generated, which are often used to produce very high magnetic fields. Usually simple metals are superconductors of type-I, whereas compounds and alloys are superconductors of type-II; the two types are drastically different from each other regarding magnetic field effects.

In type-I superconductors an external magnetic field is completely expelled from the inside of the sample by establishing circulating currents on its surface that counteract the applied field. This is the so-called Meissner effect as shown in Fig. 11.1a. Superconductivity is destroyed by the application of a field larger than a critical field $H_c(T)$. In practice, H_c is too low for useful technical applications.

The situation is very different for type-II superconductors as shown in Fig. 11.1b. In the Meissner phase, below the lower critical field $H_{c1}(T)$, the magnetic flux is completely excluded from the superconductor, whereas above the upper critical field $H_{c2}(T)$ the normal state is recovered and the magnetic field is homogeneously distributed in the inside of the material. H_{c2} is often 100 times larger than H_{c1}. Of particular interest is the mixed state between $H_{c1}(T)$ and $H_{c2}(T)$, where the magnetic field penetrates the superconductor in the form of quantized magnetic vortices, which are arranged parallel to the field direction. These vortices consist of magnetic flux lines of radius ξ (the coherence length), and each flux line carries one flux quantum $\Phi_0 = h/2e = 2.07 \cdot 10^{-15}$ Wb. The flux lines are regularly

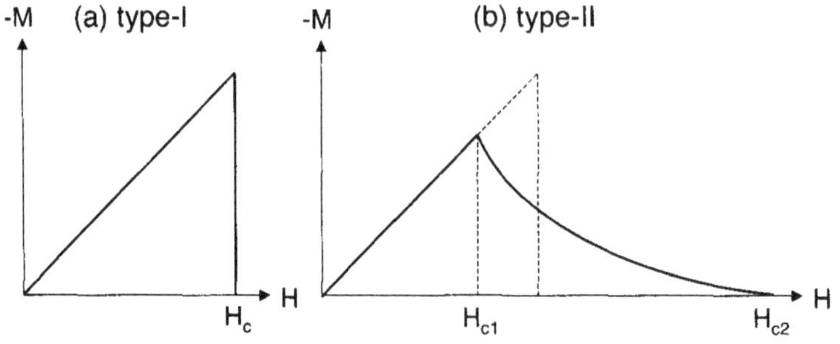

Fig. 11.1 Magnetization curves for (a) type-I and (b) type-II superconductors.

arranged and form a two-dimensional vortex lattice.

Of utmost importance was the observation of the isotope effect, i.e., the superconducting transition temperature depends on the mass m of the atom: $T_c \propto m^{-\alpha}$. This led to the conclusion that the electron-phonon interaction is the mechanism responsible for superconductivity. A quantitative understanding of superconductivity was achieved in 1957 by the theoretical concept of Bardeen, Cooper and Schrieffer (BCS). The BCS theory predicts the existence of so-called Cooper pairs (i.e., electron pairs with opposite momentum) which are bound by a weak electron-phonon interaction. As a result, an isotropic s-wave gap Δ opens in the electronic density of states, and a simple relation was established between the zero-temperature gap and T_c: $2\Delta|_{T=0} \approx 3.5\, k_B T_c$. The BCS theory also predicts the isotope coefficient to be $\alpha = \frac{1}{2}$.

The field of superconductivity was revolutionized in 1986, when Bednorz and Müller demonstrated the existence of superconductivity in copper-oxide perovskites of type $La_{2-x}(Ba,Sr)_xCuO_4$. Soon after this discovery other copper-oxide materials (cuprates, e.g., $YBa_2Cu_3O_{7-\delta}$, $Bi_2Sr_2CaCu_2O_{8+\delta}$) were discovered with T_c values breaking the temperature of liquefaction of nitrogen. It was realized that the electron-phonon interaction alone would not be sufficient to explain the high critical temperatures observed in the cuprates, and various competing theoretical models have been proposed. In particular, magnetism was recognized as one of the key elements to understand the high-T_c superconductivity of the cuprates.

Fig. 11.2 shows the generic phase diagram of high-T_c superconductors. The undoped system is antiferromagnetic and insulating. At small doping, the Néel temperature T_N decreases rapidly and vanishes. The system

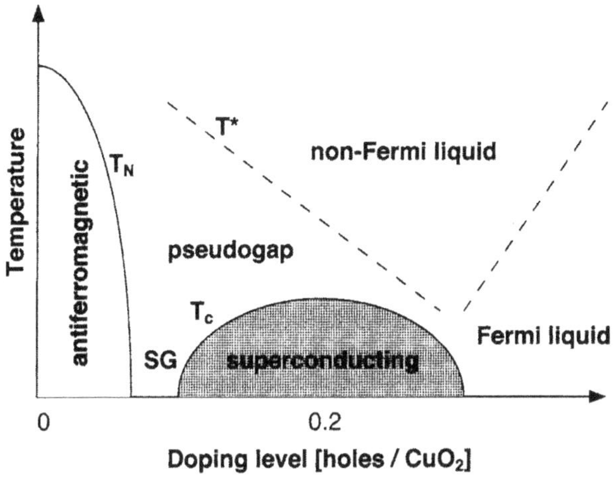

Fig. 11.2 Generic phase diagram of high-T_c superconductors vs doping.

enters then a spin-glass phase. Upon further doping superconductivity appears below a critical temperature T_c, but the system behaves like a strange metal in the underdoped regime. T_c reaches then a maximum at optimal doping and decreases again in the overdoped regime. The proximity of the antiferromagnetic and superconducting phases naturally raises the question of the importance of magnetism for the mechanism giving rise to superconductivity. One prediction of models involving magnetism was the existence of an anisotropic d-wave gap that changes sign upon 90° rotation. Indeed, a superconducting gap function with nodes was clearly identified by means of angle resolved photoemission spectroscopy (ARPES) and phase sensitive experiments. The high-T_c superconductors also differ from low-T_c materials by the observation of a so-called pseudogap well above T_c. This pseudogap closes at a temperature T^* that is increasing with decreasing doping (see Fig. 11.2) and the ratio of $\Delta_{\max}(T = 0)/T_c$ increases with underdoping.

Neutron scattering has outstanding properties in the elucidation of the basic properties of superconductors. Some of the most important achievements obtained by neutron scattering are presented in the following Sections. In particular, the flux-line lattice could be detected by SANS experiments used as a 'crystallographic' tool. Inelastic neutron scattering experiments brought unique results with respect to the lattice and spin dynamics, which yielded important information about the electron-phonon coupling and the extraordinary large scale of the spin fluctuations in the high-T_c superconductors, respectively. Finally, neutron crystal-field spectroscopy gave information on the opening of both the superconducting gap and the pseudogap.

11.2 The flux-line lattice

The diffraction of the neutrons by a magnetic vortex lattice occurs because the neutron, due to its magnetic moment, experiences a spatially varying potential. The Bragg d-spacing of a flux-line lattice is given by the condition that there is one flux quantum, Φ_0, per unit cell, and hence

$$d_{\text{tri}} = \sqrt{\frac{\sqrt{3}\Phi_0}{2B}} \tag{11.1}$$

for a triangular or hexagonal lattice and

$$d_{\text{squ}} = \sqrt{\frac{\Phi_0}{B}} \tag{11.2}$$

for a square lattice. For a magnetic field $B = \mu_0 H \approx 0.2$ T, this gives d-spacings of the order of 1000 Å, and with an incident neutron wavelength $\lambda \approx 10$ Å, Eq. (4.8) gives a Bragg angle $\Theta \approx 0.5°$. Hence neutron diffraction from a flux-line lattice is clearly a SANS experiment (see Chap. 3.3.4).

The expression for the intensity of a single (h,k) reflection is given by (Ref. [Christen *et al.* (1977)])

$$I_{hk} = \frac{\pi\gamma^2\Phi(\lambda)}{8}\frac{V\lambda^2}{\Phi_0^2|\boldsymbol{\tau}_{hk}|}|F_{hk}|^2, \tag{11.3}$$

where $\Phi(\lambda)$ is the incident neutron flux, V the sample volume, $\boldsymbol{\tau}_{hk}$ the reciprocal lattice vector associated with the appropriate d-spacing (Eq. (11.1) or Eq. (11.2)), and F_{hk} is the magnetic form factor corresponding to the Fourier inversion of the magnetic field distribution $h(\boldsymbol{r})$ inside the superconductor:

$$F_{hk} = \frac{1}{\Phi_0}\int h(\boldsymbol{r})e^{i\boldsymbol{\tau}_{hk}\cdot\boldsymbol{r}}\mathrm{d}\boldsymbol{r}. \tag{11.4}$$

In the London theory, Eq. (11.4) can be expressed by (Ref. [Forgan (1998)])

$$F_{hk} = \frac{B}{1+(\boldsymbol{\tau}_{hk}\lambda_L)^2}, \tag{11.5}$$

where λ_L is the London penetration depth. For all inductions larger than B_{c1}, the second term in the denominator of Eq. (11.5) is dominant, so that

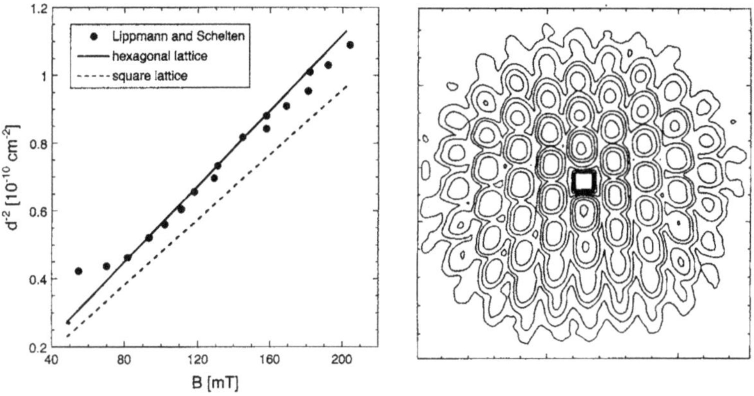

Fig. 11.3 Flux-line lattice in Nb. Left: Field dependence of the lattice parameter d. The lines denote the theoretical values of d for a hexagonal and a square flux-line lattice according to Eqs (11.1) and (11.2) (after [Lippmann and Schelten (1974)]). Right: Contour plot (logarithmic scale) of a neutron diffraction pattern from the flux-line lattice in Nb at $B = 0.1$ T and $T = 1.6$ K. The incident beam spot at the center has been masked. With kind permission from Springer+Business Media: [Forgan (1998)], Fig. 2, Copyright 1998.

the intensity given by Eq. (11.3) varies with the wavevector as $|\boldsymbol{\tau}_{hk}|^{-5}$, i.e., higher-order reflections are much weaker than low-order ones. We also see from Eqs (11.3) and (11.5) that the intensity falls off as λ_L^{-4}; this rapid variation makes the SANS signals very weak for high-T$_c$ and heavy-fermion superconductors, which have long magnetic penetration depths.

In a first example we exemplify the above considerations by the results of pioneering SANS experiments performed for a single crystal of niobium. Fig. 11.3 (left) shows the vortex lattice parameter d as a function of applied field B. A comparison with Eqs (11.1) and (11.2) clearly proves that the vortex lattice has hexagonal symmetry. More recently, with the use of two-dimensional detectors, it has been possible to image directly the hexagonal structure of the flux-line lattice in niobium (see Fig. 11.3 (right)).

With the neutron intensity being proportional to λ_L^{-4}, SANS observations of the vortex lattice in high-temperature superconductors ($\lambda_L \approx 1000 - 2000$ Å) are extremely demanding. Fig. 11.4 shows the results of SANS experiments performed for $La_{1.83}Sr_{0.17}CuO_4$ ($T_c = 37$ K). A hexagonal flux-line lattice exists for low magnetic fields (see Fig. 11.4a). However, with increasing field strength a crossover into a square coordination was observed (see Fig. 11.4b). An intrinsic fourfold symmetry is indicative of the coupling of the vortex lattice to some source of anisotropy such as the

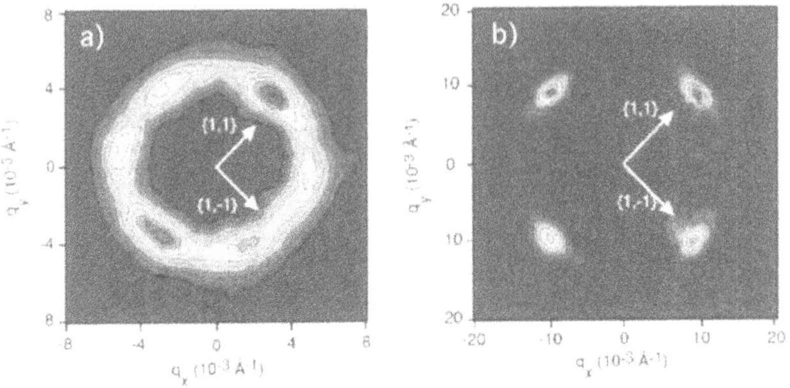

Fig. 11.4 SANS diffraction patterns obtained for $La_{1.83}Sr_{0.17}CuO_4$ at $T = 1.5$ K. (a) $B = 0.1$ T applied 10° away from the c-axis. (b) $B = 1.0$ T applied parallel to the c-axis. Reprinted from [Chang *et al.* (2006)]. Copyright 2006, with permission from Elsevier.

(anisotropic) d-wave nature of the superconducting gap function, Fermi surface/velocity anisotropies, or charge/stripe fluctuations.

11.3 Phonon density of states

In the BCS theory the superconducting transition temperature is given by the relation [Bardeen *et al.* (1957a,b)]

$$T_c = 1.14\,\hbar\omega_D\, e^{-\frac{1}{\lambda}} \tag{11.6}$$

where ω_D is the Debye frequency and λ the effective electron-phonon coupling constant. Therefore, T_c can be raised by increasing the parameters ω_D and/or λ. Equation (11.6) is valid as long as $k_BT \ll \hbar\omega_D$, which corresponds to the weak coupling limit. Both ω_D and λ are related to the phonon density of states $g(\omega)$. ω_D is an average phonon energy corresponding to the Debye temperature Θ_D which can be determined from the heat capacity:

$$\begin{aligned} c_V(T) &= k_B \int_0^\infty \left(\frac{\hbar\omega}{k_BT}\right)^2 \frac{e^{\frac{\hbar\omega}{k_BT}}\, g(\omega)}{\left(e^{\frac{\hbar\omega}{k_BT}} - 1\right)^2} \mathrm{d}\omega \\ &= \frac{12\pi^4 n N k_B}{5} \left(\frac{T}{\Theta_D}\right)^3, \end{aligned} \tag{11.7}$$

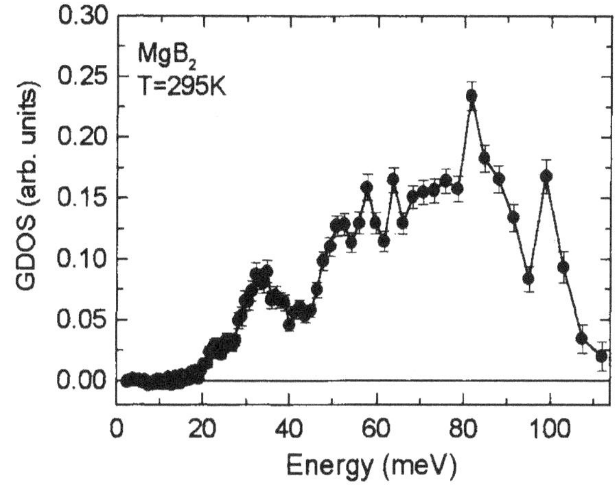

Fig. 11.5 Generalized phonon density of states in MgB_2 at $T = 295$ K. The data are corrected for multiphonon scattering, Debye-Waller attenuation and background contributions. Reprinted with kind permission from Springer Science+Business Media: [Clementyev *et al.* (2001)], Fig. 2, Copyright 2001.

where n and N denote the number of atoms per unit cell and the number of unit cells, respectively. The relation for λ reads

$$\lambda = 2 \int_0^\infty \frac{\alpha^2(\omega) g(\omega)}{\omega} \mathrm{d}\omega, \qquad (11.8)$$

where $\alpha(\omega)$ is the electron-phonon interaction which usually varies smoothly with frequency and can often be approximated by a constant. The electron-phonon spectral density, $\alpha^2(\omega)g(\omega)$, can be derived by inversion of tunneling data, while $g(\omega)$ results from inelastic neutron scattering experiments, see Chap. 5.3.

Fig. 11.5 shows the generalized phonon density of states of MgB_2 ($T_c = 39$ K) determined by inelastic neutron scattering experiments at room temperature. MgB_2 crystallizes in the hexagonal AlB_2-type structure, in which the B ions constitute graphite-like sheets in the form of primitive honeycomb lattices separated by hexagonal layers of Mg. The charge carriers reside in the boron planes which are metallic due to the covalent B-B bonds. Fig. 11.5 reflects the partial phonon density of states of both magnesium and boron. The lighter boron atoms contribute mainly to the high-energy part of the phonon spectrum, i.e., the phonon energy relevant for superconductivity is $\hbar\omega \approx 80 - 105$ meV, which means that Eq. (11.6) is no longer valid. In the intermediate- and strong-coupling limit, T_c is given by the semiempirical formula [McMillan (1968)]

$$T_c = \frac{\hbar\omega_D}{1.45k_B} \exp\left(\frac{1.04(1+\lambda)}{\lambda - \mu^*(1+0.62\lambda)}\right), \tag{11.9}$$

where μ^* is the Coulomb repulsion, which is of the order of 0.15 for all superconductors. From the McMillan formula (Eq. (11.9)) the effective electron-phonon coupling parameter $\lambda \approx 1$ is obtained, i.e., MgB_2 has to be placed into the intermediate coupling regime.

11.4 Phonon energies and linewidths

A redistribution of electronic states can result in changes of phonon peak positions and linewidths below the superconducting transition temperature due to the opening of the gap, similar to the relaxation of crystal-field transitions discussed in Chap. 11.5. An explicit relation between the phonon linewidth, $\Gamma_{\text{ep}}(\boldsymbol{q})$, due to the electron-phonon interaction, and the electron-phonon spectral density, $\alpha^2(\omega)g(\omega)$, was derived [Allen (1972)]:

$$\alpha^2(\omega)g(\omega) = \frac{2}{\pi N(E_F)\omega} \sum_{\boldsymbol{q}} \Gamma_{\text{ep}}(\boldsymbol{q})\delta(\omega - \omega(\boldsymbol{q})), \tag{11.10}$$

where $N(E_F)$ is the electronic density of states at the Fermi surface. Neutron scattering experiments are thus capable, in principle, not only of determining the phonon density of states $g(\omega)$, but $\alpha^2(\omega)g(\omega)$ as well, if $\Gamma_{\text{ep}}(\boldsymbol{q})$ is sampled over the whole Brillouin zone, thereby providing an independent access to the effective electron-phonon coupling constant λ through Eq. (11.8). However, such a program would be prohibitively lengthy, thus the following expression for the average electron-phonon linewidth $\langle\Gamma_{\text{ep}}\rangle$ could be useful [Allen (1972)]:

$$\lambda = \frac{12n\langle\Gamma_{\text{ep}}\rangle}{\pi N(E_F)\langle\omega^2\rangle}, \tag{11.11}$$

where n is the total number of atoms and $\langle\omega^2\rangle$ the mean square phonon frequency.

We exemplify the above considerations for the classical superconductor Nb_3Sn ($T_c = 18.3$ K). Fig. 11.6 (left) shows the observation of a transverse acoustic phonon (T_1 mode) with wavevector $\boldsymbol{q} = (0.18, 0.18, 0)$ and energy $\hbar\omega = 3.8$ meV both below and above the superconducting transition temperature. There is a small upward shift of the phonon energy, but the most drastic effect is the increase of the intrinsic linewidth between 6 and 26 K.

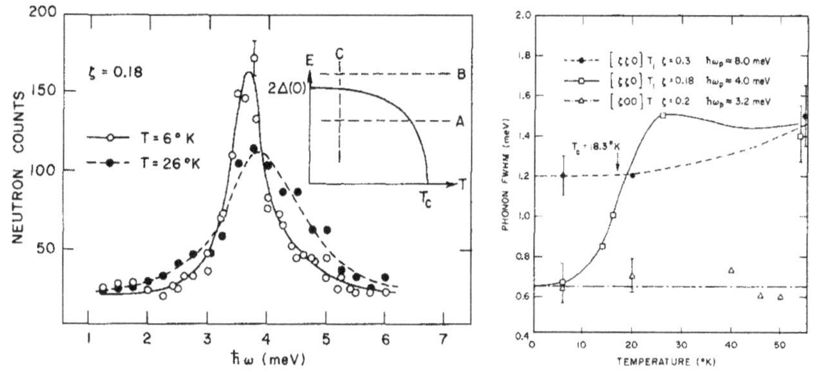

Fig. 11.6 Left: Energy spectra of neutrons scattered from Nb_3Sn below and above $T_c = 18.3$ K. The inset shows the effect of phonon damping ($\hbar\omega$ vs T) with increasing temperature (scans A and B) and at a fixed temperature (scan C). Right: Observed linewidths of three phonon modes with different wavevector $\boldsymbol{q}$. Reprinted with permission from [Axe and Shirane (1973)]. Copyright 1973 by the American Physical Society.

Obviously phonons with energy $\hbar\omega < 2\Delta$ do not have enough energy to span the gap Δ, and consequently there is no interaction with the charge carriers. For an isotropic gap, the intrinsic linewidth in the superconducting state is then given by

$$\Gamma_s(T) = \Gamma_n(T)e^{-\frac{\Delta}{k_B T}}, \tag{11.12}$$

where $\Gamma_n(T)$ is the linewidth in the normal state. This means that $\Gamma_s(T \ll T_c) \approx 0$, and line broadening sets in just below T_c where the superconducting gap opens. This behavior was clearly observed as shown in Fig. 11.6 (right) which summarizes the observed linewidth vs temperature. Fig. 11.6 (right) also includes data for two other phonon modes. The linewidth observed for the same T_1 mode with $\boldsymbol{q} = (0.3, 0.3, 0)$ remains unaffected at T_c, since its energy $\hbar\omega = 8.0$ meV obviously exceeds the gap energy. Similarly, the transverse acoustic phonon with $\boldsymbol{q} = (0.2, 0, 0)$ does not exhibit a linewidth change across T_c. Such mode-selective linewidth experiments are highly useful to determine the gap size through Eq. (11.12) as well as to identify the particular phonon modes that contribute to the electron-phonon interaction responsible for the BCS mechanism giving rise to superconductivity in Nb_3Sn.

More recently, experiments on the superconducting compounds RNi_2B_2C (R=Y,Lu) showed not only an effect on the phonon linewidth, but also a dramatic alteration of the phonon line shape when the temperature is reduced below T_c. The temperature dependence of the phonon

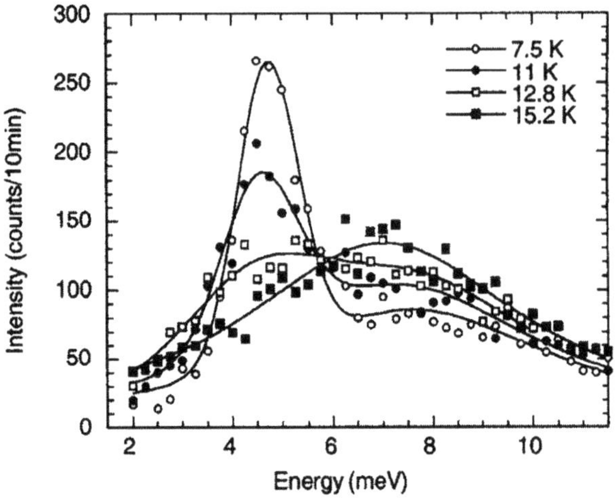

Fig. 11.7 Temperature dependence of the phonon profile observed for YNi_2B_2C at $Q = (0.55, 0, 8)$. Reprinted with permission from [Kawano *et al.* (1996)]. Copyright 1996 by the American Physical Society.

spectra observed for YNi_2B_2C ($T_c = 14.2$ K) at $Q = (0.55, 0, 8)$ is shown in Fig. 11.7. Above T_c the phonon response corresponds to a broad line centered at 7 meV. With decreasing the temperature below T_c a new peak emerges at 5 meV. The intensity of the new peak shows a clear onset at T_c, and its temperature dependence is akin to the order parameter of superconductivity. The new peak truly reflects the superconducting state, since its intensity starts to drop continuously above H_{c1} and completely disappears at H_{c2} [Kawano *et al.* (1996)]. It was shown theoretically that the two-component response of phonon scattering occurs under the condition $\omega < 2(1 + 2\Gamma_{ep}/\omega)\Delta$ [Allen *et al.* (1997)].

11.5 Relaxation effects of crystal-field transitions

In Chap. 9.4 we discussed the line broadening of crystal-field transitions caused by the interaction of the rare-earth ions with the charge carriers. The situation is anomalous for superconducting compounds, because the Cooper pairing of the charge carriers creates an energy gap Δ below the superconducting transition temperature T_c, thus crystal-field excitations with energy $\hbar\omega < 2\Delta$ do not have enough energy to span the gap, similar as for the case of phonons discussed in Chap. 11.4. For an isotropic gap,

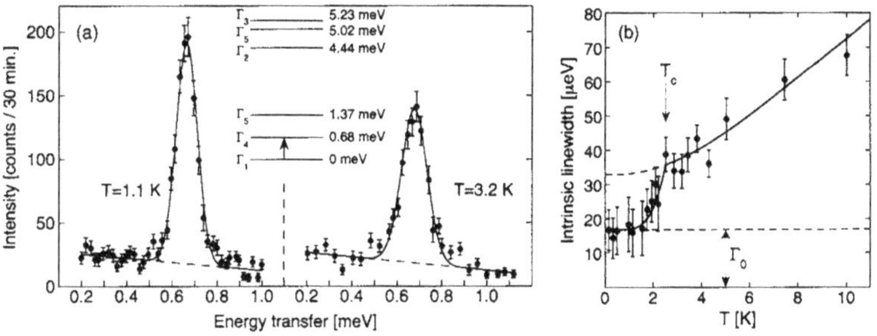

Fig. 11.8 (a) Energy spectra of neutrons scattered from $La_{0.997}Tb_{0.003}Al_2$ ($x = 0.003$) below and above $T_c = 2.6$ K. The insert shows the sequence of crystal-field levels for the Tb^{3+} ions in cubic symmetry. (b) Observed linewidths of the crystal-field transitions in $La_{0.997}Tb_{0.003}Al_2$ vs temperature. The solid curve is calculated from Eqs (9.21) and (11.12). The calculations assume a constant additive contribution Γ_0 due to local structural distortions induced by the different ionic radii of the La^{3+} and Tb^{3+} ions (after [Feile *et al.* (1981)]).

the intrinsic linewidth $\Gamma_s(T)$ in the superconducting state is then given by Eq. (11.12) with $\Gamma_n(T)$ being the linewidth in the normal state defined by Eq. (9.21). The exponential temperature dependence of $\Gamma_s(T)$ predicted by Eq. (11.12) was nicely demonstrated for the first time by a neutron spectroscopic study of the classical superconductor $La_{1-x}Tb_xAl_2$. The experiments were performed for a low Tb concentration ($x = 0.001$ and 0.003) in order to avoid line broadening due to the exchange interaction between the Tb ions. Fig. 11.8a shows the observation of the lowest crystal-field transition at $\hbar\omega = 0.68$ meV both below and above the superconducting transition temperature. The observed linewidths corrected for instrumental resolution are shown in Fig. 11.8b. The crossover from the normal state behavior governed by Eq. (9.21) and the exponential decrease of the linewidth below T_c is in striking agreement with the theoretical expectations.

The situation is different for high-T_c superconductors, where the pairing of the charge carriers occurs at the pseudogap temperature T^*, but long-range phase coherence sets in only at $T_c < T^*$. The gap opening at T^* was clearly observed in neutron spectroscopic experiments performed for $La_{1.96-x}Ho_{0.04}Sr_xCuO_4$. Fig. 11.9 shows the temperature dependence of the linewidth Γ observed for the lowest crystal-field transition at $\hbar\omega = \pm 0.2$ meV at different doping levels ($x = 0.11, 0.15, 0.20$). All data exhibit a qualitatively similar behavior: the linewidth is rather small at low temperatures, then it raises with increasing slope $d\Gamma/dT$ up to T^*; from thereon the slope $d\Gamma/dT$ is reduced and remains constant according to the

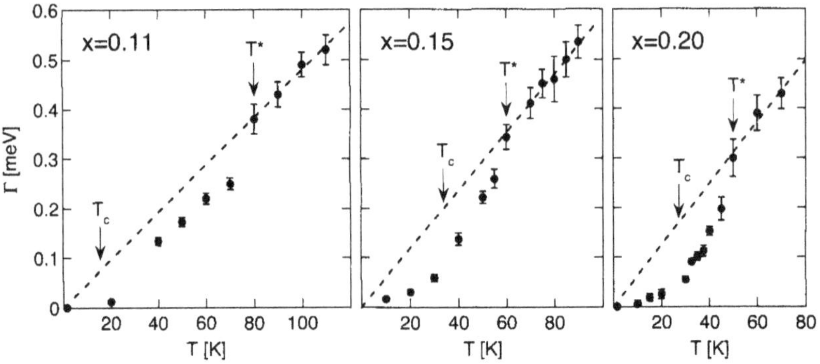

Fig. 11.9 Temperature dependence of the intrinsic linewidth (HWHM) corresponding to the lowest ground-state crystal-field transition in $La_{1.96-x}Ho_{0.04}Sr_xCuO_4$. The lines denote the linewidth in the normal state calculated from Eq. (9.20) (after [Häfliger *et al.* (2006)]).

Korringa law Eq. (9.20). The doping dependence of T^* established from these experiments is in good agreement with magnetic susceptibility, resistivity, heat capacity, NQR ([Häfliger *et al.* (2006)] and references therein) and ARPES experiments [Norman and Pépin (2003)].

11.6 Spin fluctuations in high-temperature superconductors

Inelastic neutron scattering experiments revealed the existence of strong magnetic fluctuations in high-T_c superconductors. In the undoped parent compounds the magnetic excitations can be explained by standard spin-wave theory as exemplified for La_2CuO_4 in Chap. 8.2.2. The situation is more complex for doped materials; not only are the magnetic excitations different in the normal state, but they also renormalize in a spectacular way through the superconducting transition temperature T_c. The cartoon shown in Fig. 11.10 illustrates the main features observed for practically all cuprate compounds around optimal doping. In the normal state, some broad excitations exist around the antiferromagnetic reciprocal lattice vector (π, π). In $La_{2-x}Sr_xCuO_4$, these excitations are peaked at some incommensurate wavevector $(\pi + \delta, \pi)$ [Mason *et al.* (1992)], and the incommensurability δ is enhanced with increasing Sr doping x.

The situation changes drastically as one enters the superconducting

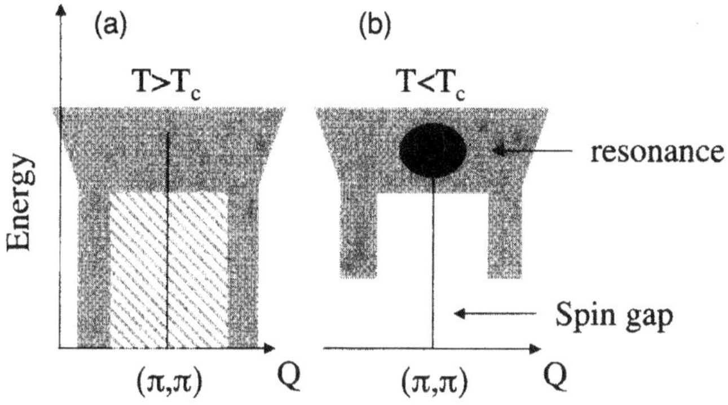

Fig. 11.10 Cartoon of the spin excitations in high-temperature superconductors (a) in the normal state and (b) in the superconducting state.

state. A strong renormalization of the spectral weight distribution is observed with the following salient features: (i) In the vicinity of the (π, π) point, a depletion of the spectral weight at low energies gives rise to the opening of a spin gap. (ii) In cuprates having values of T_c higher than about 80 K a so-called resonance peak is observed, i.e., a strong build-up of intensity at the (π, π) point at energies larger than the spin gap [Bourges (1998)]. This is illustrated in Fig. 11.11 for the compound $YBa_2Cu_3O_{6+x}$. More recent experiments on $YBa_2Cu_3O_{6.6}$ gave evidence for a square-shaped continuum of excitations peaked at incommensurate positions and having spectral weight far exceeding that of the resonance [Hayden *et al.* (2004)].

At optimal doping, both the spin gap and the resonance vanish above T_c. This fact, together with the observation that the energy E_r of the resonance scales with T_c as $E_r \approx 5\, k_B T_c$, indicates that a tight link exists between the magnetic and electronic degrees of freedom in the cuprate materials. Based on itinerant magnetism approaches, many models have been proposed to explain the observed magnetic excitations [Norman and Pépin (2003)]. Basically, these models start from a Fermi liquid picture where the Lindhard function of the spin susceptibility χ^0 can be written as

$$\chi^0(\boldsymbol{q}, \omega) \approx \sum_{\boldsymbol{k}} \frac{f_{\boldsymbol{k}} - f_{\boldsymbol{k}+\boldsymbol{q}}}{\hbar\omega - (\epsilon_{\boldsymbol{k}+\boldsymbol{q}} - \epsilon_{\boldsymbol{k}}) + i\delta}, \tag{11.13}$$

with $\epsilon_{\boldsymbol{k}}$ and $f_{\boldsymbol{k}}$ being the electronic band dispersion and the Fermi function, respectively. In the superconducting phase the Cooper pairs have to be

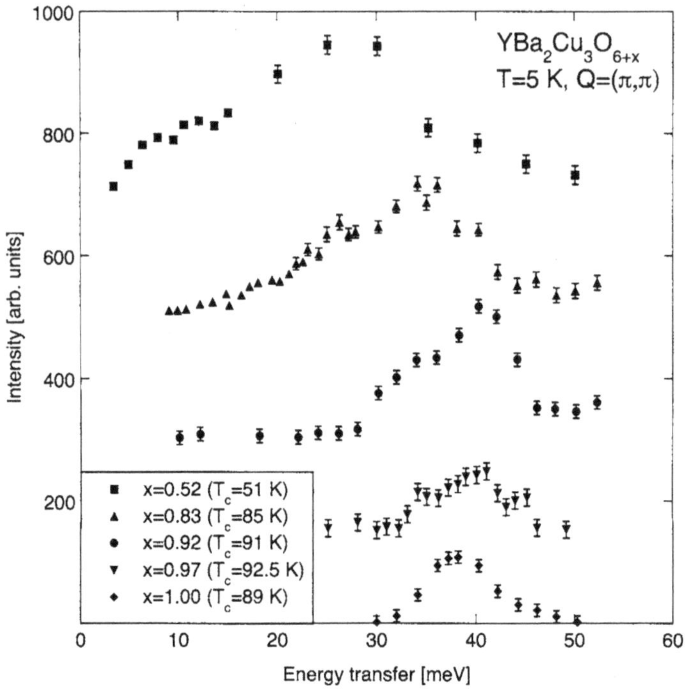

Fig. 11.11 Imaginary part of the spin susceptibility observed for $YBa_2Cu_3O_{6+x}$ at $T = 5$ K and $\boldsymbol{Q} = (\pi, \pi)$ for various doping levels (after [Bourges (1998)]). The data for $x = 0.97$, 0.92, 0.83, and 0.52 are shifted upwards by 150, 300, 500, and 700 counts, respectively.

considered, and χ^0 becomes at $T = 0$:

$$\chi^0(\boldsymbol{q},\omega) \approx \sum_{\boldsymbol{k}} \left(1 - \frac{\Delta_{\boldsymbol{k}}\Delta_{\boldsymbol{k}+\boldsymbol{q}} + \epsilon_{\boldsymbol{k}}\epsilon_{\boldsymbol{k}+\boldsymbol{q}}}{E_{\boldsymbol{k}}E_{\boldsymbol{k}+\boldsymbol{q}}}\right) \times \frac{f_{\boldsymbol{k}} + f_{\boldsymbol{k}+\boldsymbol{q}} - 1}{\hbar\omega - (E_{\boldsymbol{k}} + E_{\boldsymbol{k}+\boldsymbol{q}}) + i\delta}, \tag{11.14}$$

where $\Delta_{\boldsymbol{k}}$ and $E_{\boldsymbol{k}} = \sqrt{\epsilon_{\boldsymbol{k}}^2 + \Delta_{\boldsymbol{k}}^2}$ and are the superconducting gap function and the quasi-particle energy, respectively. The term in the brackets is called the coherence factor which plays a crucial role to explain the existence of spin fluctuations in high-temperature superconductors. In the limit of $\epsilon \to 0$ and for an isotropic gap function $\Delta_{\boldsymbol{k}} = |\Delta|$, the coherence factor vanishes. However, for a d-wave gap function we have $\Delta_{\boldsymbol{k}} = -\Delta_{\boldsymbol{k}+(\pi,\pi)}$, thus the coherence factor is equal to 2. Within such a fermionic model, it follows from Eq. (11.14) that inelastic neutron scattering is also probing, indirectly, the symmetry of the superconducting gap function.

The susceptibility as expressed by Eq. (11.14) is not sufficient to explain the measured inelastic neutron scattering data, but one has to consider an interacting susceptibility $\chi(\boldsymbol{q}, \omega)$ as defined in the RPA model by Eq. (8.41). The Fourier transformed exchange function for cuprates is then of the form of Eq. (8.24). Within the RPA treatment most features (spin gap, magnetic resonance, and incommensurate excitations) of the measured susceptibility $\chi''(\boldsymbol{q}, \omega) = \mathrm{Im}\chi(\boldsymbol{q}, \omega)$ can be reproduced at optimal doping. Nevertheless, it should be mentioned that other scenarios such as those involving the presence of stripes [Tranquada *et al.* (2004)] or SO(5) super-symmetry [Zhang (1997)] have also been invoked to explain the neutron scattering data.

11.7 Further reading

- J. F. Annett, *Superconductivity, superfluids and condensates* (Oxford University Press, Oxford, 2004)
- A. Furrer, *Neutron scattering in layered copper-oxide superconductors* (Kluwer Academic Publishers, Dordrecht, 1998)
- M. R. Norman and C. Pépin, Rep. Prog. Phys. 66, 1547 (2003): *The electronic nature of high temperature cuprate superconductors*
- D. R. Tilley and J. Tilley, *Superfluidity and superconductivity* (Hilger, Bristol, 1990)

Chapter 12

Superfluidity

12.1 Introduction

The phenomenon of superfluidity is observed only at low temperatures. The best known superfluids are the two isotopes of helium, ^{3}He and ^{4}He. Evidence of superfluidity has also been found in cold atomic gases, and it is believed that superfluidity occurs inside neutron stars. Superfluidity was discovered in 1937 by Kapitza, Allen and Misener. Upon cooling ^{4}He to a critical temperature of $T_\lambda = 2.17$ K, they observed a remarkable discontinuity in heat capacity, and a fraction of the liquid becomes a zero viscosity superfluid. It is called the lambda point, because the shape of the specific heat curve is like the Greek letter λ.

Although the phenomenologies of the superfluid states of ^{4}He and ^{3}He are very similar, the microscopic details of the transition are very different. ^{4}He atoms are bosons, and their superfluidity can be understood in terms of the Bose statistics that they obey. Specifically, the superfluidity of ^{4}He can be regarded as a consequence of Bose-Einstein condensation in an interacting system (see Chap. 10.5). On the other hand, ^{3}He atoms are fermions, and the superfluid transition in this system is described by a generalization of the Bardeen-Cooper-Schrieffer theory of superconductivity (see Chap. 11.1). However, Cooper pairing takes place between atoms rather than electrons, and the attractive interaction between them is mediated by spin fluctuations rather than phonons.

Superfluids exhibit many unusual properties. A superfluid acts as if it were a mixture of a normal component, with all the properties associated with a normal fluid, and a superfluid component. The latter has zero viscosity, zero entropy, and infinite thermal conductivity. It is thus impossible to set up a temperature gradient in a superfluid, much as it is impossible to set up a voltage difference in a superconductor. An interesting property becomes visible if the superfluid is placed in a rotating container. Instead of rotating uniformly with the container like a normal fluid, the rotating state of a superfluid consists of quantized vortices. The number of vortex lines depends on the constant h/m, where h is the Planck's constant and m the mass of the atoms. ^{3}He supports also spin vortices, where the spin-up condensate circles in opposite direction than the spin-down condensate. Up to the present, vortex lines in a superfluid – being analogous to flux lines in a type-II superconductor (see Chap. 11.2) – have not been observed by neutron scattering.

The present chapter will discuss some important neutron scattering results obtained for ^{4}He and ^{3}He in the liquid state, which provides the best opportunity for experimental tests for the theories of Bose and Fermi liquids.

12.2 Liquid ^{4}He

12.2.1 *Phase diagram*

The fact that ^{4}He does not solidify down to absolute zero is a pure quantum effect. The weak van der Waals forces between the atoms are not strong enough to overcome the very large zero-point motion (resulting from the small mass) to confine a helium atom to a lattice site. It may be solidified only under the application of pressure. The phase diagram of ^{4}He is shown in Fig. 12.1. At zero pressure the liquid phases above and below T_λ are designated ^{4}He-I and ^{4}He-II, respectively. In ^{4}He-II only a small fraction of the atoms form the condensate, thus superfluid helium is a mixture of two fluids. Neutron scattering experiments have provided important information on the size of the superfluid fraction as well as on the nature of the excitation spectrum as outlined below.

12.2.2 *Elementary excitations*

Unlike other liquids, the elementary excitations in ^{4}He-II are sharp and well defined as shown in Fig. 12.2. At low wavevectors, there is an almost linear dispersion

$$\hbar\omega(q) = c \cdot q, \tag{12.1}$$

similar to the longitudinal phonon modes in a solid (see Chap. 5.2). With

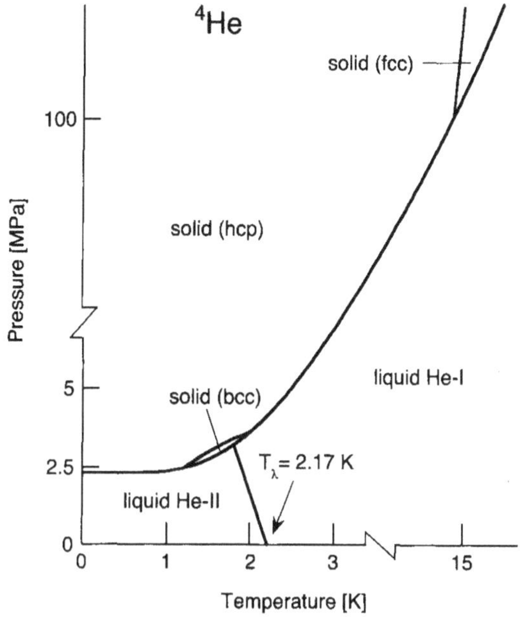

Fig. 12.1 Phase diagram of ^{4}He.

increasing wavevector the dispersion curve exhibits a maximum, which is called the maxon, followed by a minimum, which is called the roton, since this part of the dispersion curve was originally believed to carry all the rotational properties of the liquid. In the region of the roton minimum the dispersion is parabolic in shape and very accurately modeled by the Landau expression

$$\hbar\omega(q) = \Delta_R + \frac{\hbar^2(Q - Q_R)^2}{2\mu_R}, \tag{12.2}$$

where Δ_R denotes the roton energy gap at Q_R and μ_R is the roton effective mass. The presence of the roton gap has thus come to be associated with the superfluid component of ^{4}He-II, in analogy to the gap in superconductors (see Chap. 11).

Above T_λ the phonon is still a reasonably sharp, well-defined excitation, whereas the roton, extremely sharp in the superfluid, collapses into a broad, ill-defined excitation due to roton-roton interactions. The temperature dependence of the roton energy Δ_R and roton half-width Γ_R was calculated

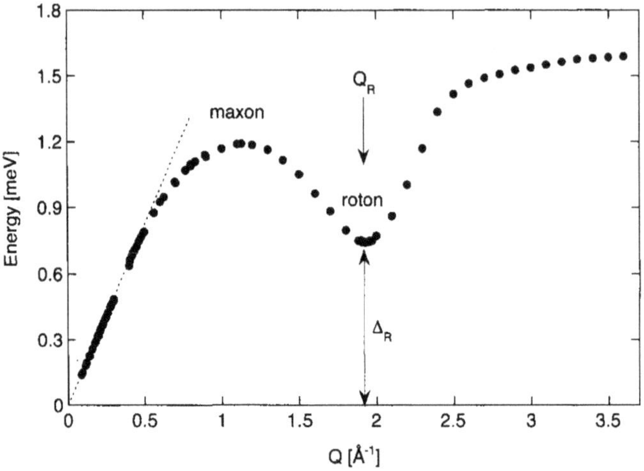

Fig. 12.2 Dispersion curve observed for ^{4}He at $T \leq 1.2$ K (after [Donnelly *et al.* (1981)]).

to be (see [Bedell *et al.* (1984)]).

$$\Delta_R(T) = \Delta_R(0) + P_\Delta \cdot (1 + R\sqrt{T})\sqrt{T} \cdot e^{-\frac{\Delta_R(T)}{k_B T}}, \tag{12.3}$$
$$\Gamma_R(T) = P_\Gamma \cdot (1 + R\sqrt{T})\sqrt{T} \cdot e^{-\frac{\Delta_R(T)}{k_B T}}.$$

Figure 12.3 summarizes a series of experimental data which are well explained with the model parameters $P_\Delta = 24.72$, $P_\Gamma = 41.6$, and $R = 0.0603$.

12.2.3 *The condensate fraction*

At sufficiently high moduli of the scattering vector $\boldsymbol{Q}$ for the impulse approximation to be valid, $S(\boldsymbol{Q}, \omega)$ will directly reflect the momentum distribution $n(\boldsymbol{p})$ of the atoms as indicated by Eq. (G.11). If a finite fraction n_0 of the atoms in the superfluid phase ^{4}He-II are condensed into the zero-momentum state as originally proposed by London, then $n(\boldsymbol{p})$ can be written as

$$n(\boldsymbol{p}) = n_0 \delta(\boldsymbol{p}) + (1 - n_0) n^*(\boldsymbol{p}), \tag{12.4}$$

where $n^*(\boldsymbol{p})$ is the momentum distribution for the noncondensate atoms. Substituting Eq. (12.4) into Eq. (G.11) then gives an $S(\boldsymbol{Q}, \omega)$, which consists of a δ-function condensate component at the recoil energy $E_r = \frac{\hbar^2 Q^2}{2M}$, sitting on top of a broad noncondensate peak, which is symmetric about

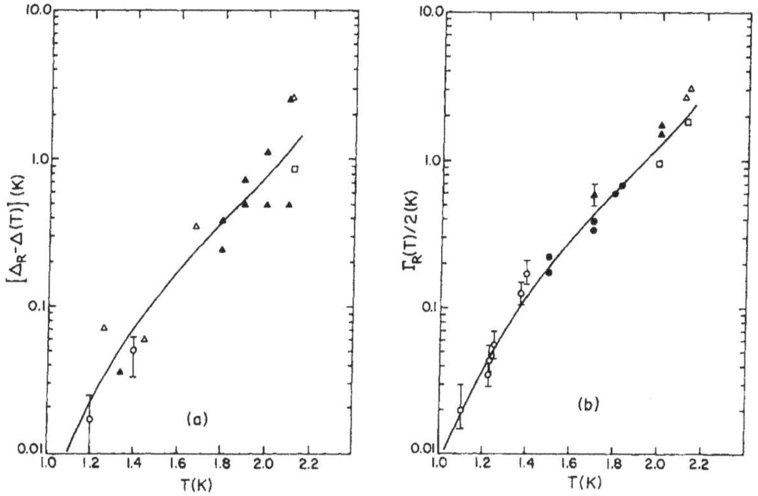

Fig. 12.3 Temperature dependence of the roton energy Δ_R (left) and linewidth Γ_R (right). The lines correspond to the theoretical prediction based on Eq. (12.3). Reprinted with permission from [Bedell *et al.* (1984)]. Copyright 1984 by the American Physical Society.

E_r. This is demonstrated by experimental data taken at different temperatures in Fig. 12.4 (left), which clearly shows the extra scattering due to the condensate component below T_λ. The relative weight of the δ-function component gives directly the condensate fraction n_0 as shown in Fig. 12.4 (right). Analyzing the temperature dependence according to

$$n_0(T) = n_0(0)\left(1 - \left(\frac{T}{T_\lambda}\right)^\alpha\right) \tag{12.5}$$

yields $n_0(0) = 0.139(23)$ and $\alpha = 3.6(1.4)$.

12.2.4 *Static structure factor*

Measurements of the static structure factor $S(\boldsymbol{Q})$ for liquid ^{4}He above and below T_λ are shown in Fig. 12.5 (left). As a result of the low density and poorly defined short-range order due to the large zero-point motion, the oscillations in $S(\boldsymbol{Q})$ have small amplitudes that decrease rapidly with increasing Q. The infinite-Q limit has effectively been reached beyond 6 Å^{-1}, which allows a very accurate Fourier analysis to determine the pair

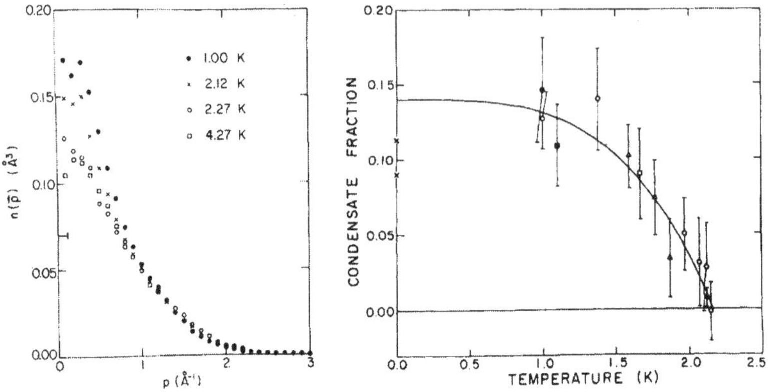

Fig. 12.4 Left: Temperature dependence of the momentum distribution $n(\boldsymbol{p})$ for liquid ^{4}He. Right: Experimentally based values for the condensate fraction $n_0(T)$ in superfluid ^{4}He. The line is a fit to Eq. (12.5). Reprinted with permission from [Sears *et al.* (1982)]. Copyright 1982 by the American Physical Society.

correlation function $g(r)$ via the relationship (see Chap. 6.2)

$$g(r) = \frac{1}{2\pi^2 \rho r} \int_0^\infty Q\,(S(Q) - 1) \sin(Qr)\,\mathrm{d}Q, \tag{12.6}$$

where ρ is the number density. The temperature dependence of various features of $g(r)$ derived from the results of Fig. 12.5 (left) are shown in Fig. 12.5 (right). The amplitudes of the oscillations about the line $g(r) = 1$, which are a direct measure of the correlations between the atoms, exhibit the normal increase with decreasing temperature above T_λ, but then there is a drastic reversal and the amplitudes decrease continuously with further cooling in the superfluid phase. This anomalous loss of spatial order – being unique to superfluid ^{4}He – is a direct consequence of the formation of the zero-momentum Bose condensate below T_λ. This offers the possibility to determine the condensate fraction n_0, in analogy to Eq. (12.4), via the relationship

$$g(r) - 1 = (1 - n_0)^2\,(g_n(r) - 1), \tag{12.7}$$

where $g_n(r)$ is the pair correlation function for the noncondensate atoms. The open circles in Fig. 12.4 (right) show values of n_0 obtained by application of Eq. (12.7). They are obviously in excellent agreement with the values derived from the analysis of the momentum distribution $n(\boldsymbol{p})$ discussed in Chap. 12.2.3.

Upon decreasing the temperature, ^{4}He contracts as a normal liquid, thus the maxima of $g(r)$ shift to lower values of r as shown by the intersections

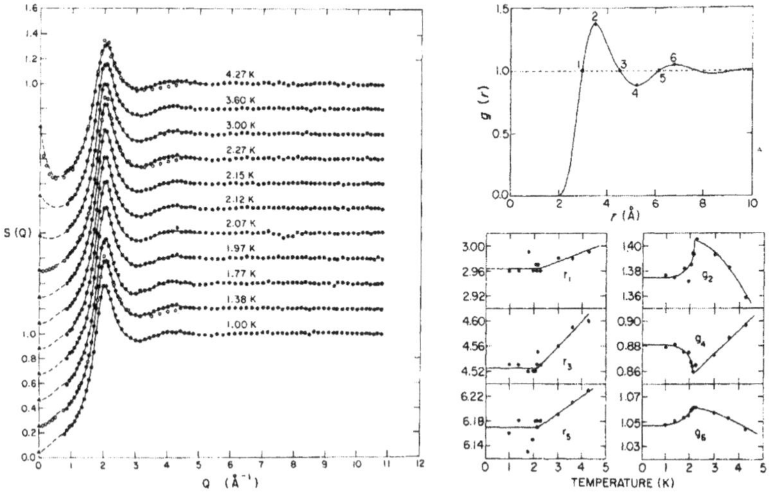

Fig. 12.5 Left: Static structure factor $S(Q)$ of liquid ^{4}He observed for various temperatures (after [Svensson *et al.* (1980)]). Right: Pair correlation function $g(r)$ for liquid ^{4}He at $T = 1.00$ K and the temperature variation of selected amplitudes g_i and intersections r_i at $g(r) = 1$. Reprinted from [Glyde and Svensson (1987)]. Copyright 1987, Academic Press.

r_i in Fig. 12.5 (right). However, a spontaneous expansion takes place at T_λ, which continues down to about $T = 1$ K, from thereon the contraction sets in again. These unusual features below T_λ, however, have not been observed in the data of Fig. 12.5 (right).

12.3 Liquid ^{3}He

12.3.1 *Phase diagram*

The superfluidity of ^{3}He was discovered in 1972 by Osheroff, Richardson, Lee and Leggett at temperatures below 3 mK. The phase diagram of ^{3}He is shown in Fig. 12.6. Its overall features are very similar to that of ^{4}He, with the exception of the onset of superfluidity which occurs at very low temperatures. This has been taken as a proof of the theory of superfluidity in ^{4}He, since ^{3}He is governed by Fermi-Dirac statistics which does not allow the condensation of a macroscopic number of atoms in the state of lowest energy.

Neutron scattering experiments on ^{3}He are hampered by the enormously

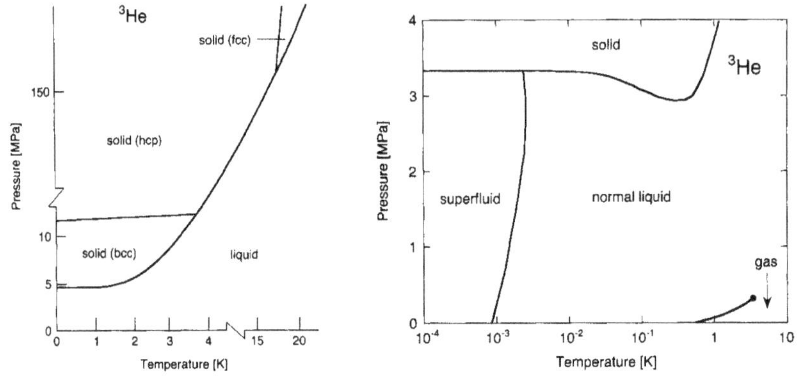

Fig. 12.6 Phase diagram of ^{3}He.

large absorption cross section (see Appendix B), resulting in a striking lack of experimental data. Moreover, neutron experiments in the mK region are extremely demanding; to date there are still no neutron measurements available on superfluid ^{3}He, thus we just summarize some results obtained in the normal liquid state.

12.3.2 *Elementary excitations*

Since ^{3}He has a nuclear spin $I = \frac{1}{2}$, the scattering law $S(\boldsymbol{Q},\omega)$ is the sum of coherent scattering (S_{coh}) due to the density excitations as in ^{4}He and of spin-dependent scattering (S_{inc}) due to the spin-density excitations:

$$S(\boldsymbol{Q},\omega) = S_{\text{coh}}(\boldsymbol{Q},\omega) + \frac{\sigma_{\text{inc}}}{\sigma_{\text{coh}}} S_{\text{inc}}(\boldsymbol{Q},\omega), \tag{12.8}$$

where σ_{coh} and σ_{inc} denote the coherent and incoherent scattering cross-sections defined by Eqs (2.30) and (2.31), respectively. Indeed, the energy spectra taken at $T = 15$ mK exhibit two peaks for $Q \leq 1.2$ Å^{-1} as shown in Fig. 12.7. The peak at lower energies is attributed to the spin fluctuation resonance occurring in the component S_{inc} as predicted by Leggett, whereas the peak at higher energy is interpreted as the zero-sound mode in the component S_{coh}. This interpretation is confirmed by the Q-dependence of the intensities based on $S_{\text{inc}} \propto F^2(Q)$ (see Chap. 2.6), $S_{\text{coh}} \propto Q^2$ (see Chap. 5.1), $\sigma_{\text{coh}} = 4.4$ barn (see Appendix B), and $\sigma_{\text{inc}} \approx 1.1$ barn (adjusted to the experimental data). The energy of the spin fluctuation is roughly independent of Q, whereas the zero-sound mode evolves up to $Q \approx 1$ Å^{-1} with a linear dispersion according to Eq. (12.1).

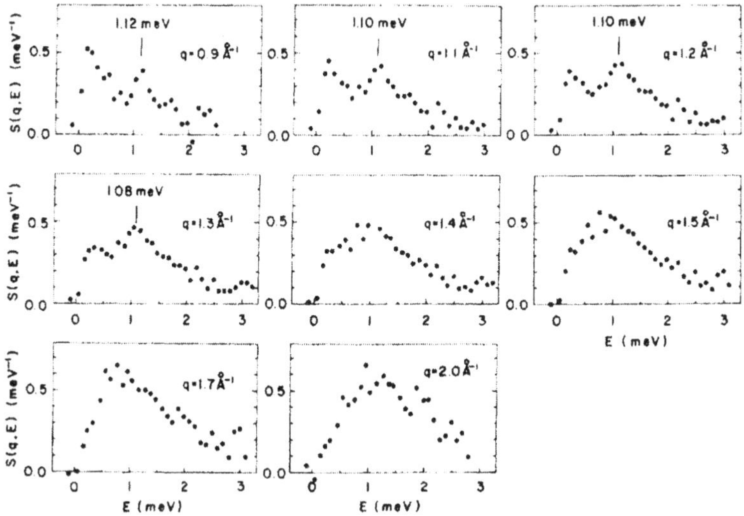

Fig. 12.7 $S(Q,\omega)$ observed for liquid ^{3}He at $T = 15$ mK. Reprinted with permission from [Sköld *et al.* (1976)]. Copyright 1976 by the American Physical Society.

For $Q \geq 1.3$ Å^{-1} a distinct separation between the S_{coh} and S_{inc} component is prevented by line broadening, which is presumably due to interactions with single particle-hole (p-h) excitations or multiples of p-h excitations (Landau damping). In particular, the decay of a phonon into a single p-h pair is possible if these modes directly overlap energetically. This is sketched in Fig. 12.8, which displays the observed dispersion of the zero-sound mode together with the p-h band calculated for an effective mass $m^* = 3m_0$.

12.4 Further reading

- M. A. Adams, in *Neutron scattering in novel materials*, ed. by A. Furrer (World Scientific, Singapore, 2000), p. 289: *Superfluid ^{4}He – a very novel material*
- J. F. Annett, *Superconductivity, superfluids and condensates* (Oxford University Press, Oxford, 2004)
- B. Fåk, in *Cold neutrons: large scales – high resolution*, ed. by A. Furrer (Proc. 97-01, ISSN 1019-6447, PSI Villigen, 1997), p. 47: *Excitations in normal liquid ^{3}He and superfluid ^{4}He*

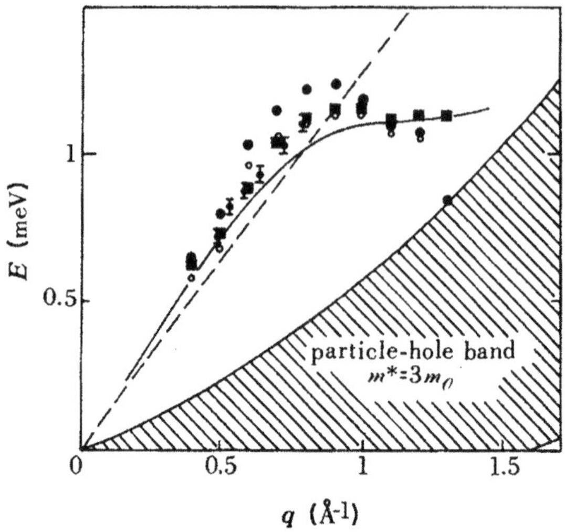

Fig. 12.8 Dispersion of the zero-sound mode in ^{3}He at $T = 15$ mK (after [Sköld *et al.* (1976)]). The shaded area shows the single particle-hole band, assuming a Q-independent effective mass $m^* = 3m_0$. Reprinted from [Glyde and Svensson (1987)]. Copyright 1987, Academic Press.

- H. Glyde, in *Methods of experimental physics*, Vol. 23, Part B, ed. by D. L. Price and K. Sköld (Academic Press, London, 1987), p. 303: *Solid and liquid helium*
- W. G. Stirling and H. R. Glyde, Phys. Rev. B 41, 4224 (1990): *Temperature dependence of the phonon and roton excitations in liquid* ^{4}*He*
- D. R. Tilley and J. Tilley, *Superfluidity and superconductivity* (Hilger, Bristol, 1990)

Chapter 13

Defects in Solids

13.1 Introduction

The considerations in the previous chapters were based on the concept of an ideal periodic lattice, i.e., we assumed the materials to be perfect. However, real materials are never perfect, but they may be disturbed by lattice defects, dislocations, impurities, density gradients, etc. If the concentration of such imperfections is of a marginal size, say less than 10^{-3}, they have usually no impact on the materials properties. Moreover, the neutrons are not a very sensitive tool to unravel such imperfections, so that we can use the ideal concepts in terms of an average response. For higher concentrations, typically for 10^{-2} and above, the imperfections can be important and must be properly accounted for. Then the imperfections are not only associated with local structural and dynamical effects, but they may also have an impact on the surrounding matrix and in addition they may interact with each other. Therefore, neutron scattering studies are preferably performed for detailed concentration series.

While the coherent elastic scattering of a perfect crystal is restricted to the Bragg peaks (see Chap. 4), real crystals show additional scattering around and between Bragg peaks. The elastic part of this scattering is called diffuse scattering and contains information on the occupation of the lattice sites and on static local displacements.

The simplest imperfections are point defects such as vacancies (called Schottky defect), substitutional atoms, self interstitials (called Frenkel defect), and foreign interstitials as illustrated in Fig. 13.1. Point defects give rise to local structural distortions which may have a considerable effect for instance on the properties of magnetic materials, since the exchange coupling is strongly dependent on the distance between the magnetic ions.

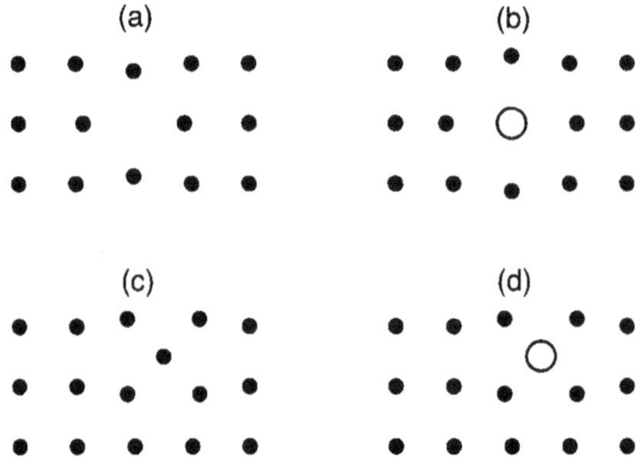

Fig. 13.1 Different types of point defects. (a) Vacancy; (b) substitutional atom; (c) self interstitial; (d) foreign interstitial.

The present Chapter concentrates on a few types of imperfections whose elucidation by neutron scattering has been useful. This includes firstly the static properties of point defects which often have the tendency towards clustering in pairs or even in large aggregates to form macro-defects. Secondly, point defects have an influence on the dynamical properties of the crystalline host, giving rise to so-called resonant and local modes.

13.2 Short-range order of point defects

Let us consider a compound composed of two types of atoms A and B sharing randomly the same crystallographic positions. The relative concentrations are p_A and p_B with $p_A + p_B = 1$. According to Eq. (2.32) the average scattering lengths are given by

$$\langle b \rangle = p_A b_A + p_B b_B, \qquad (13.1)$$
$$\langle b^2 \rangle = p_A b_A^2 + p_B b_B^2.$$

Insertion of Eq. (13.1) into the incoherent part of the scattering law defined by Eq. (2.31) yields the diffuse cross-section

$$\left(\frac{d\sigma}{d\Omega}\right)_{\text{diff}} \sim \langle b^2 \rangle - \langle b \rangle^2 = p_A p_B (b_A - b_B)^2, \qquad (13.2)$$

which just corresponds to a structureless background. From Eq. (13.2), we can directly infer the diffuse cross-section from vacancies by setting for example $b_B = 0$.

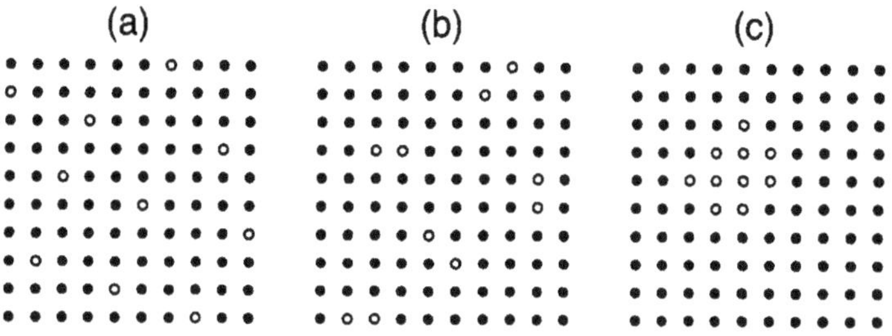

Fig. 13.2 Atomic site distribution of a binary alloy $A_{0.9}B_{0.1}$. (a) Statistical distribution of the minority atoms B; (b) preference for nearest-neighbor and next-nearest-neighbor pair formation of atoms B; (c) clustering of atoms B to form a large aggregate.

We now consider a binary alloy with $p_A \gg p_B$, i.e., the atoms of type B are introduced as point defects substituting some of the atoms of type A of the host lattice. In principle, point defects are randomly distributed over the lattice sites, but they often have the tendency to build clusters or even aggregates as illustrated in Fig. 13.2. Then there exist well defined correlations between the positions of the defects. Under the assumption that the defects are pairwise correlated, the diffuse scattering can be calculated from Eq. (4.4):

$$\left(\frac{d\sigma}{d\Omega}\right)_{\text{diff}} = p_B b_B^2 \sum_{|\boldsymbol{R}_j - \boldsymbol{R}_{j'}|=r} \alpha(r)\, e^{-i\boldsymbol{Q}\cdot(\boldsymbol{R}_{j'}-\boldsymbol{R}_j)}, \tag{13.3}$$

where $\alpha(r)$ is the probability that two point defects are separated by a distance $r = |\boldsymbol{R}_j - \boldsymbol{R}_{j'}|$. The $\boldsymbol{Q}$ dependence of Eq. (13.3) is similar to the cross section derived for spin dimer systems, see Eqs (8.9) and (8.10). For polycrystalline material Eq. (13.3) transforms to

$$\left(\frac{d\sigma}{d\Omega}\right)_{\text{diff}} = 2p_B b_B^2 \sum_r \alpha(r) \left(1 - \frac{\sin(Qr)}{Qr}\right), \tag{13.4}$$

i.e., the pair correlations give rise to an oscillatory modulation of the intensity vs Q.

The preferential pair formation of atoms was nicely observed in the oxygen-deficient compound $Zr(Y)O_{2-x}$ as illustrated in Fig. 13.3. The neutron diffraction data exhibit - besides the dominant Bragg peaks - a sinusoidally modulated diffuse background which could be interpreted according to Eq. (13.4) by pair formation due to the relaxation of nearest-neighbor oxygen ions towards the vacancy.

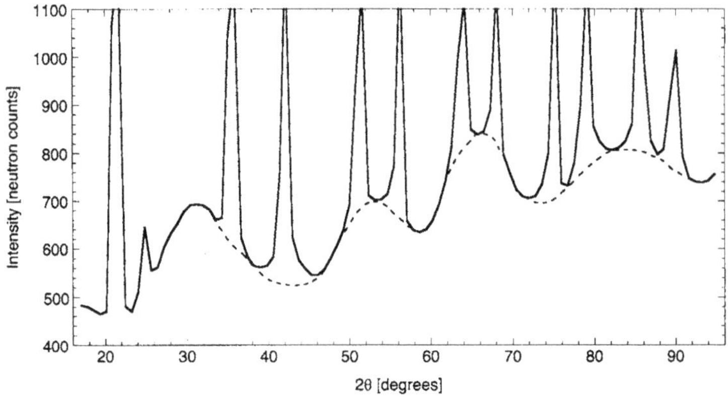

Fig. 13.3 Neutron diffraction pattern of the solid solution $Zr(Y)O_{2-x}$ showing the heavily modulated diffuse background (after [Steele and Fender (1974)]).

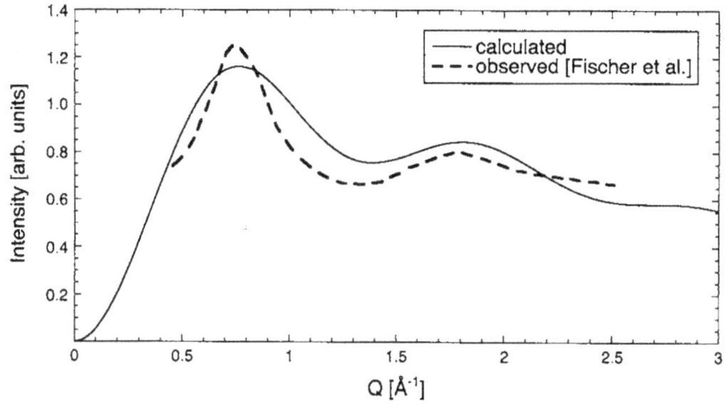

Fig. 13.4 Observed diffuse scattering from non-stoichiometric $TbSe_{1-x}$ at $T = 4.2$ K (dashed line) (after [Fischer *et al.* (1976)]). The solid line corresponds to short-range correlations of next-nearest neighboring Tb pairs, calculated from Eq. (13.5).

Other examples concern the antiferromagnetic rare-earth monochalcogenides which crystallize in the NaCl structure. Small deviations from the optimal stoichiometry give rise to considerable changes of the magnetic properties. For instance, for the slightly non-stoichiometric compound $TbSe_{1-x}$ an unusual coexistence of long-range antiferromagnetic order below $T_N = 52$ K and short-range magnetic order down to 4.2 K was observed. The latter is due to Schottky defects in the Se sublattice (see Fig. 13.1a) which results in a build-up of strong local correlations between neighboring

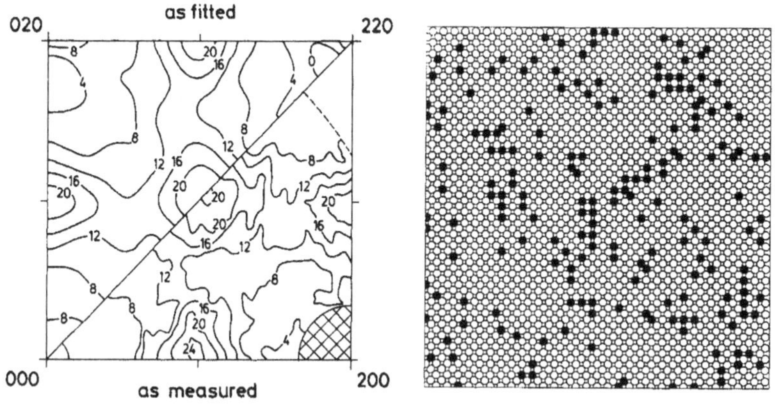

Fig. 13.5 Left: Diffuse elastic scattering observed for $Ni_{0.9}Al_{0.1}$ in the reciprocal (100) plane. The lines correspond to diffuse scattering with equal intensity. Experimental results (as measured) are compared with model calculations (as fitted). Right: Reconstructed atom distribution in real space. Reprinted from [Schönfeld *et al.* (1997)]). Copyright 1997, with permission from Elsevier.

pairs of Tb^{3+} ions as illustrated in Fig. 13.4. For the data interpretation the prefactors of Eq. (13.4) have to be adjusted to the case of magnetic scattering defined by Eq. (7.3) as follows:

$$\left(\frac{d\sigma}{d\Omega}\right)_{\text{diff.magn.}} \propto F^2(Q)\,\langle\hat{S}^z\rangle^2\left(1-\frac{\sin(Qr)}{Qr}\right), \tag{13.5}$$

where r is the (reduced) distance between two Tb^{3+} ions at the site of the Schottky defect.

In general, short-range order is not necessarily restricted to nearest-neighbor pairs as in the two examples discussed above, but we have to deal with a variety of probabilities $\alpha(r)$ in Eqs (13.3) and (13.4). In addition, a complete evaluation of diffuse scattering has to take into account displacements of the individual atoms from the ideal lattice sites as sketched in Fig. 13.1. This is a rather complex problem, thus simulations based on Monte Carlo or cluster variation algorithms are used to calculate the diffuse scattering, which is then compared to the measured data. The experimental strategy is to collect a complete set of diffuse scattering data, preferably covering the irreducible part of the Brillouin zone, as exemplified in Fig. 13.5 for the alloy $Ni_{0.9}Al_{0.1}$.

13.3 Macro-defects

As illustrated in Fig. 13.2c, point defects may have the tendency to build large aggregates. For instance, vacancies (see Fig. 13.1a) can accumulate to create extended voids, or substitutional atoms (see Fig. 13.1b) can form islands within the matrix of the host atoms. Aggregates with diameters exceeding about 50 Å are called macro-defects. In order to describe the scattering from macro-defects, we use the cross-section Eq. (6.13) derived for liquids:

$$\frac{\mathrm{d}\sigma}{\mathrm{d}\Omega} \propto S(Q) = \int \left(\rho(\boldsymbol{r}) - \rho_h\right) e^{\imath \boldsymbol{Q}\cdot\boldsymbol{r}} \mathrm{d}\boldsymbol{r}, \tag{13.6}$$

where $\rho(\boldsymbol{r})$ and ρ_h denote the scattering density of the macro-defects and of the host lattice, respectively (the scattering density involves the square of the neutron scattering lengths of the corresponding atoms). Due to the large size of the macro-defects, the relevant scattering will occur for small moduli of the scattering vector $\boldsymbol{Q}$, thus we expand Eq. (13.6) as a Taylor series:

$$\frac{\mathrm{d}\sigma}{\mathrm{d}\Omega} \propto S(Q) = \int \left(\rho(\boldsymbol{r}) - \rho_h\right) \left(1 + \imath \boldsymbol{Q}\cdot\boldsymbol{r} - \frac{1}{2}\left(\boldsymbol{Q}\cdot\boldsymbol{r}\right)^2 + \dots\right) \mathrm{d}\boldsymbol{r}. \tag{13.7}$$

If we assume inversion symmetry for $\rho(\boldsymbol{r})$, the integration of the linear term vanishes; furthermore, we replace the quadratic term by its spatial average:

$$\frac{\mathrm{d}\sigma}{\mathrm{d}\Omega} \propto S(Q) = \int \left(\rho(\boldsymbol{r}) - \rho_h\right) \left(1 - \frac{1}{2}\cdot\frac{2}{3}Q^2 r^2\right) \mathrm{d}\boldsymbol{r}. \tag{13.8}$$

We define the contrast scattering density by

$$\bar{\rho} = \frac{1}{V}\int \rho(\boldsymbol{r})\mathrm{d}\boldsymbol{r} - \rho_h \tag{13.9}$$

and insert this into Eq. (13.8):

$$\frac{\mathrm{d}\sigma}{\mathrm{d}\Omega} \propto S(Q) = \bar{\rho}V\left(1 - \frac{Q^2}{3}\cdot\frac{1}{\bar{\rho}V}\int \left(\rho(\boldsymbol{r}) - \rho_h\right)\boldsymbol{r}^2 \mathrm{d}\boldsymbol{r}\right) \tag{13.10}$$

where V is the volume of the sample. The last term of Eq. (13.10) corresponds to the second moment of the contrast density and therefore describes the mean square radius of the macro-defects:

$$R_g^2 = \frac{1}{\bar{\rho}V}\int \left(\rho(\boldsymbol{r}) - \rho_h\right) r^2 \mathrm{d}\boldsymbol{r}. \tag{13.11}$$

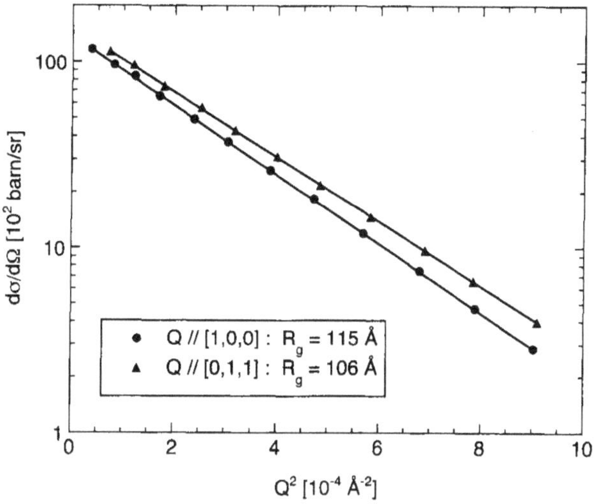

Fig. 13.6 Results of SANS experiments performed for a Cu single crystal with Co impurities. The data indicate a slight anisotropy of the Co aggregate which forms an ellipsoid with its principal axis along [100] (after [Abersfelder *et al.* (1980)]).

R_g is called radius of gyration. Inserting Eq. (13.11) back into Eq. (13.10) yields

$$\frac{d\sigma}{d\Omega} \propto S(Q) = \bar{\rho} V \left(1 - \frac{Q^2 R_g^2}{3}\right) \approx \bar{\rho} V e^{-Q^2 R_g^2/3} \tag{13.12}$$

where the bracket, corresponding to a first-order Taylor series, is transformed back to an exponential. Eq. (13.12) is known as the Guinier approximation for the analysis of elastic neutron scattering data taken at small moduli of the scattering vector $\boldsymbol{Q}$.

The Guinier approximation has been widely used to solve a variety of problems involving large scale structures, notably in biology (e.g., determination of the size of proteins, colloids or viruses in solution). Its application, however, relies on a sufficiently strong scattering contrast defined by Eq. (13.9). It is one of the outstanding properties of neutron scattering (contrary to x-ray experiments) to provide such a strong contrast often through isotope substitution (see Fig. 1.1). For biological objects, the contrast is obvious for hydrogen and its isotope deuterium.

Figure 13.6 shows the results of a SANS experiment performed for a single crystal of the alloy $Cu_{1-x}Co_x$ ($x \ll 1$), in which the Co atoms form

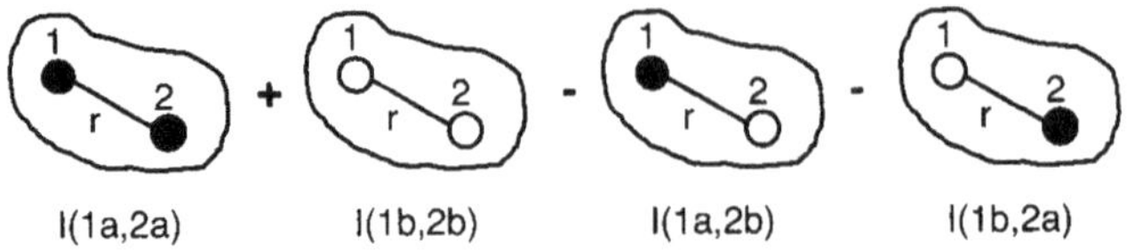

Fig. 13.7 Illustration of the triangulation method.

large aggregates. As listed in the Appendix B, natural Cu and Co have drastically different scattering lengths, thus a strong scattering contrast is guaranteed. According to Eq. (13.12), the intensity is logarithmically plotted vs Q^2 in Fig. 13.6, thus the radius of gyration R_g can be derived from the slope of the experimental data.

13.4 The triangulation method

If the interest in macro-defects or more generally in a large scale structure is restricted to explore only the location of certain subunits, then the triangulation method is a useful tool. The idea stems from classical cartography, where mutual distances between certain points of interest are successively determined to produce a final triangulation network. The operation principle for neutron scattering is illustrated in Fig. 13.7. The subunits 1 and 2 separated by the distance r are marked by substitution with isotopes either of type a or b, which offers four different structures. Thereby a pair correlation is introduced into the system which is reflected in the total sum of intensities:

$$I(1,2) = I(1a,2a) + I(1b,2b) - I(1a,2b) - I(1b,2a). \tag{13.13}$$

In analogy to the pair correlation of defects discussed in Chap. 13.2, Eq. (13.13) exhibits a Q dependence according to

$$I(1,2) \propto \frac{\sin(Qr)}{Qr}, \tag{13.14}$$

from which the mutual distance between the subunits 1 and 2 can be derived. We are not aware of triangulation experiments in condensed matter physics, but the method (which is extremely time consuming concerning the sample preparation) was successfully used in biology as illustrated in Fig. 13.8.

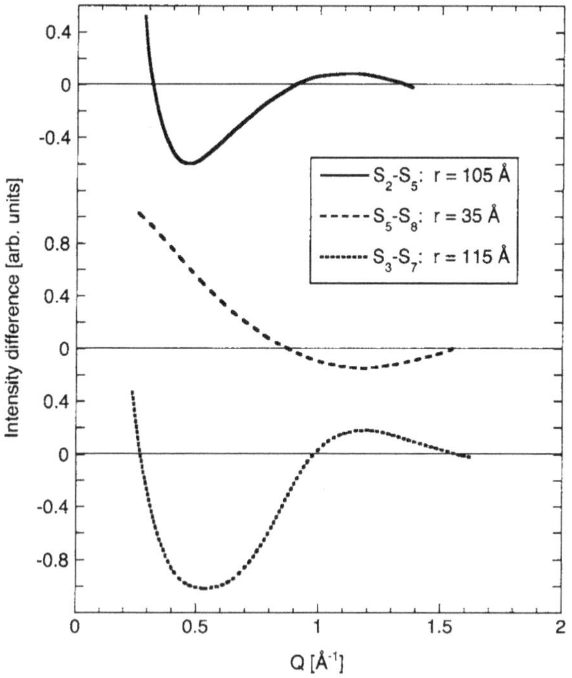

Fig. 13.8 SANS results for the ribosome unit 30S. The curves represent the intensity differences according to Eq. (13.13). With use of Eq. (13.14) the following distances between the subunits S_j and $S_{j'}$ were determined: (a) S_2- S_5: $r = 105$ Å; (b) S_5- S_8: $r = 35$ Å; (a) S_3- S_7: $r = 115$ Å. The subunits were marked by H and D (after [Engelman *et al.* (1975)]).

13.5 Resonant and local modes

The main difficulty encountered in the theoretical description of the vibrational properties of a crystal containing defects is the lack of exact periodicity. A simple model for the scattering law $S(\boldsymbol{Q}, \omega)$ has been worked out for very low concentration c of impurities that differ from the host atoms they replace only in mass (after [Nicklow (1983)]):

$$S(\boldsymbol{Q}, \omega) \propto \frac{c \operatorname{Im} T(\omega)}{\left(\omega^2 - \omega_j^2(\boldsymbol{q}) - c \operatorname{Re} T(\omega)\right)^2 + \left(c \operatorname{Im} T(\omega)\right)^2} \tag{13.15}$$

with

$$T(\omega) = \frac{M \epsilon \omega^2}{1 - (1-c) M \epsilon \omega^2 P(\omega)}, \tag{13.16}$$

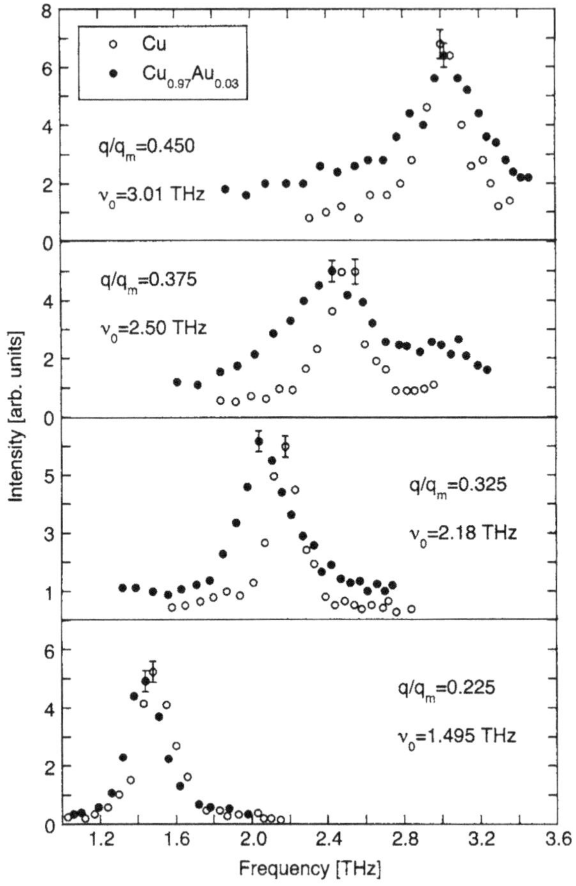

Fig. 13.9 Constant-Q scans measured for T_1 phonons with wavevector $\boldsymbol{q} = (0, x, x)$ in Cu and $Cu_{0.97}Au_{0.03}$. The full lines are calculations based on Eq. (13.15). q_m denotes the wavevector at the reciprocal (011) point (after [Nicklow (1983)]).

where $\epsilon = (M - M')/M$, M and M' denote the mass of the host and defect atoms, respectively, and $P(\omega)$ is the Green function of the host with poles occurring at the frequency $\omega_j(\boldsymbol{q})$ of the perfect crystal. Consequently the δ-function peak in the scattering cross section for the perfect crystal, Eq. (5.23), is replaced by a broadened scattering function that is shifted in frequency from $\omega_j(\boldsymbol{q})$ and is nearly Lorentzian if the ω dependence of $\mathrm{Re}\,T(\omega)$ and $\mathrm{Im}\,T(\omega)$ is not strong.

For small concentrations of heavy or light substitutional atoms, inelas-

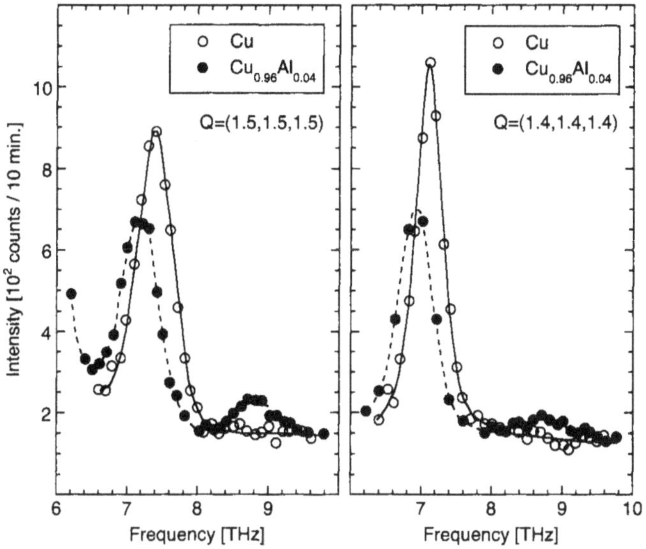

Fig. 13.10 Constant-Q scans performed for Cu and $Cu_{0.96}Al_{0.04}$ (after [Nicklow (1983)]). The lines are Gaussian least-squares fits to the data.

tic neutron scattering experiments give rise to so-called resonant or local modes, respectively. For heavy impurities with $M' > M$, the resonant modes appear within the band of frequencies for the pure host. This is illustrated in Fig. 13.9 for a Cu crystal doped with 3 at. % Au. Around the resonant mode of frequency $\nu \approx 2.4$ THz, the peaks observed for the Cu-Au sample are significantly broader than those for Cu. In addition, they are shifted in frequency from the positions ν_0 measured for Cu. Note that this shift is large and negative for $\nu_0 < 2.4$ THz, and it is small but positive for $\nu_0 > 2.5$ THz, consistent with Eq. (13.15).

Impurities with mass $M' < M$ give rise to vibrational modes greater than the maximum frequency of the host lattice. However, the impurity mode does not propagate far into the host lattice, because its frequency is not a normal-mode frequency of that lattice. It is therefore localized in space. This is illustrated in Fig. 13.10 for a Cu crystal doped with 4 at. % Al, showing results at and just off the [111]-zone boundary. In addition to the longitudinal phonon mode with maximum frequency of about 7 THz, there is a peak near 8.8 THz. The local character of the latter is demonstrated by the decrease of the intensity as $\boldsymbol{Q}$ is moved away from the zone boundary. It can be shown that for the analysis of local modes Eq. (13.15) applies by setting $\omega^2 - \omega_j^2(q) - c\,\mathrm{Re}T(\omega) = 0$, thus the local mode is given by a δ-function as observed in the experiments, see Fig. 13.10 (the Gaussian shape is due to the limited instrumental resolution).

13.6 Further reading

- T. Egami and S. Billinge, *Underneath the Bragg peaks: Structural analysis of complex materials*, Pergamon Materials Series, Vol. 7 (Pergamon, New York, 2003)
- G. Kostorz, in *Neutron scattering*, ed. by A. Furrer (Proc. 93-01, ISSN 1019-6447, PSI Villigen, 1993), p. 273: *Wide- and small-angle diffuse scattering from metals and alloys*
- G. Kostorz, in *Introduction to neutron scattering*, ed. by A. Furrer (Proc. 96-01, ISSN 1019-6447, PSI Villigen, 1996), p. 129: *Diffuse scattering*
- M. A. Krivoglaz, *X-ray and neutron diffraction in nonideal crystals* (Springer, Berlin, 1996)
- R. M. Nicklow, in *Methods of experimental physics*, Vol. 21, ed. by J. N. Mundy, S. J. Rothman, M. J. Fluss and L. C. Smedskjaer (Academic Press, Orlando, 1983), p. 172: *Dynamic properties of defects*
- W. Schmatz, in *Methods of experimental physics*, Vol. 21, ed. by J. N. Mundy, S. J. Rothman, M. J. Fluss, and L. C. Smedskjaer (Academic Press, Orlando, 1983), p. 147: *Static properties of defects*
- W. Schmatz, in *Methods of experimental physics*, Vol. 23, Part B, ed. by D. L. Price and K. Sköld (Academic Press, London, 1987), p. 85: *Defects in solids*
- B. Schönfeld, Progr. Mater. Sci. 44, 435 (1999): *Local atomic arrangements in binary alloys*
- B. Schönfeld, in *Neutron scattering in novel materials*, ed. by A. Furrer (World Scientific, Singapore, 2000), p. 79: *Scattering between Bragg peaks*

Chapter 14

Surfaces and Interfaces

14.1 Introduction

Advanced technologies developed for the deposition of thin films and for the production of multilayer systems brought about a myriad of new artificial materials with novel physical properties in both fundamental and applied physics, thus there is an obvious need to characterize the surfaces and interfaces in these new materials by suitable experimental methods. Surfaces and interfaces correspond to very small quantities of matter of the order of a few micrograms, thus at first sight neutron scattering being an intensity-limited technique does not appear to be a useful tool. However, neutron reflection at grazing incidence increases the interaction with the sample, thus neutron reflectometers are the appropriate instruments for such studies (see Chap. 3.3.5).

In neutron reflectometry we are dealing with elastic scattering, thus the moduli of the wavevectors $\boldsymbol{k}_0 \equiv \boldsymbol{k}$ and $\boldsymbol{k}_1 \equiv \boldsymbol{k}$ of the incident and scattered neutrons, respectively, are equal: $|\boldsymbol{k}_0| = |\boldsymbol{k}_1|$. For the geometry sketched in Fig. 14.1, the scattering vector $\boldsymbol{Q} = \boldsymbol{k}_0 - \boldsymbol{k}_1$ has the following components:

$$\boldsymbol{Q} = k_0 \begin{pmatrix} \cos\theta_0 - \cos\theta_1 \cos\chi \\ -\sin\chi \\ -\sin\theta_0 - \sin\theta_1 \cos\chi \end{pmatrix} . \tag{14.1}$$

Three different types of reflectometry measurements have been developed which probe different length scales ξ and directions as visualized in Fig. 14.1:

(i) Specular reflection corresponds to scattering in the incidence plane with $\theta_0 = \theta_1$ and $\chi = 0$. As a consequence, the scattering vector $\boldsymbol{Q} =$

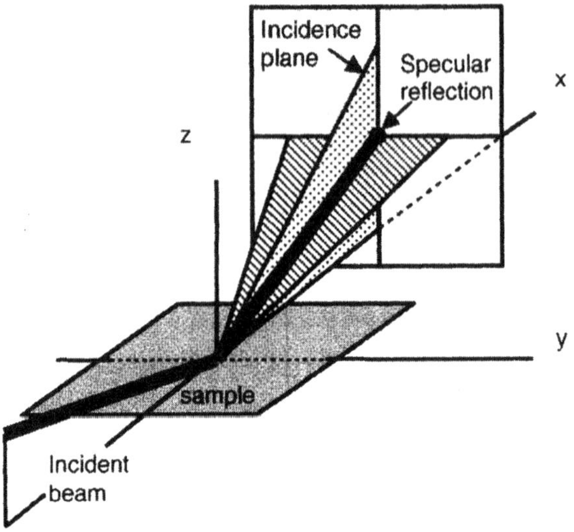

Fig. 14.1 Different scattering geometries for reflectivity measurements. Specular reflection is marked by the thick line. The dotted and hatched planes refer to off-specular reflection and grazing incidence scattering. Reprinted from [Ott (2007)]. Copyright 2007, with permission from Elsevier.

$(0, 0, Q_z)$ is perpendicular to the surface, thus the structure perpendicular to the surface is probed on a length scale 2 nm $< \xi <$ 100 nm.

(ii) Off-specular reflection corresponds to scattering in the incidence plane with $\theta_0 \neq \theta_1$ and $\chi = 0$. This introduces an in-plane component Q_x to the scattering vector, $\boldsymbol{Q} = (Q_x, 0, Q_z)$, with $Q_x \ll Q_z$ for geometrical reasons, thus in-plane structures (e.g., the surface roughness) are probed on a rather large length scale 500 nm $< \xi <$ 50 μm.

(iii) Grazing incidence scattering corresponds to scattering perpendicular to the incidence plane with $\theta_0 = \theta_1$ and $\chi \neq 0$, thus the scattering vector is $\boldsymbol{Q} = (0, Q_y, Q_z)$. It is an extension of off-specular reflection to probe surface features on a smaller length scale 2 nm $< \xi <$ 100 nm.

The length scales estimated above result directly from the Q-ranges expressed in Eq. (14.1) for neutrons in a wavelength band 2 Å $< \lambda_0 <$ 20 Å ($\lambda_0 = 2\pi/k_0$).

For the description of neutron reflectivity we refer to Chap. 2.8. In order to include scattering from magnetic layers we add a magnetic term

Eq. (2.39) to the interaction potential U:

$$U = \frac{2\pi\hbar^2 \rho\langle b\rangle}{m} - \boldsymbol{\mu}\cdot\boldsymbol{H}. \tag{14.2}$$

The refractive index n of neutrons with energy E (defined by Eq. (1.5)), entering a medium with potential $U \ll E$ is given by

$$n = \sqrt{1 - \frac{U}{E}} \approx 1 - \frac{U}{2E}. \tag{14.3}$$

The magnetic part of U gives rise to refractive indices $n^{\pm}$ that depend on the relative orientation of the spin of the incident neutron and the internal field of the magnetic layer:

$$n^{\pm} = 1 - \delta \pm \delta_M = 1 - \frac{\rho\lambda^2\langle b\rangle}{2\pi} \pm \frac{2m\lambda^2\mu H}{h^2}, \tag{14.4}$$

where δ and δ_M are the nuclear and magnetic contributions, respectively, whose sizes are of the same order of magnitude. According to Eq. (2.63) there is total reflection up to a critical angle $\gamma_c = \arccos(n^{\pm})$. Above γ_c the wavefunctions of the reflected and transmitted neutrons are described by

$$\begin{aligned} \psi_0 &= e^{ik_0 z} + r_0 \cdot e^{-ik_0 z}, \\ \psi_1 &= t_0 \cdot e^{ik_1 z}, \end{aligned} \tag{14.5}$$

where the z-axis is perpendicular to the reflecting surface and the wavenumber of the transmitted neutron is given by

$$k_1 = \sqrt{k_0^2 - \frac{2mU}{\hbar^2}}. \tag{14.6}$$

As mentioned in Chap. 2.8, the wavefunctions Eq. (14.5) have to be continuous and continuously differentiable at the surface:

$$\begin{aligned} \psi_0(z)|_{z=z_0} &= \psi_1(z)|_{z=z_0}, \\ \frac{\partial}{\partial z}\psi_0(z)|_{z=z_0} &= \frac{\partial}{\partial z}\psi_1(z)|_{z=z_0}. \end{aligned} \tag{14.7}$$

With use of Eq. (14.7) the reflection and transmission coefficients introduced in Eq. (14.5) turn out to be:

$$r_0 = \frac{k_0 - k_1}{k_0 + k_1} e^{i2k_0 z_0}, \quad t_0 = \frac{2k_0}{k_0 + k_1} e^{i(k_0 - k_1) z_0}. \tag{14.8}$$

For an n-layer system the calculations become increasingly involved, since the neutron wave transmitted at the ith interface will partially be reflected at the $(i+1)$th interface, thus Eq. (14.5) has to be extended:

$$\begin{aligned} \psi_i &= e^{ik_i z} + r_i \cdot e^{-ik_i z}, \\ \psi_{i+1} &= t_i \cdot e^{ik_{i+1} z} + r_{i+1} \cdot e^{-ik_{i+1} z}. \end{aligned} \tag{14.9}$$

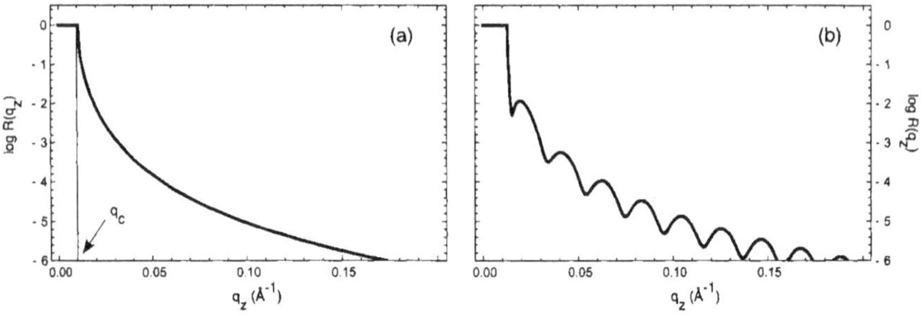

Fig. 14.2 (a) Calculated reflectivity from the surface of a substrate with scattering length density $\rho\langle b\rangle = 2.387 \cdot 10^{-6}$ Å^{-2}. (b) Calculated reflectivity from a monolayer system with scattering length density $\rho\langle b\rangle = 2.387 \cdot 10^{-6}$ Å^{-2} and thickness $d = 30$ nm deposited on a substrate with scattering length density $\rho\langle b\rangle = 3.183 \cdot 10^{-6}$ Å^{-2} (after [Hoppler (2005)]).

For the calculation of the effective reflectivity and transmission, Eq. (14.9) has to be successively applied, which is carried out with the help of a multiplication scheme based on matrices of the form

$$\mathcal{Q}_{i-1,i} = \frac{1}{2}\begin{pmatrix} \left(1+\frac{k_i}{k_{i-1}}\right) e^{i(k_i-k_{i-1})z_i} & \left(1-\frac{k_i}{k_{i-1}}\right) e^{-i(k_i+k_{i-1})z_i} \\ \left(1-\frac{k_i}{k_{i-1}}\right) e^{i(k_i+k_{i-1})z_i} & \left(1+\frac{k_i}{k_{i-1}}\right) e^{-i(k_i-k_{i-1})z_i} \end{pmatrix}, \tag{14.10}$$

which results in the final expression [Blundell and Bland (1992)]

$$\begin{pmatrix} 1 \\ r \end{pmatrix} = \prod_{i=1}^{n} \mathcal{Q}_{i-1,i} \begin{pmatrix} t \\ 0 \end{pmatrix} = \begin{pmatrix} M_{11} & M_{12} \\ M_{21} & M_{22} \end{pmatrix} \begin{pmatrix} t \\ 0 \end{pmatrix}. \tag{14.11}$$

The reflection and transmission coefficients $r = M_{21}/M_{11}$ and $t = 1/M_{11}$, respectively, correspond to the familiar Fresnel coefficients. Calculations based on Eq. (14.11) show that the total effective reflectivity $R = r^2$ rapidly decreases as Q_z^{-4} above γ_c.

14.2 Specular reflection

For specular reflection, the modulus of the scattering vector $\boldsymbol{Q}$ corresponds to Q_z, see Eq. (14.1). We start by considering the reflectivity from the surface of a substrate with scattering length density $\rho\langle b\rangle = 3.183 \cdot 10^{-6}$ Å^{-2}. The calculated reflectivity nicely exhibits the Q_z^{-4} decrease as shown in Fig. 14.2a.

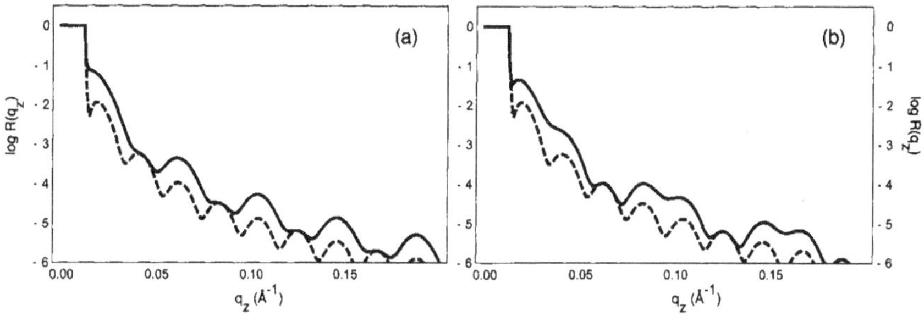

Fig. 14.3 (a) Calculated reflectivity from a bilayer system with scattering length densities $\rho\langle b\rangle = 2.387 \cdot 10^{-6}$ Å^{-2} (dashed) and $\rho\langle b\rangle = 3.979 \cdot 10^{-6}$ Å^{-2} (solid) and equal thickness $d = 15$ nm deposited on a substrate with $\rho\langle b\rangle = 3.183 \cdot 10^{-6}$ Å^{-2}. (b) The same as (a), but with a 2:1 ratio of the layer thicknesses, i.e., $d_1 = 20$ nm and $d_2 = 10$ nm. The results for the monolayer are indicated by the broken lines (after [Hoppler (2005)]).

Figure 14.2b displays the calculated reflectivity from a monolayer system with scattering length density $\rho\langle b\rangle = 2.387 \cdot 10^{-6}$ Å^{-2} and thickness $d = 30$ nm deposited on the substrate described in Fig. 14.2a. The reflectivity exhibits modulations (superimposed on the overall Q_z^{-4} decrease), which correspond to constructive and destructive interferences of the neutron waves scattered from the two interfaces (vacuum-monolayer and monolayer-substrate). The periodicity of the oscillating pattern is related to the thickness d of the monolayer through the Bragg condition Eq. (4.8): $d = 2\pi/Q_z$.

Figure 14.3a shows the reflectivity from a bilayer system with equal layer thickness $d = 15$ nm and scattering length densities $\rho\langle b\rangle = 2.387 \cdot 10^{-6}$ Å^{-2} (as in Fig. 14.2b) and $\rho\langle b\rangle = 3.979 \cdot 10^{-6}$ Å^{-2}. In this case there are three interfaces giving rise to interference effects. A comparison with Fig. 14.2b shows that every second maximum of the oscillating reflectivity pattern is suppressed. By choosing a 2:1 ratio of the layer thickness, i.e., $d_1 = 20$ nm and $d_2 = 10$ nm, every third maximum is suppressed, see Fig. 14.3b.

We can now generalize these results for a system containing n bilayers. The reflectivity will exhibit Bragg reflections at scattering angles corresponding to the thickness of the bilayer. Between the Bragg positions there will be (n-1) weak oscillations reflecting the total thickness of the system. These oscillations are called Kiessig fringes. This is illustrated in Fig. 14.4 for a system of eight bilayers being composed of nickel (7 nm) and titanium (7 nm).

The above considerations refer to ideal interfaces, but in reality interdiffusion and roughness exists at interfaces as shown in Fig. 14.5a, thus

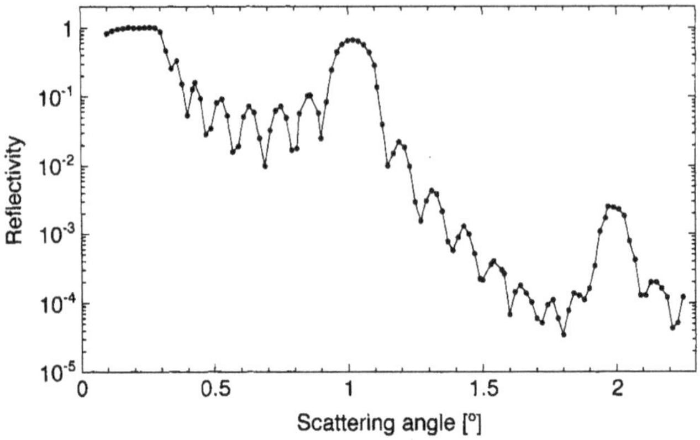

Fig. 14.4 Reflectivity observed for a system of eight bilayers of composition Ni (7 nm) / Ti (7 nm) deposited on float glass. The wavelength of the incident neutrons was $\lambda = 4.7$ Å (by courtesy of SwissNeutronics AG, CH-5313 Klingnau).

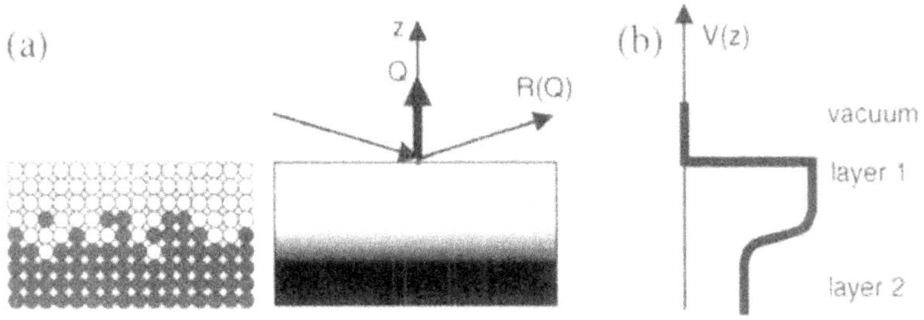

Fig. 14.5 (a) Interface between two layers. (b) Variation of the interaction potential along the z-axis. Reprinted from [Ott (2007)]. Copyright 2007, with permission from Elsevier.

numerical models have to be applied to reconstruct all the characteristics of layered systems. The statistical imperfection of interfaces is usually described by a Gaussian distribution whose variance corresponds to the roughness of the interface. Thereby the interface is approximated by a continuous medium, and the interaction potential is smoothed out accordingly, see Fig. 14.5b. It should be noted, however, that detailed information on the in-plane roughness requires off-specular reflectivity or grazing incidence scattering measurements.

In polarized reflectivity experiments it is possible to measure four cross sections, two non-spin-flip cross sections R^{++} and R^{--} and two spin-

Table 14.1 Scattering length densities for GaMnAs and GaAs in 10^{-6} Å^{-2} units. b and p refer to the nuclear and magnetic scattering amplitudes introduced in Chap. 7.8.

GaMnAs: R^{++} $\rho \cdot (b-p)$	GaMnAs: R^{--} $\rho \cdot (b+p)$	GaMnAs: $R^{+-} = R^{+-}$ $\rho \cdot p$	GaAs: R (unpolarized) $\rho \cdot p$
2.713	3.067	0.177	3.070

flip process cross-sections R^{+-} and R^{-+}. According to Eq. (2.53), the spin-flip and non-spin-flip cross sections are sensitive to the component of the magnetization parallel and perpendicular to the applied field, respectively. This is demonstrated for a polarized neutron reflectometry study of a GaMnAs/GaAs superlattice [Kepa *et al.* (2002)] which is a good semiconductor and exhibits ferromagnetism, thus it is of potential interest as a solid-state electronic device in spintronics. The magnetically active GaMnAs layer had a thickness of 50 monolayers and a Mn concentration of 6%, and the thickness of the GaAs spacer was 6 monolayers. The number of repeats for the total superlattice was 50, and the reflecting surface area was of the order of 1 cm^2. As probed by magnetization measurements, the sample was ferromagnetic below $T_c = 40$ K. The calculated cross sections are listed in Table 14.1. Due to the negative sign of the scattering length of Mn (see Appendix B) and the positive magnetic scattering amplitude for the Mn spins, there is practically no contrast between GaAs and GaMnAs for the R^{--} cross section. Consequently, experiments probing R^{--} should not produce superlattice peaks below T_c.

The observed reflectivity profiles around the first-order superlattice peak at $Q_z = 0.041$ Å^{-1} are summarized in Fig. 14.6. The measurements below T_c and in a small stray field of 2 Gauss (originating from the guide field, see Chap. 3.2.7) are shown in Fig. 14.6a. The presence of the superlattice peak in the R^{--} data and its absence in the R^{++} data means that the layer moments are aligned in the direction opposite to the stray field. Furthermore, the spin-flip reflectivities R^{+-} and R^{-+} are essentially zero, thus there is no magnetization component perpendicular to the stray field. When a field opposite to the stray field is applied, see Fig. 14.6b, the superlattice peak appears in the R^{++} data and is absent in the R^{--} data, consistent with the scattering length densities listed in Table 14.1. Above T_c (Fig. 14.7c) the sample is paramagnetic, thus we have $R^{++} = R^{--}$ since only nuclear scattering is present. All these observations lead to the conclusion that each GaMnAs layer forms a single magnetic domain. Moreover, the magneti-

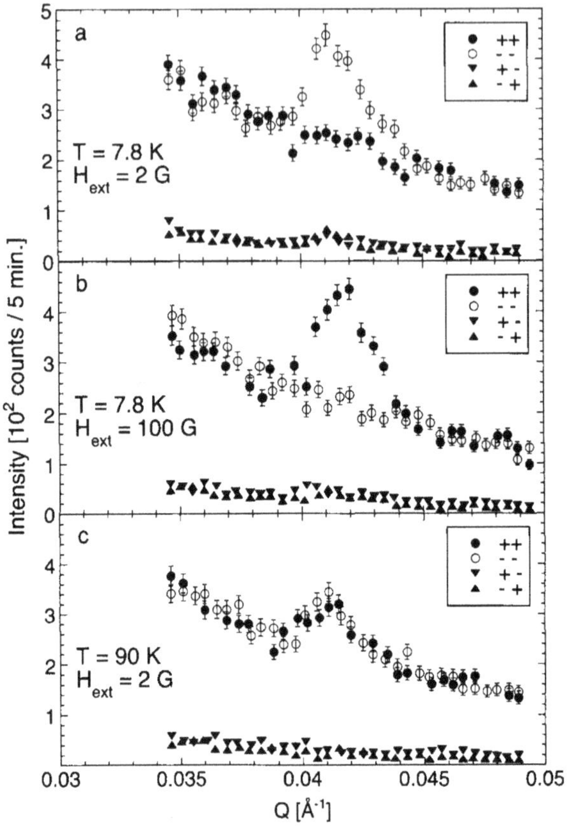

Fig. 14.6 Polarized neutron reflectivity profiles observed for the GaMnAs/GaAs superlattice described in the text around the first-order superlattice peak (after [Kepa *et al.* (2002)]).

zation of all layers is parallel, characteristic of ferromagnetically coupled superlattices.

14.3 Off-specular reflection

In reality, surfaces and interfaces may be disturbed by roughness (see Fig. 14.5a) or domain formation, giving rise to diffuse scattering (see Chap. 14). In specular reflectivity measurements, such in-plane imperfections are averaged out, since the sample is probed along its depth only. It is the purpose of off-specular reflection to probe the in-plane structure by

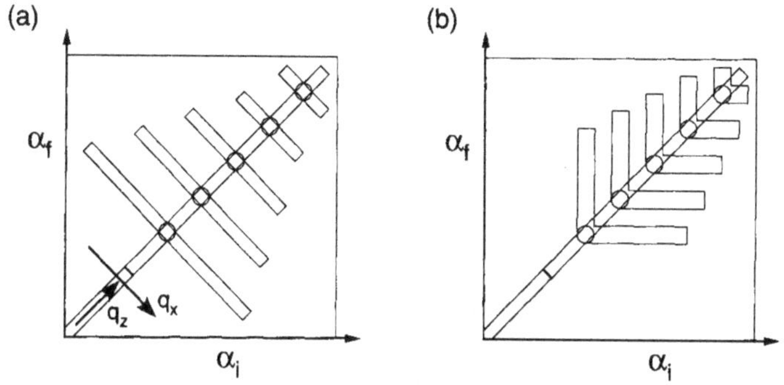

Fig. 14.7 Schematic representation of off-specular reflectivity. (a) Observation of Bragg sheets. (b) Observation of Yoneda wings (after [Hoppler (2005)]).

introducing a small Q_x component to the scattering vector $\boldsymbol{Q} = (Q_x, 0, Q_z)$, see Eq. (14.1). In particular, the length scale for off-specular reflectivity (which is in the μm region) matches the typical sizes of magnetic micro-domain structures.

The effects to be observed by off-specular reflectivity are twofold. Lateral in-plane correlations of imperfections due to roughness or domain formation will produce so-called Bragg sheets, i.e., line broadening of the Bragg reflections along Q_x as shown in Fig. 14.7a. On the other hand, if there are both lateral correlations and correlations perpendicular to the surface, then the diffuse pattern will exhibit Yoneda-like wings as visualized in Fig. 14.7b. (Strictly speaking, the term Yoneda wings should only be used at the edge of total reflection). In both cases, the correlation range is inversely proportional to the width of the observed patterns.

The modeling of off-specular reflectivity data is quite involved, so that we refer to the pertinent literature [Pietsch *et al.* (2004)] for further details. For illustrative purposes only, Fig. 14.8 shows a two-dimensional intensity map observed for the multilayer $[^{57}\mathrm{Fe}(68\text{Å})/\mathrm{Cr}(9\text{Å})]_{12}/\mathrm{Cr}(68\text{Å})$ grown on a sapphire substrate [Lauter *et al.* (2002)]. Nearly no off-specular scattering is visible around the first-order Bragg peak situated on the reflectivity line at $Q_x = Q_z = 0.041$ Å^{-1}. However, the Bragg sheets running through the $\frac{1}{2}$- and $\frac{3}{2}$-order Bragg peak positions at $Q_x = Q_z = 0.020$ Å^{-1} and $Q_x = Q_z = 0.063$ Å^{-1}, respectively, show that this scattering is due to lateral magnetic fluctuations which are correlated antiferromagnetically in neighboring layers, involving a doubling of the unit cell perpendicular to the surface.

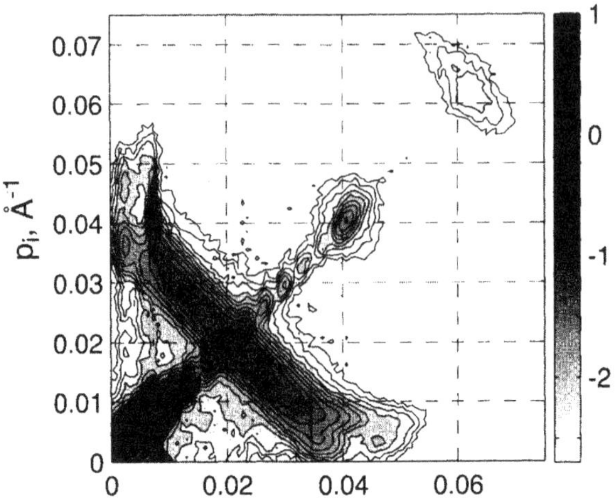

Fig. 14.8 Two-dimensional reflectivity map observed for the Fe/Cr multilayer described in the text. Reprinted with kind permission from Springer Science+Business Media: [Lauter *et al.* (2002)], Fig. 1a, copyright 2002.

14.4 Grazing incidence scattering

Since the Q_x component is limited in off-specular reflectivity measurements due to geometrical reasons, the off-specular technique has been extended to the pin-hole SANS geometry (see Chap. 3.3.4) to probe scattering vectors $\boldsymbol{Q} = (0, Q_y, Q_z)$ in the range $10^{-4} < Q_y < 3$ nm^{-1}. Fig. 14.9 shows the results of the first grazing incidence scattering experiments performed for a 50 nm thick film of alternating Fe and Pd monolayers [Fermon *et al.* (1999)]. After magnetizing the sample along the easy axis, a magnetic stripe domain pattern is formed as shown at the bottom of the figure. This stripe pattern was studied with use of the SANS technique under grazing incidence ($\theta = 0.7°$) of the incoming neutrons, the magnetic domains being parallel to the incidence plane. The diffracted neutron beam was detected by a two-dimensional counter as shown in the upper part of the figure. One can observe a bright specular spot and two weaker (10^{-3}) off-specular peaks, whose positions reflect the periodicity of the stripe domains (100 nm).

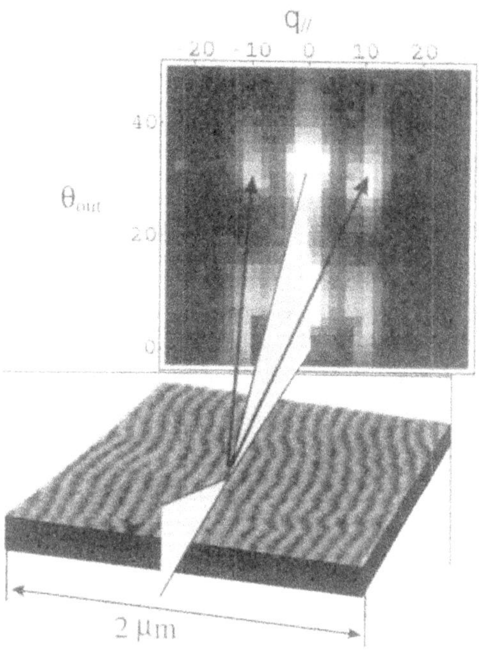

Fig. 14.9 Diffraction geometry of the grazing incidence SANS measurements performed for a thin film of FePd. Bottom: Magnetic force microscopy picture of the magnetic stripe pattern. Top: Multidetector showing the specular and off-specular peaks. The bottom signal is due to the refracted wave. Reprinted from [Fermon *et al.* (1999)]. Copyright 1999, with permission from Elsevier.

14.5 Further reading

- S. J. Blundell and J. A. C. Bland, Phys. Rev. B 46, 3391 (1992): *Polarized neutron reflection as a probe of magnetic films and multilayers*
- G. P. Felcher, Physica B 267-268, 154 (1999): *Polarized neutron reflectometry a historical perspective*
- C. F. Majkrzak, in *Magnetic neutron scattering*, ed. by A. Furrer (World Scientific, Singapore, 1995), p. 78: *Neutron scattering studies of magnetic superlattices*
- C. F. Majkrzak, K. V. O'Donovan and N. F .Berk, in *Neutron scattering from magnetic materials*, ed. by T. Chatterji (Elsevier, Amsterdam, 2006), p. 397: *Polarized neutron reflectometry*
- D. F. McMorrow, in *Complementarity between neutron and synchrotron x-ray scattering*, ed. by A. Furrer (World Scientific, Singapore, 1998),

p. 3: *From thin films to superlattices studied with x-rays and neutrons*

- F. Ott, C.R. Physique 8, 763 (2007): *Neutron scattering on magnetic surfaces*
- J. R. P. Webster, in *Cold neutrons: large scales – high resolution*, ed. by A. Furrer (Proc. 97-01, ISSN 1019-6447, PSI Villigen, 1997), p. 143: *The potential of neutron off-specular reflectivity studies*
- H. Zabel, in *Frontiers of neutron scattering*, ed. by A. Furrer (World Scientific, Singapore, 2000), p. 210: *Future trends in heterostructure research with neutron scattering*
- H. Zabel and K. Theis-Bröhl, J. Phys.: Condens. Matter 15, S505 (2003): *Polarized neutron reflectivity and scattering studies of magnetic heterostructures*

Chapter 15

Hydrogen Dynamics

15.1 Introduction

Hydrogen is the most abundant element in the universe, making up 75% of normal matter by mass and over 90% by number of atoms. Hydrogen gas is very rare in the earth's atmosphere, but it is the third most abundant element on the earth's surface, occurring mostly in the form of water, chemical compounds and biological objects. Hydrogen can form compounds with most elements and is highly soluble in many metal compounds, thereby forming materials with novel properties and applications. As the only neutral atom for which the Schrödinger equation can be solved analytically, the hydrogen atom has played a key role in the development of quantum mechanics, also supported by detailed neutron scattering studies of the energetics and bonding. In some of the preceding chapters, the important role of hydrogen was highlighted for the case of isotope substitution (Chap. 2.4 and 13.4), metal hydride formation (Chap. 4.3), hydrogen diffusion (Chap. 6.3) and phase transitions in ice (Chap. 10.3). Furthermore, hydrogen atoms give rise to local modes (Chap. 13.5) and are the best suited example for the application of the impulse approximation (Appendix G). In the present chapter, we will discuss in more detail the quantum mechanical aspects of hydrogen bonds including hydrogen tunnelling.

15.2 Dynamics of the hydrogen bond

Hydrogen bonds are important and ubiquitous; they are responsible for the unique properties of water, they provide biological activity through the imposition of tertiary and quaternary structures in large molecules, and they are the physical basis for proton conductivity. A hydrogen bond

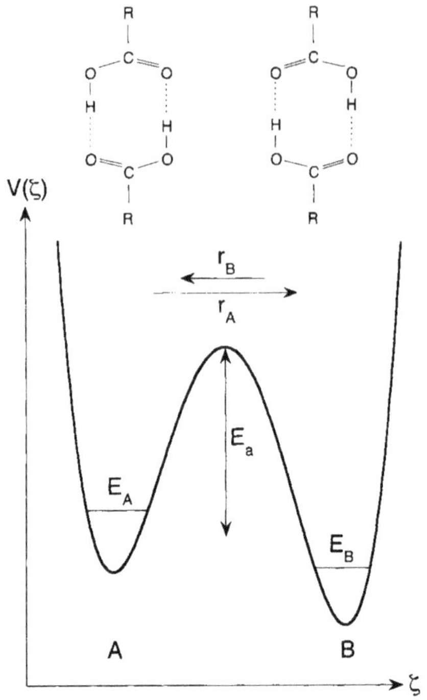

Fig. 15.1 Schematic sketch of the double-proton exchange (the example shown is a carboxylic acid). V is the asymmetric double-minimum potential. E_A and E_B denote the lowest energy levels, E_a the activation energy, and r_A and r_B the transition rates.

is conventionally written as A–H$\cdots$B to indicate the strong bond to the atom A and the weaker bond to the atom B. The distance A–H is usually shorter than the distance H$\cdots$B. As a consequence, the energy of the H atom is described by an asymmetric double-minimum potential as sketched in Fig. 15.1. For some special cases the double-minimum potential may be symmetric. The hydrogen atom can change its configuration from A–H$\cdots$B to A$\cdots$H–B with a transition rate r_A, and from there back to A–H$\cdots$B with a transition rate r_B. We denote the probabilities for the realization of the configurations A and B by $p_A(t)$ and $p_B(t)$, respectively, with $p_A(t) + p_B(t) = 1$. The equation of motion for the hydrogen atom then reads

$$\frac{\partial}{\partial t}p_A = -\frac{\partial}{\partial t}p_B = -r_A p_A(t) + r_B p_B(t) = -(r_A + r_B)p_A(t) + r_B, \quad (15.1)$$

which has the following solution:

$$p_A(t) = \left(p_A(0) - \frac{r_B}{r_A + r_B}\right) e^{-(r_A+r_B)t} + \frac{r_B}{r_A + r_B},$$
$$p_B(t) = 1 - p_A(t). \tag{15.2}$$

Thermal equilibrium is reached for $t \to \infty$. Inserting $t = \infty$ in Eq. (15.2) yields $p_A(\infty) = r_B/(r_A + r_B)$ and $p_B(\infty) = r_A/(r_A + r_B)$, thus Eq. (15.2) can be rewritten as

$$p_A(t) = (p_A(0) - p_A(\infty))\, e^{-(r_A+r_B)t} + p_A(\infty), \tag{15.3}$$
$$p_B(t) = (p_B(0) - p_B(\infty))\, e^{-(r_A+r_B)t} + p_B(\infty).$$

The population factors $p_{A,B}(\infty)$ are governed by Boltzmann statistics:

$$p_A(\infty) = \frac{1}{Z}, \quad p_B(\infty) = \frac{1}{Z} e^{-\frac{E_B - E_A}{k_B T}}, \quad Z = 1 + e^{-\frac{E_B - E_A}{k_B T}}. \tag{15.4}$$

where $E_{A,B}$ are the lowest energy levels of the double-minimum potential.

In order to calculate the neutron cross-section for the hydrogen-bond dynamics, we apply the jump-diffusion model discussed in Chap. 6.3. The hydrogen atom jumps from the initial position $\boldsymbol{l}_\alpha$ to the final position $\boldsymbol{l}_\beta = \boldsymbol{l}_\alpha + \boldsymbol{R}$, where $\boldsymbol{R}$ is the jump vector. We introduce these facts into Eq. (5.2)

$$\frac{d^2\sigma}{d\Omega d\omega} \propto \sum_{\alpha,\beta} e^{\imath \boldsymbol{Q}\cdot(\boldsymbol{l}_\alpha - \boldsymbol{l}_\beta)} \int_{-\infty}^{\infty} \langle e^{-\imath \boldsymbol{Q}\cdot \boldsymbol{u}_\alpha(0)} e^{\imath \boldsymbol{Q}\cdot \boldsymbol{u}_\beta(t)} \rangle e^{-\imath\omega t} dt \tag{15.5}$$

by setting $\boldsymbol{l}_\alpha = \boldsymbol{l}_\beta$ (i.e., incoherent scattering, see Chap. 2.4), $\boldsymbol{u}_\alpha(0) = 0$, and $\boldsymbol{u}_\beta(t) = \boldsymbol{R}$ (note that we lose the Debye-Waller factor Eq. (5.12) by identifying the displacement vectors $\boldsymbol{u}$ by the jump vector $\boldsymbol{R}$). The expectation value of Eq. (15.5) is then given by

$$\sum_{\alpha,\beta} \langle e^{-\imath \boldsymbol{Q}\cdot \boldsymbol{u}_\alpha(0)} e^{\imath \boldsymbol{Q}\cdot \boldsymbol{u}_\beta(t)} \rangle =$$
$$= p_A(\infty)\left(P(A,A,t) + P(A,B,t) e^{\imath \boldsymbol{Q}\cdot\boldsymbol{R}}\right) +$$
$$+ p_B(\infty)\left(P(B,A,t) e^{-\imath \boldsymbol{Q}\cdot\boldsymbol{R}} + P(B,B,t)\right), \tag{15.6}$$

where $P(\alpha, \beta, t)$ denotes the probability for the hydrogen atom being in the state β at the time t, if it was in the state α at the time $t = 0$. The probabilities $P(A, B, t)$ are calculated from Eq. (15.3) by setting $p_A(0) = 1$

and $p_B(0) = 0$ for $\alpha = A$ as well as $p_A(0) = 0$ and $p_B(0) = 1$ for $\alpha = B$:

$$\begin{aligned}
p(A,A,t) = p_A(t) &= (1 - p_A(\infty))\, e^{-(r_A+r_B)t} + p_A(\infty) \\
&= p_B(\infty) e^{-(r_A+r_B)t} + p_A(\infty), \\
p(A,B,t) = p_B(t) &= (0 - p_B(\infty))\, e^{-(r_A+r_B)t} + p_B(\infty) \\
&= -p_B(\infty) e^{-(r_A+r_B)t} + p_B(\infty), \\
p(B,A,t) = p_A(t) &= (0 - p_A(\infty))\, e^{-(r_A+r_B)t} + p_A(\infty) \\
&= -p_A(\infty) e^{-(r_A+r_B)t} + p_A(\infty), \\
p(B,B,t) = p_B(t) &= (1 - p_B(\infty))\, e^{-(r_A+r_B)t} + p_B(\infty) \\
&= p_A(\infty) e^{-(r_A+r_B)t} + p_B(\infty).
\end{aligned} \tag{15.7}$$

Combining Eq. (15.6) and Eq. (15.7) yields

$$\begin{aligned}
\sum_{\alpha,\beta} \langle e^{-\imath \boldsymbol{Q}\cdot \boldsymbol{u}_\alpha(0)} e^{\imath \boldsymbol{Q}\cdot \boldsymbol{u}_\beta(t)} \rangle &= 1 - 4\,p_A(\infty) p_B(\infty) \sin^2\left(\frac{\boldsymbol{Q}\cdot\boldsymbol{R}}{2}\right) \\
&\quad + 4\,p_A(\infty) p_B(\infty) \sin^2\left(\frac{\boldsymbol{Q}\cdot\boldsymbol{R}}{2}\right) e^{-(r_A+r_B)t}.
\end{aligned} \tag{15.8}$$

When introducing the expectation value Eq. (15.8) back into Eq. (15.5) and performing the integration, the time-independent term yields a δ-function, whereas the time-dependent term yields a Lorentzian. The final neutron cross-section turns out to be

$$\begin{aligned}
\frac{\mathrm{d}^2\sigma}{\mathrm{d}\Omega \mathrm{d}\omega} &= N \frac{k'}{k} \left(\langle b_H^2\rangle - \langle b_H\rangle^2\right) e^{-2W(Q)} \left(S_0(Q)\delta(\omega) + S_1(Q) \frac{1}{\pi} \cdot \frac{\tau}{1+(\omega\tau)^2} \right), \\
S_0(Q) &= 1 - 4\,p_A(\infty) p_B(\infty) \sin^2\left(\frac{\boldsymbol{Q}\cdot\boldsymbol{R}}{2}\right), \\
S_1(Q) &= 4\,p_A(\infty) p_B(\infty) \sin^2\left(\frac{\boldsymbol{Q}\cdot\boldsymbol{R}}{2}\right),
\end{aligned} \tag{15.9}$$

where $\tau = 1/(r_A + r_B)$ corresponds to the average relaxation time of the configurations A and B. For experiments with polycrystalline materials, Eq. (15.9) has to be averaged in $\boldsymbol{Q}$ space:

$$\begin{aligned}
S_0(Q) &= 1 - 2p_A(\infty) p_B(\infty) \left(1 - \frac{\sin(QR)}{QR}\right), \\
S_1(Q) &= 2p_A(\infty) p_B(\infty) \left(1 - \frac{\sin(QR)}{QR}\right).
\end{aligned} \tag{15.10}$$

Both contributions $S_0(Q)$ and $S_1(Q)$ are centered at zero-energy transfer. $S_0(Q)$ corresponds to pure elastic scattering, whereas $S_1(Q)$ corresponds to quasielastic scattering with a finite energy width.

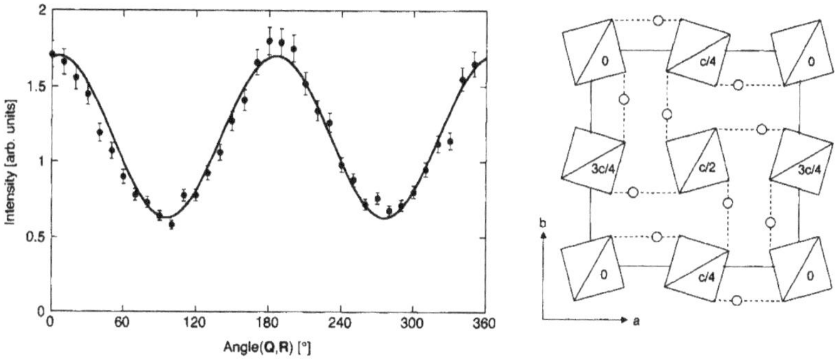

Fig. 15.2 Left: Incoherent elastic peak intensity observed for a single crystal of KH_2PO_4 as a function of the angle between $\boldsymbol{Q}$ and $\boldsymbol{R}$. The line is a fit to $S_0(\boldsymbol{Q})$ defined in Eq. (15.9). Right: Projection of the unit cell of KH_2PO_4 onto the (a, b)-plane. There are four PO_4 tetrahedra ($\square$) and eight H atoms ($\circ$) in the units cell. The heights of the centers of the PO_4 tetrahedra are also denoted.

In a first example we consider the ferroelectric compound KH_2PO_4 in which the H atoms form a bond between the corners of the PO_4 tetrahedra as visualized in Fig. 15.2. Each hydrogen atom can jump between two sites corresponding to O−H···O and O···H−O in the paraelectric phase, but is ordered in the ferroelectric phase. Neutron scattering experiments were performed on a single crystal above the Curie temperature to probe the elastic component $S_0(Q)$ of the cross section Eq. (15.9). The measurements were carried out by scanning the angle between the scattering vector $\boldsymbol{Q}$ and the average jump vector $\boldsymbol{R}$ from 0° to 360°. The observed modulation of the intensities is in excellent agreement with the model prediction as shown in Fig. 15.2.

Many carboxylic acids of chemical composition HOOC−R−COOH are known to form dimeric hydrogen-bonded systems as shown in Fig. 15.1 (e.g., R=C_2 and R=C_6H_4 for dicarboxylic acid and terephthalic acid, respectively). There had been some controversy in the literature about the mechanism of proton dynamics in these systems. One view was that the protons jump directly along the hydrogen bond direction; alternatively, it was suggested that the proton exchange occurs through a torsional motion of the COOH groups implying a 180° rotation. This question was clarified by neutron scattering experiments, out of which we present here the results obtained for terephthalic acid. The polycrystalline sample was ring-deuterated (C_6D_4) in order to avoid significant contributions from hydrogen atoms not belonging to the H bond. The measurements were carried out

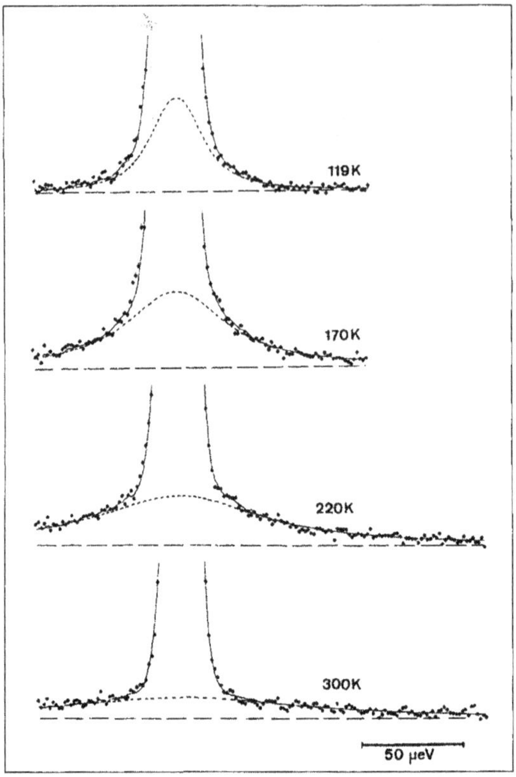

Fig. 15.3 Energy spectra of neutrons scattered from ring-deuterated terephthalic acid. The full line is a least-squares fit as explained in the text. The dotted and dashed lines denote the background and the quasielastic contribution, respectively (after [Meier (1984)]).

with use of a backscattering spectrometer (see Chap. 3.3.8) with an energy resolution of about 10 μeV, which allowed an unambiguous separation of the two components $S_0(Q)$ and $S_1(Q)$ as shown in Fig. 15.3. The signal centered at $\omega = 0$ is well described by the superposition of a δ-function and a Lorentzian, both being folded with the Gaussian instrumental resolution function. The Q-dependence of the intensity of the Lorentzian observed at $T = 170$ K is shown in Fig. 15.4a and compared with the two jump models mentioned above. The direct mechanism involves a jump distance $R = 0.70$ Å, whereas the jump distance is much longer for the torsional mechanism with $R = 2.23$ Å. The results of least-squares fits based on $S_1(Q)$ defined by Eq. (15.10) clearly support the direct jump of the H

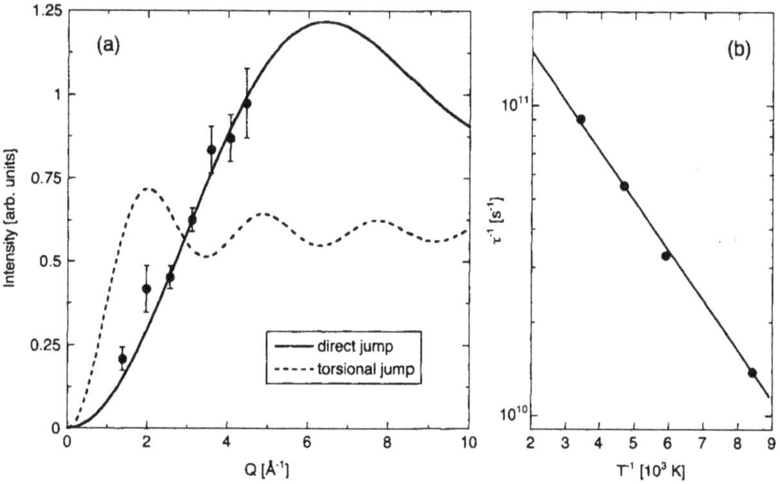

Fig. 15.4 (a) Q-dependence of the quasielastic neutron scattering intensity observed for ring-deuterated terephthalic acid. The full and broken lines denote least-squares fits to the direct and torsional H jump mechanisms, respectively (after [Meier *et al.* (1984)]). (b) Logarithmic Arrhenius plot of the inverse relaxation rate τ vs the inverse temperature (after [Meier (1984)]).

atoms along the hydrogen bond.

As can be seen from Fig. 15.3, the width of the Lorentzian increases considerably with increasing temperature. According to Eq. (15.9), the linewidth parameter of the Lorentzian is the relaxation time τ, whose temperature dependence may be described by an Arrhenius law

$$\frac{1}{\tau} = \frac{1}{\tau_0} e^{-\frac{E_a}{k_B T}}, \tag{15.11}$$

where E_a is the activation energy as indicated in Fig. 15.1. The inverse relaxation time is logarithmically plotted vs the inverse temperature in Fig. 15.4b. A least-squares fit to Eq. (15.11) yields $1/\tau_0 = 1.5 \cdot 10^{11}$ s^{-1} and $E_a = 50$ meV.

15.3 Hydrogen tunnelling

Tunnelling is a quantum-mechanical phenomenon in which a particle violates the principles of classical mechanics by penetrating or passing through a potential barrier higher than the kinetic energy of the particle. This process is known since 1928, when Gamov solved the theory of the α-decay of

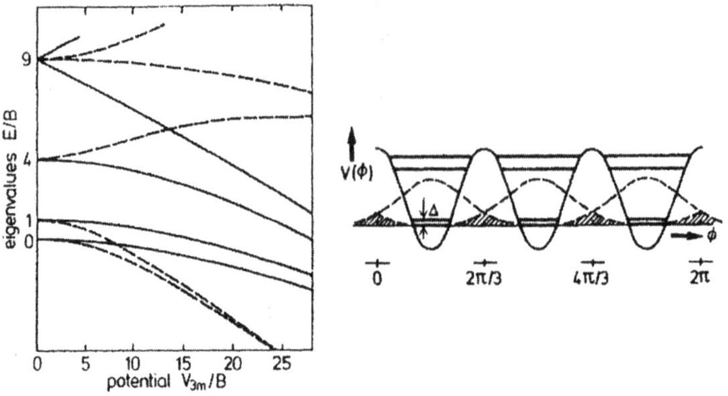

Fig. 15.5 (a) Eigenvalues E of a CH_3 group in a potential $V(\phi)$ with $n = 3$. Both E and V are normalized to $B = \hbar^2/2I$. (b) Schematic sketch representing the tunnel splitting of the rotational states in a potential $V(\phi)$ with $n = 3$. Reprinted with permission from [Asmussen and Press (1994)]. Copyright 1994, World Scientific.

a nucleus via tunnelling. Born realized shortly afterwards that tunnelling is not restricted to nuclear physics, but is a general phenomenon that applies to many different systems. Here we will discuss tunnelling for the case of quantum rotations of molecules, whose detailed understanding has greatly benefitted from neutron scattering experiments.

The discussion of quantum rotations is easiest with just one rotational degree of freedom. The potential created by the environment of the molecule is given by

$$V(\phi) = \frac{V_n}{2}\left(1 - \cos(n\phi)\right), \tag{15.12}$$

where the index n denotes the number of barriers for a full rotation. The Schrödinger equation has the form

$$\frac{\hbar^2}{2I}\nabla^2\psi + \left(E - V(\phi)\right)\psi = 0, \tag{15.13}$$

where I is the moment of inertia (e.g., $I = 5.31 \cdot 10^{-47}$ kg m^2 for CH_3). The simplest case is a dumbbell type of molecule experiencing a potential $V(\phi)$ with $n = 2$, but so far no example for this case has been found in nature. The simplest examples known are rotors with threefold symmetry ($n = 3$) confined to rotations around the σ-bond or the molecular dipole moment. Typical examples are CH_3 and NH_3 groups.

Solutions of the Schrödinger equation (15.13) can be obtained by the related Mathieu equation as exemplified in Fig. 15.5a for a potential $V(\phi)$

with $n = 3$. The free rotor limit of the gas phase is obtained for $V(\phi) = 0$. With increasing strength of the potential, the eigenstates with energy E may be considered as torsional oscillators within the potential wells as shown in Fig. 15.5b. If the wave functions of a single proton in neighboring potential wells overlap, a tunnel splitting $\hbar\omega_t$ is caused as illustrated in Fig. 15.5b. For a more detailed treatment of the tunnel splitting we refer to the exercise No. 15.1.

The tunnel splitting is extremely sensitive to weak changes of the potential $V(\phi)$. For tetrahedral molecules the asymptotic behavior at large potentials $V > 30$ meV can be described by an exponential decrease [Voll and Hüller (1988)]

$$\hbar\omega_t = \beta e^{-\alpha\sqrt{\frac{V(\phi)}{B}}}, \tag{15.14}$$

where $B = \hbar^2/2I$. The values of α (normally ≈ 1) and β (normally $\approx B$) are characteristic of the type of rotor and the shape of the potential. Since the barrier is narrower for an excited torsional state, the tunnel splitting is usually much larger than that in the ground state. Equation (15.14) clearly shows that there is a strong isotope effect for tunnel splittings. While in the harmonic approximation the energy of librational modes for CH_3 groups is reduced by $\sqrt{2}$ on deuteration, the tunnel splittings may decrease by an order of magnitude.

The scattering law $S(\boldsymbol{Q},\omega)$ for a polycrystalline sample of isolated, hydrogen containing rotors with threefold symmetry (e.g., CH_3) and for equally populated tunnel states is given by [Hüller (1977)]

$$S(Q,\omega) = \frac{1}{3}\left[\left(5 + 4\frac{\sin(Qr)}{Qr}\right)\delta(\hbar\omega) + 2\left(1 - \frac{\sin(Qr)}{Qr}\right)\delta(\hbar\omega \pm \hbar\omega_t)\right],$$

where r is the proton-proton distance. Consequently the neutron scattering spectrum consists of an elastic line and a Stokes and anti-Stokes pair of inelastic peaks at energy transfers corresponding to the tunnel splitting $\hbar\omega_t$. The ratio of the elastic line to the inelastic lines is determined by the scattering vector Q, thus in cases where there is more than one type of rotor it may be possible to establish the relative number of rotors from the observed intensities.

Figure 15.6 shows tunnelling spectra of methyl groups in acetamide CH_3CONH_2, recorded on backscattering spectrometers (see Chap. 3.3.8). For the protonated compound (Fig. 15.6a), the tunnelling transitions are clearly revealed as peaks in the observed energy spectrum at ± 32.0 μeV.

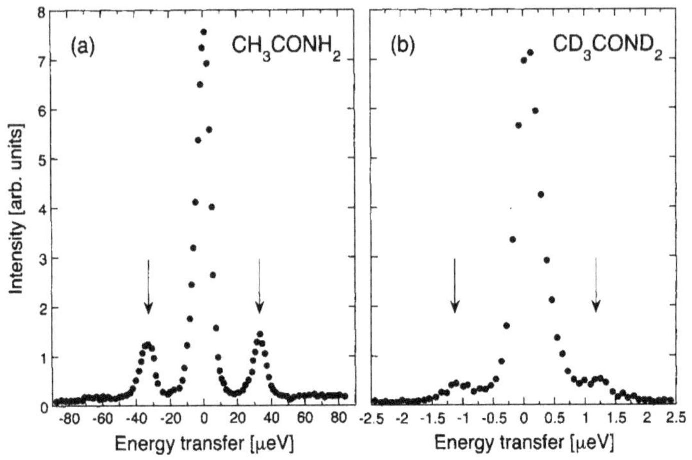

Fig. 15.6 Energy spectra of neutrons scattered from both protonated (a) and deuterated (b) acetamide at $T = 4$ K. The instrumental energy resolutions for (a) and (b) are 5.0 and 0.5 μeV, respectively (after [Heidemann *et al.* (1989b)]).

The tunnel splittings are drastically reduced to ± 1.18 μeV for the deuterated compound as visualized in Fig. 15.6b.

So far we have only considered isolated rotors and described them in terms of a single-particle potential (Eq. (15.12)). In reality, this concept is disputable, when neighboring molecules have orientational degrees of freedom, so that coupling effects have to be taken into account. The appropriate Hamiltonian for CH_3 dimers to be used in the Schrödinger equation Eq. (15.13) is then described by

$$\begin{aligned} H &= H_1 + H_2 + H_{12}, \\ H_i &= -\frac{\hbar^2}{2I}\nabla^2 + \frac{V_3}{2}\left(1 - \cos(3\phi_i)\right), \quad i = 1, 2, \\ H_{12} &= \frac{W}{2}\left(1 - \cos(3(\phi_1 - \phi_2))\right). \end{aligned} \tag{15.15}$$

The effect of coupling is to remove certain degeneracies of the uncoupled rotor states, so that the tunnelling states split up into several components. This was nicely observed for lithium acetate dihydrate $CH_3COOLi{\cdot}2D_2O$ as shown in Fig. 15.7. Eq. (15.15) can of course be extended to three and even more interacting molecules [Häusler (1992)].

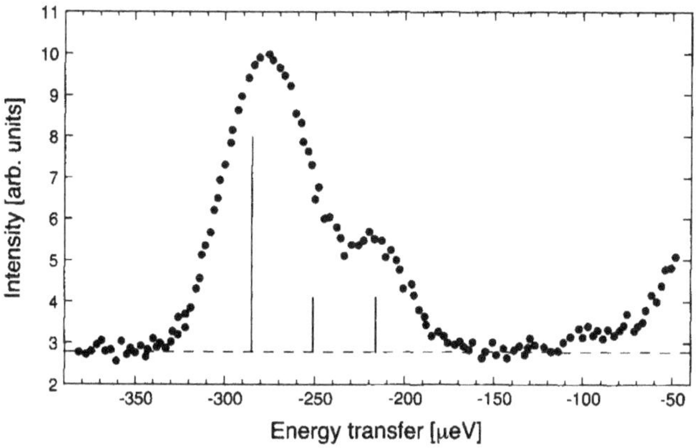

Fig. 15.7 Energy spectrum of neutrons scattered from $CH_3COOLi{\cdot}2D_2O$ at $T = 4$ K and $Q = 1.7$ Å^{-1}. The vertical bars mark the three components of the tunneling transitions (after [Heidemann *et al.* (1989a)]).

15.4 Further reading

- B. Asmussen and W. Press, in *Neutron scattering from hydrogen in materials*, ed. by A. Furrer (World Scientific, Singapore, 1994), p. 184: *Rotational dynamics of molecular groups*
- C. Carlile and M. Prager, Int. J. Mod. Phys. B 7, 3113 (1993): *Rotational tunnelling spectroscopy with neutrons*
- A. J. Horsewill, Spectrochimica Acta 48 A, 379 (1992): *Rotational tunnelling in organic molecules*
- A. Hüller, Phys. Rev. B 16, 1844 (1977): *Rotational tunnelling in solids: The theory of neutron scattering*
- P. C. H. Mitchell, S. F. Parker, A. J. Ramirez-Cuesta and J. Tomkinson, in *Vibrational spectroscopy with neutrons* (World Scientific, Singapore, 2005), p. 393: *Hydrogen bonding*
- W. Press, *Single-particle rotations in molecular crystals* (Springer, Berlin, 1981)
- T. Springer and D. Richter, in *Methods of experimental physics*, Vol. 23, Part B, ed. by D. L. Price and K. Sköld (Academic Press, London, 1987), p. 131: *Hydrogen in metals.*
- J. Tomkinson, in *Neutron scattering from hydrogen in materials*, ed. by A. Furrer (World Scientific, Singapore, 1994), p. 168: *The inelastic neutron scattering spectroscopy of hydrogen bonds*

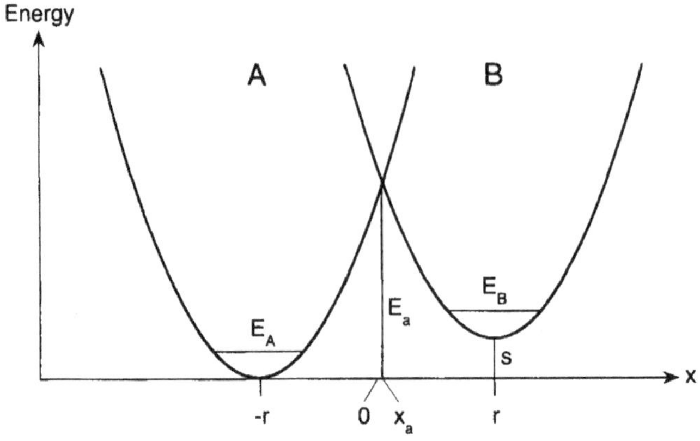

Fig. 15.8 Schematic sketch of an idealized double-minimum potential.

15.5 Exercises

Exercise No. 15.1

Figure 15.8 shows an idealized double-minimum potential described by two parabola

$$V_A = \frac{m\omega^2}{2}(x+r)^2 \quad \text{and} \quad V_B = s + \frac{m\omega^2}{2}(x-r)^2, \tag{15.16}$$

where m is the mass of the atom oscillating with frequency ω.

(a) Calculate the activation energy E_a.

(b) If we consider the two potentials separately, we have the situation of two harmonic oscillators, with ground-state energy $\hbar\omega$ for the potential A. Determine the amplitude σ of the zero-point motion of the harmonic oscillator.

(c) Determine the wavefunction of the harmonic oscillator in the potential A by solving the Schrödinger equation

$$\frac{\hbar^2}{2m}\frac{\partial^2\psi}{\partial x^2} + (E - V(x))\,\psi = 0 \tag{15.17}$$

with the ansatz

$$\psi(x) = a \cdot e^{-bx^2/2}. \tag{15.18}$$

Since the shape of the two potentials is identical, the wavefunction of the harmonic oscillator in the potential B will be same as that in the potential A.

(d) We now express the wavefunctions for the double-minimum potential by a linear combination of the wavefunctions found above:

$$\begin{aligned} |A\rangle &= \alpha|\psi_A\rangle + \beta|\psi_B\rangle \\ |B\rangle &= -\beta|\psi_A\rangle + \alpha|\psi_B\rangle \end{aligned} \tag{15.19}$$

where $\alpha > 0$, $\beta > 0$, and $\alpha^2 + \beta^2 = 1$. Determine the ground-state energies E_A and E_B as well as the components α and β from the Hamiltonian

$$\hat{\mathcal{H}} = -\frac{\hbar^2}{2m}\frac{\partial^2}{\partial x^2} + V(x). \tag{15.20}$$

(e) Discuss the results obtained for E_A and E_B and derive a criterion for the occurrence of tunnelling.

15.6 Solutions

Exercise No. 15.1

(a) The two parabola defined by Eq. (15.16) exhibit a crossing at

$$x_a = \frac{s}{2mr\omega^2}.$$

Inserting x_a into Eq. (15.16) yields the activation energy

$$E_a = \frac{m\omega^2}{2}\left(r + \frac{s}{2mr\omega^2}\right).$$

(b) Inserting σ into Eq. (15.16) yields

$$\frac{m\omega^2}{2}\sigma^2 = \frac{1}{2}\hbar\omega \quad \rightarrow \quad \sigma = \sqrt{\frac{\hbar}{m\omega}}.$$

(c) The solution of the Schrödinger equation Eq. (15.17) yields

$$b = \frac{m\omega}{\hbar}, \qquad a = \left(\frac{m\omega}{\pi\hbar}\right)^{1/4},$$

the latter resulting from the normalization condition $\int |\psi(x)|^2 \mathrm{d}x = 1$.

(d) The ground-state energies E_A and E_B are the solutions of the secular determinant

$$\begin{vmatrix} H_{AA} - E & H_{AB} \\ H_{BA} & H_{BB} - E \end{vmatrix} = 0,$$

with

$$H_{ij} = \int_{-\infty}^{x_a} \psi_i H \psi_j \mathrm{d}x + \int_{x_a}^{\infty} \psi_i H \psi_j \mathrm{d}x, \quad i, j = A, B.$$

ψ_i and H are defined by Eqs (15.19) and (15.20), respectively. After some lengthy calculations, the following result is obtained:

$$E_{A,B} = \frac{s}{2} \pm \sqrt{\left(\frac{s}{2}\right)^2 + t^2}, \quad t = \frac{2}{\sqrt{\pi}} u \cdot e^{-u^2} \hbar\omega, \quad u = \frac{2r}{\sigma},$$

where σ corresponds to the result obtained in (b). The components of the wavefunctions (Eq. (15.19)) turn out to be

$$\alpha = \sqrt{\frac{\left(\frac{s}{2}\right)^2 + t^2}{\left(\frac{s}{2}\right)^2 + 2t^2}}, \quad \beta = \sqrt{\frac{t^2}{\left(\frac{s}{2}\right)^2 + 2t^2}}.$$

(e) A nearly symmetric double-minimum potential is realized for $s \ll t$, thus

$$E_A - E_B \approx 2t \quad \text{and} \quad \alpha \approx \beta \approx \frac{1}{\sqrt{2}},$$

In this case, the wavefunctions $|A\rangle$ and $|B\rangle$ are delocalized and tunnelling occurs with the tunnelling energy $2t$. On the other hand, for an asymmetric double-minimum potential the condition $s \gg t$ holds, thus

$$E_A - E_B \approx s \quad \text{and} \quad \alpha \approx 1, \quad \beta \approx 0,$$

i.e., there is no overlap of the localized wavefunctions $|A\rangle$ and $|B\rangle$ and tunnelling is not possible.

Appendix A

Dirac δ-function and Lattice Sums

The Dirac δ-function

Definition:

$$\delta(x) = \begin{cases} 0, & \text{for } x \neq 0 \\ \infty, & \text{for } x = 0 \end{cases} \tag{A.1}$$

$$\int_{-\infty}^{\infty} \delta(x)\mathrm{d}x = 1$$

Consequently:

$$\int_{-\infty}^{\infty} f(x)\delta(x-a)\mathrm{d}x = f(a)$$

$$\delta(c \cdot x) = \tfrac{1}{c} \cdot \delta(x), \quad \text{for} \quad c > 0 \tag{A.2}$$

$$\delta(-x) = \delta(x)$$

$\delta(x)$ is not a proper mathematical function, however, it can be represented as the limit of a mathematical function in terms of an infinite integral. Consider the function

$$f(x) = \int_{-k_0}^{k_0} e^{ikx}\mathrm{d}k = \frac{1}{ix}\left(e^{ik_0x} - e^{-ik_0x}\right) = \frac{2}{x}\sin(k_0x), \tag{A.3}$$

which has a peak for $x = 0$. Upon increasing the parameter k_0 the height of the peak increases and its width decreases. For $k_0 \to \infty$ Eq. (A.3) has the shape of a δ-function. Since the integral

$$\int_{-\infty}^{\infty} f(x)\mathrm{d}x = 2\int_{-\infty}^{\infty} \frac{1}{x}\sin(k_0x)\mathrm{d}x = 2\pi \tag{A.4}$$

is independent of the value of k_0, we arrive at the following integral representation:

$$\delta(x) = \frac{1}{2\pi}\int_{-\infty}^{\infty} e^{ikx}\mathrm{d}k. \tag{A.5}$$

The following application is very useful for calculations of neutron cross-sections:

$$\delta(\hbar\omega) = \frac{1}{\hbar}\delta(\omega) = \frac{1}{2\pi\hbar}\int_{-\infty}^{\infty} e^{i\omega t}\mathrm{d}t. \tag{A.6}$$

In three dimensions Eqs (A.1) and (A.2) are rewritten as

$$\delta(\boldsymbol{r}) = \begin{cases} 0 & \text{for } |\boldsymbol{r}| \neq 0 \\ \infty & \text{for } |\boldsymbol{r}| = 0 \end{cases}$$

$$\int_{-\infty}^{\infty} \delta(\boldsymbol{r})\mathrm{d}\boldsymbol{r} = 1 \tag{A.7}$$

$$\int f(\boldsymbol{r})\delta(\boldsymbol{r}-\boldsymbol{r}_0)\mathrm{d}\boldsymbol{r} = f(\boldsymbol{r}_0) \tag{A.8}$$

$$\delta(\boldsymbol{r}) = \frac{1}{(2\pi)^3}\int e^{i\boldsymbol{Q}\cdot\boldsymbol{r}}\mathrm{d}\boldsymbol{Q} \tag{A.9}$$

Lattice sums and lattice integrals

Let us define the lattice vectors in real space and reciprocal space by $\boldsymbol{l} = l_1\boldsymbol{a}_1 + l_2\boldsymbol{a}_2 + l_3\boldsymbol{a}_3$ and $\boldsymbol{\tau} = t_1\boldsymbol{\tau}_1 + t_2\boldsymbol{\tau}_2 + t_3\boldsymbol{\tau}_3$, respectively. Then the following relations hold:

$$\sum_{\boldsymbol{l}} e^{i\boldsymbol{Q}\cdot\boldsymbol{l}} = \frac{(2\pi)^3}{v_0}\sum_{\boldsymbol{\tau}}\delta(\boldsymbol{Q}-\boldsymbol{\tau}) \tag{A.10}$$

$$\sum_{\boldsymbol{l}} e^{i(\boldsymbol{Q}-\boldsymbol{Q}')\cdot\boldsymbol{l}} = N\cdot\delta_{\boldsymbol{Q}\boldsymbol{Q}'} \tag{A.11}$$

$$\int e^{i(\boldsymbol{\tau}-\boldsymbol{\tau}')\cdot\boldsymbol{r}}\mathrm{d}\boldsymbol{r} = v_0\delta_{\boldsymbol{\tau}\boldsymbol{\tau}'} \tag{A.12}$$

$$\int e^{i\boldsymbol{Q}\cdot(\boldsymbol{l}-\boldsymbol{l}')}\mathrm{d}\boldsymbol{Q} = \frac{(2\pi)^3}{v_0}\delta_{\boldsymbol{l}\boldsymbol{l}'} \tag{A.13}$$

where the integration is carried out over the unit cell of volume v_0 in real space and where N denotes the number of unit cells of the system.

Appendix B

Neutron Scattering Lengths and Cross Sections

Table B.1: Neutron scattering lengths and cross sections (in barn) of selected elements and corresponding isotopes.

Isotope	conc (%)	b_{coh} (fm)	b_{inc} (fm)	σ_{coh}	σ_{inc}	σ_{tot}	σ_{abs}
H	—	-3.7390	—	1.7568	80.26	82.02	0.3326
^{1}H	99.985	-3.7406	25.274	1.7583	80.27	82.03	0.3326
^{2}H	0.015	6.671	4.04	5.592	2.05	7.64	0.000519
He	—	3.26(3)	—	1.34	0	1.34	0.00747
^{3}He	0.00014	5.74-1.483i	-2.5+2.568i	4.42	1.6	6	5333.(7.)
^{4}He	99.99986	3.26	0	1.34	0	1.34	0
Li	—	-1.90	—	0.454	0.92	1.37	70.5
^{6}Li	7.5	2.00-0.261i	-1.89+0.26i	0.51	0.46	0.97	940.(4.)
^{7}Li	92.5	-2.22	-2.49	0.619	0.78	1.4	0.0454
Be	100	7.79	0.12	7.63	0.0018	7.63	0.0076
B	—	5.30-0.213i	—	3.54	1.7	5.24	767.(8.)
^{10}B	20	-0.1-1.066i	-4.7+1.231i	0.144	3	3.1	3835.(9.)
^{11}B	80	6.65	-1.3	5.56	0.21	5.77	0.0055
C	—	6.6460	—	5.551	0.001	5.551	0.0035
^{12}C	98.9	6.6511	0	5.559	0	5.559	0.00353
^{13}C	1.1	6.19	-0.52	4.81	0.034	4.84	0.00137
N	—	9.36	—	11.01	0.5	11.51	1.9
^{14}N	99.63	9.37	2.0	11.03	0.5	11.53	1.91
^{15}N	0.37	6.44	-0.02	5.21	0.00005	5.21	0.000024
O	—	5.803	—	4.232	0.0008	4.232	0.00019
^{16}O	99.762	5.803	0	4.232	0	4.232	0.0001
^{17}O	0.038	5.78	0.18	4.2	0.004	4.2	0.236
^{18}O	0.2	5.84	0	4.29	0	4.29	0.00016
F	100	5.654	-0.082	4.017	0.0008	4.018	0.0096
Na	100	3.63	3.59	1.66	1.62	3.28	0.53
Mg	—	5.375	—	3.631	0.08	3.71	0.063
^{24}Mg	78.99	5.66	0	4.03	0	4.03	0.05
^{25}Mg	10	3.62	1.48	1.65	0.28	1.93	0.19
^{26}Mg	11.01	4.89	0	3	0	3	0.0382
Al	100	3.449	0.256	1.495	0.0082	1.503	0.231
Si	—	4.1491	—	2.163	0.004	2.167	0.171
^{28}Si	92.23	4.107	0	2.12	0	2.12	0.177
^{29}Si	4.67	4.70	0.09	2.78	0.001	2.78	0.101
^{30}Si	3.1	4.58	0	2.64	0	2.64	0.107
P	100	5.13	0.2	3.307	0.005	3.312	0.172

(note: 1 fm = 10^{-15} m, 1 barn = 10^{-24} cm^{2})

Table B.1: (continued)

Isotope	conc (%)	b_{coh} (fm)	b_{inc} (fm)	σ_{coh}	σ_{inc}	σ_{tot}	σ_{abs}
S	—	2.847	—	1.0186	0.007	1.026	0.53
^{32}S	95.02	2.804	0	0.988	0	0.988	0.54
^{33}S	0.75	4.74	1.5	2.8	0.3	3.1	0.54
^{34}S	4.21	3.48	0	1.52	0	1.52	0.227
^{36}S	0.02	3.(1.)	0	1.1	0	1.1	0.15
Cl	—	9.5770	—	11.5257	5.3	16.8	33.5
^{35}Cl	75.77	11.65	6.1	17.06	4.7	21.8	44.1
^{37}Cl	24.23	3.08	0.1	1.19	0.001	1.19	0.433
Ar	—	1.909	—	0.458	0.225	0.683	0.675
^{36}Ar	0.337	24.90	0	77.9	0	77.9	5.2
^{38}Ar	0.063	3.5	0	1.5(3.1)	0	1.5(3.1)	0.8
^{40}Ar	99.6	1.830	0	0.421	0	0.421	0.66
K	—	3.67	—	1.69	0.27	1.96	2.1
^{39}K	93.258	3.74	1.4	1.76	0.25	2.01	2.1
^{40}K	0.012	3.(1.)	—	1.1	0.5	1.6	35.(8.)
^{41}K	6.73	2.69	1.5	0.91	0.3	1.2	1.46
Ca	—	4.70	—	2.78	0.05	2.83	0.43
^{40}Ca	96.941	4.80	0	2.9	0	2.9	0.41
^{42}Ca	0.647	3.36	0	1.42	0	1.42	0.68
^{43}Ca	0.135	-1.56	—	0.31	0.5	0.8	6.2
^{44}Ca	2.086	1.42	0	0.25	0	0.25	0.88
^{46}Ca	0.004	3.6	0	1.6	0	1.6	0.74
^{48}Ca	0.187	0.39	0	0.019	0	0.019	1.09
Ti	—	-3.438	—	1.485	2.87	4.35	6.09
^{46}Ti	8.2	4.93	0	3.05	0	3.05	0.59
^{47}Ti	7.4	3.63	-3.5	1.66	1.5	3.2	1.7
^{48}Ti	73.8	-6.08	0	4.65	0	4.65	7.84
^{49}Ti	5.4	1.04	5.1	0.14	3.3	3.4	2.2
^{50}Ti	5.2	6.18	0	4.8	0	4.8	0.179
V	—	-0.3824	—	0.0184	5.08	5.1	5.08
^{50}V	0.25	7.6	—	7.3(1.1)	0.5	7.8(1.0)	60.(40.)
^{51}V	99.75	-0.402	6.35	0.0203	5.07	5.09	4.9
Cr	—	3.635	—	1.66	1.83	3.49	3.05
^{50}Cr	4.35	-4.50	0	2.54	0	2.54	15.8
^{52}Cr	83.79	4.920	0	3.042	0	3.042	0.76
^{53}Cr	9.5	-4.20	6.87	2.22	5.93	8.15	18.1(1.5)
^{54}Cr	2.36	4.55	0	2.6	0	2.6	0.36
Mn	100	-3.73	1.79	1.75	0.4	2.15	13.3
Fe	—	9.45	—	11.22	0.4	11.62	2.56
^{54}Fe	5.8	4.2	0	2.2	0	2.2	2.25
^{56}Fe	91.7	9.94	0	12.42	0	12.42	2.59
^{57}Fe	2.2	2.3	—	0.66	0.3	1	2.48
^{58}Fe	0.3	15.(7.)	0	28	0	28.(26.)	1.28
Co	100	2.49	-6.2	0.779	4.8	5.6	37.18
Ni	—	10.3	—	13.3	5.2	18.5	4.49
^{58}Ni	68.27	14.4	0	26.1	0	26.1	4.6
^{60}Ni	26.1	2.8	0	0.99	0	0.99	2.9
^{61}Ni	1.13	7.60	±3.9	7.26	1.9	9.2	2.5
^{62}Ni	3.59	-8.7	0	9.5	0	9.5	14.5
^{64}Ni	0.91	-0.37	0	0.017	0	0.017	1.52
Cu	—	7.718	—	7.485	0.55	8.03	3.78
^{63}Cu	69.17	6.43	0.22	5.2	0.006	5.2	4.5
^{65}Cu	30.83	10.61	1.79	14.1	0.4	14.5	2.17
Zn	—	5.680	—	4.054	0.077	4.131	1.11
^{64}Zn	48.6	5.22	0	3.42	0	3.42	0.93

(note: 1 fm = 10^{-15} m, 1 barn = 10^{-24} cm^2)

Table B.1: (continued)

Isotope	conc (%)	b_{coh} (fm)	b_{inc} (fm)	σ_{coh}	σ_{inc}	σ_{tot}	σ_{abs}
^{66}Zn	27.9	5.97	0	4.48	0	4.48	0.62
^{67}Zn	4.1	7.56	-1.50	7.18	0.28	7.46	6.8
^{68}Zn	18.8	6.03	0	4.57	0	4.57	1.1
^{70}Zn	0.6	6.(1.)	0	4.5	0	4.5(1.5)	0.092
Ga	—	7.288	—	6.675	0.16	6.83	2.75
^{69}Ga	60.1	7.88	-0.85	7.8	0.091	7.89	2.18
^{71}Ga	39.9	6.40	-0.82	5.15	0.084	5.23	3.61
Ge	—	8.185	—	8.42	0.18	8.6	2.2
^{70}Ge	20.5	10.0	0	12.6	0	12.6	3
^{72}Ge	27.4	8.51	0	9.1	0	9.1	0.8
^{73}Ge	7.8	5.02	3.4	3.17	1.5	4.7	15.1
^{74}Ge	36.5	7.58	0	7.2	0	7.2	0.4
^{76}Ge	7.8	8.2	0	8.(3.)	0	8.(3.)	0.16
As	100	6.58	-0.69	5.44	0.06	5.5	4.5
Se	—	7.970	—	7.98	0.32	8.3	11.7
^{74}Se	0.9	0.8	0	0.1	0	0.1	51.8(1.2)
^{76}Se	9	12.2	0	18.7	0	18.7	85.(7.)
^{77}Se	7.6	8.25	±0.6(1.6)	8.6	0.05	8.65	42.(4.)
^{78}Se	23.5	8.24	0	8.5	0	8.5	0.43
^{80}Se	49.6	7.48	0	7.03	0	7.03	0.61
^{82}Se	9.4	6.34	0	5.05	0	5.05	0.044
Br	—	6.795	—	5.8	0.1	5.9	6.9
^{79}Br	50.69	6.80	-1.1	5.81	0.15	5.96	11
^{81}Br	49.31	6.79	0.6	5.79	0.05	5.84	2.7
Sr	—	7.02	—	6.19	0.06	6.25	1.28
^{84}Sr	0.56	7.(1.)	0	6.(2.)	0	6.(2.)	0.87
^{86}Sr	9.86	5.67	0	4.04	0	4.04	1.04
^{87}Sr	7	7.40	—	6.88	0.5	7.4	16.(3.)
^{88}Sr	82.58	7.15	0	6.42	0	6.42	0.058
Y	100	7.75	1.1	7.55	0.15	7.7	1.28
Zr	—	7.16	—	6.44	0.02	6.46	0.185
^{90}Zr	51.45	6.4	0	5.1	0	5.1	0.011
^{91}Zr	11.32	8.7	-1.08	9.5	0.15	9.7	1.17
^{92}Zr	17.19	7.4	0	6.9	0	6.9	0.22
^{94}Zr	17.28	8.2	0	8.4	0	8.4	0.0499
^{96}Zr	2.76	5.5	0	3.8	0	3.8	0.0229
Nb	100	7.054	-0.139	6.253	0.0024	6.255	1.15
Mo	—	6.715	—	5.67	0.04	5.71	2.48
^{92}Mo	14.84	6.91	0	6	0	6	0.019
^{94}Mo	9.25	6.80	0	5.81	0	5.81	0.015
^{95}Mo	15.92	6.91	—	6	0.5	6.5	13.1
^{96}Mo	16.68	6.20	0	4.83	0	4.83	0.5
^{97}Mo	9.55	7.24	—	6.59	0.5	7.1	2.5
^{98}Mo	24.13	6.58	0	5.44	0	5.44	0.127
^{100}Mo	9.63	6.73	0	5.69	0	5.69	0.4
Pd	—	5.91	—	4.39	0.093	4.48	6.9
^{102}Pd	1.02	7.7(7)	0	7.5(1.4)	0	7.5(1.4)	3.4
^{104}Pd	11.14	7.7(7)	0	7.5(1.4)	0	7.5(1.4)	0.6
^{105}Pd	22.33	5.5	-2.6(1.6)	3.8	0.8	4.6(1.1)	20.(3.)
^{106}Pd	27.33	6.4	0	5.1	0	5.1	0.304
^{108}Pd	26.46	4.1	0	2.1	0	2.1	8.55
^{110}Pd	11.72	7.7(7)	0	7.5(1.4)	0	7.5(1.4)	0.226
Ag	—	5.922	—	4.407	0.58	4.99	63.3
^{107}Ag	51.83	7.555	1.00	7.17	0.13	7.3	37.6(1.2)
^{109}Ag	48.17	4.165	-1.60	2.18	0.32	2.5	91.0(1.0)

(note: 1 fm = 10^{-15} m, 1 barn = 10^{-24} cm^2)

Table B.1: (continued)

Isotope	conc (%)	b_{coh} (fm)	b_{inc} (fm)	σ_{coh}	σ_{inc}	σ_{tot}	σ_{abs}
Cd	—	4.87-0.70i	—	3.04	3.46	6.5	2520.(50.)
^{106}Cd	1.25	5.(2.)	0	3.1	0	3.1(2.5)	1
^{108}Cd	0.89	5.4	0	3.7	0	3.7	1.1
^{110}Cd	12.51	5.9	0	4.4	0	4.4	11
^{111}Cd	12.81	6.5	—	5.3	0.3	5.6	24
^{112}Cd	24.13	6.4	0	5.1	0	5.1	2.2
^{113}Cd	12.22	-8.0-5.73i	—	12.1	0.3	12.4	20600.(400.)
^{114}Cd	28.72	7.5	0	7.1	0	7.1	0.34
^{116}Cd	7.47	6.3	0	5	0	5	0.075
Sn	—	6.225	—	4.871	0.022	4.892	0.626
^{112}Sn	1	6.(1.)	0	4.5(1.5)	0	4.5(1.5)	1
^{114}Sn	0.7	6.2	0	4.8	0	4.8	0.114
^{115}Sn	0.4	6.(1.)	—	4.5(1.5)	0.3	4.8(1.5)	30.(7.)
^{116}Sn	14.7	5.93	0	4.42	0	4.42	0.14
^{117}Sn	7.7	6.48	—	5.28	0.3	5.6	2.3
^{118}Sn	24.3	6.07	0	4.63	0	4.63	0.22
^{119}Sn	8.6	6.12	—	4.71	0.3	5	2.2
^{120}Sn	32.4	6.49	0	5.29	0	5.29	0.14
^{122}Sn	4.6	5.74	0	4.14	0	4.14	0.18
^{124}Sn	5.6	5.97	0	4.48	0	4.48	0.133
Cs	100	5.42	1.29	3.69	0.21	3.9	29.0(1.5)
Ba	—	5.07	—	3.23	0.15	3.38	1.1
^{130}Ba	0.11	-3.6	0	1.6	0	1.6	30.(5.)
^{132}Ba	0.1	7.8	0	7.6	0	7.6	7
^{134}Ba	2.42	5.7	0	4.08	0	4.08	2.0(1.6)
^{135}Ba	6.59	4.67	—	2.74	0.5	3.2	5.8
^{136}Ba	7.85	4.91	0	3.03	0	3.03	0.68
^{137}Ba	11.23	6.83	—	5.86	0.5	6.4	3.6
^{138}Ba	71.7	4.84	0	2.94	0	2.94	0.27
La	—	8.24	—	8.53	1.13	9.66	8.97
^{138}La	0.09	8.(2.)	—	8.(4.)	0.5	8.5(4.0)	57.(6.)
^{139}La	99.91	8.24	3.0	8.53	1.13	9.66	8.93
Ce	—	4.84	—	2.94	0.001	2.94	0.63
^{136}Ce	0.19	5.80	0	4.23	0	4.23	7.3(1.5)
^{138}Ce	0.25	6.70	0	5.64	0	5.64	1.1
^{140}Ce	88.48	4.84	0	2.94	0	2.94	0.57
^{142}Ce	11.08	4.75	0	2.84	0	2.84	0.95
Pr	100	4.58	-0.35	2.64	0.015	2.66	11.5
Nd	—	7.69	—	7.43	9.2	16.6	50.5(1.2)
^{142}Nd	27.16	7.7	0	7.5	0	7.5	18.7
^{143}Nd	12.18	14.(2.)	±21.(1.)	25.(7.)	55.(7.)	80.(2.)	337.(10.)
^{144}Nd	23.8	2.8	0	1	0	1	3.6
^{145}Nd	8.29	14.(2.)	—	25.(7.)	5.(5.)	30.(9.)	42.(2.)
^{146}Nd	17.19	8.7	0	9.5	0	9.5	1.4
^{148}Nd	5.75	5.7	0	4.1	0	4.1	2.5
^{150}Nd	5.63	5.3	0	3.5	0	3.5	1.2
Sm	—	0.80-1.65i	—	0.422	39.(3.)	39.(3.)	5922.(56.)
^{144}Sm	3.1	-3.(4.)	0	1.(3.)	0	1.(3.)	0.7
^{147}Sm	15.1	14.(3.)	±11.(7.)	25.(11.)	143(19.)	39.(16.)	57.(3.)
^{148}Sm	11.3	-3.(4.)	0	1.(3.)	0	1.(3.)	2.4
^{149}Sm	13.9	-19.2-11.7i	±31.4-10.3i	63.5	137.(5.)	200.(5.)	42080.(400.)
^{150}Sm	7.4	14.(3.)	0	25.(11.)	0	25.(11.)	104.(4.)
^{152}Sm	26.6	-5.0	0	3.1	0	3.1	206.(6.)
^{154}Sm	22.6	9.3	0	11.(2.)	0	11.(2.)	8.4
Gd	—	6.5-13.82i	—	29.3	151.(2.)	180.(2.)	49700.(125.)
^{152}Gd	0.2	10.(3.)	0	13.(8.)	0	13.(8.)	735.(20.)

(note: 1 fm = 10^{-15} m, 1 barn = 10^{-24} cm^2)

Table B.1: (continued)

Isotope	conc (%)	b_{coh} (fm)	b_{inc} (fm)	σ_{coh}	σ_{inc}	σ_{tot}	σ_{abs}
^{154}Gd	2.1	10.(3.)	0	13.(8.)	0	13.(8.)	85.(12.)
^{155}Gd	14.8	6.0-17.0i	±5.(5.)-13.16i	40.8	25.(6.)	66.(6.)	61100.(400.)
^{156}Gd	20.6	6.3	0	5	0	5	1.5(1.2)
^{157}Gd	15.7	-1.14-71.9i	±5.(5.)-55.8i	650.(4.)	394.(7.)	1044.(8.)	259000.(700.)
^{158}Gd	24.8	9.(2.)	0	10.(5.)	0	10.(5.)	2.2
^{160}Gd	21.8	9.15	0	10.52	0	10.52	0.77
Tb	100	7.38	-0.17	6.84	0.004	6.84	23.4
Dy	—	16.9-0.276i	—	35.9	54.4(1.2)	90.3	994.(13.)
^{156}Dy	0.06	6.1	0	4.7	0	4.7	33.(3.)
^{158}Dy	0.1	6.(4.)	0	5.(6.)	0	5.(6.)	43.(6.)
^{160}Dy	2.34	6.7	0	5.6	0	5.6	56.(5.)
^{161}Dy	19	10.3	±4.9	13.3	3.(1.)	16.(1.)	600.(25.)
^{162}Dy	25.5	-1.4	0	0.25	0	0.25	194.(10.)
^{163}Dy	24.9	5.0	1.3	3.1	0.21	3.3	124.(7.)
^{164}Dy	28.1	49.4-0.79i	0	307.(3.)	0	307.(3.)	2840.(40.)
Ho	100	8.01	-1.70	8.06	0.36	8.42	64.7(1.2)
Er	—	7.79	—	7.63	1.1	8.7	159.(4.)
^{162}Er	0.14	8.8	0	9.7	0	9.7	19.(2.)
^{164}Er	1.56	8.2	0	8.4	0	8.4	13.(2.)
^{166}Er	33.4	10.6	0	14.1	0	14.1	19.6(1.5)
^{167}Er	22.9	3.0	1.0	1.1	0.13	1.2	659.(16.)
^{168}Er	27.1	7.4	0	6.9	0	6.9	2.74
^{170}Er	14.9	9.6	0	11.6	0	11.6(1.2)	5.8
Tm	100	7.07	0.9	6.28	0.1	6.38	100.(2.)
Ta	—	6.91	—	6	0.01	6.01	20.6
^{180}Ta	0.012	7.(2.)	—	6.2	0.5	7.(4.)	563.(60.)
^{181}Ta	99.988	6.91	-0.29	6	0.011	6.01	20.5
W	—	4.86	—	2.97	1.63	4.6	18.3
^{180}W	0.1	5.(3.)	0	3.(4.)	0	3.(4.)	30.(20.)
^{182}W	26.3	6.97	0	6.1	0	6.1	20.7
^{183}W	14.3	6.53	—	5.36	0.3	5.7	10.1
^{184}W	30.7	7.48	0	7.03	0	7.03	1.7
^{186}W	28.6	-0.72	0	0.065	0	0.065	37.9
Au	100	7.63	-1.84	7.32	0.43	7.75	98.65
Tl	—	8.776	—	9.678	0.21	9.89	3.43
^{203}Tl	29.524	6.99	1.06	6.14	0.14	6.28	11.4
^{205}Tl	70.476	9.52	-0.242	11.39	0.007	11.4	0.104
Pb	—	9.405	—	11.115	0.003	11.118	0.171
^{204}Pb	1.4	9.90	0	12.3	0	12.3	0.65
^{206}Pb	24.1	9.22	0	10.68	0	10.68	0.03
^{207}Pb	22.1	9.28	0.14	10.82	0.002	10.82	0.699
^{208}Pb	52.4	9.50	0	11.34	0	11.34	0.00048
Bi	100	8.532	—	9.148	0.0084	9.156	0.0338
U	—	8.417	—	8.903	0.005	8.908	7.57
^{234}U	0.005	12.4	0	19.3	0	19.3	100.1(1.3)
^{235}U	0.72	10.47	±1.3	13.78	0.2	14	680.9(1.1)
^{238}U	99.275	8.402	0	8.871	0	8.871	2.68

(note: 1 fm = 10^{-15} m, 1 barn = 10^{-24} cm^2)

Further reading

- L. Koester, *Neutron scattering lengths and fundamental neutron interactions*, Springer tracts in modern physics, Vol. 80 (Springer, Berlin, 1977)
- V. F. Sears, in *Methods of experimental physics*, Vol. 23, Part A, ed. by D. L. Price and K. Sköld (Academic Press, London, 1986), p. 521: *Neutron scattering lengths and cross sections*

Appendix C

Pauli Spin Operators

The Pauli spin operators are identical to the $S = 1/2$ case of the general spin operators. In that case one has to consider the following two states $|S, M\rangle$:

$$u = |\frac{1}{2}, \frac{1}{2}\rangle \qquad (\text{"spin up"});$$
$$v = |\frac{1}{2}, -\frac{1}{2}\rangle \qquad (\text{"spin down"}). \tag{C.1}$$

These states are normalized and orthogonal:

$$\langle u|u\rangle = \langle v|v\rangle = 1,$$
$$\langle u|v\rangle = \langle v|u\rangle = 0. \tag{C.2}$$

Application of the usual operators $\hat{S}^{\pm} = \hat{S}^x \pm \imath\hat{S}^y$ and $\hat{S}^z$ on these states yields

$$\begin{array}{lll} \hat{S}^+u = 0 & \hat{S}^-u = v & \hat{S}^zu = u \\ \hat{S}^+v = u & \hat{S}^-v = 0 & \hat{S}^zv = -v \end{array} \tag{C.3}$$

The Pauli operators are defined as follows:

$$\begin{aligned} \hat{\sigma}_x &= 2\hat{S}^x = \hat{S}^+ + \hat{S}^- \\ \hat{\sigma}_y &= 2\hat{S}^y = -\imath(\hat{S}^+ - \hat{S}^-) \\ \hat{\sigma}_z &= \hat{S}^z \end{aligned} \tag{C.4}$$

From Eq. (C.3) and Eq. (C.4) follows:

$$\begin{array}{ll} \hat{\sigma}_xu = v & \hat{\sigma}_xv = u \\ \hat{\sigma}_yu = \imath v & \hat{\sigma}_yv = -\imath u \\ \hat{\sigma}_zu = u & \hat{\sigma}_zv = -v \end{array} \tag{C.5}$$

Appendix D

Cross Section for Magnetic Neutron Scattering

The calculation of the matrix element $\langle \boldsymbol{k}', \boldsymbol{\sigma}', \lambda' | \hat{U} | \boldsymbol{k}, \boldsymbol{\sigma}, \lambda \rangle$ of Eq. (2.4) with the operator $\hat{U}$ defined by Eqs (2.39) - (2.41) can be performed separately for the momentum and spin states $|\boldsymbol{k}\rangle$ and $|\boldsymbol{\sigma}\rangle$, respectively. First we calculate the matrix element $\langle \boldsymbol{k}' | \hat{U} | \boldsymbol{k} \rangle$ by making use of the following identities:[1]

$$\frac{\boldsymbol{R}}{R^3} = -\nabla \left(\frac{1}{R} \right) \tag{D.1}$$

$$\mathrm{curl}(\boldsymbol{v}) \equiv \nabla \wedge \boldsymbol{v} \tag{D.2}$$

$$\frac{1}{R} = \frac{1}{2\pi^2} \int \mathrm{d}\boldsymbol{q} \frac{1}{q^2} e^{\imath \boldsymbol{q} \cdot \boldsymbol{R}} \tag{D.3}$$

$$\nabla e^{\imath \boldsymbol{q} \cdot \boldsymbol{r}} = \imath \boldsymbol{q} e^{\imath \boldsymbol{q} \cdot \boldsymbol{r}} \tag{D.4}$$

$$(\nabla \wedge (\boldsymbol{s} \wedge \nabla))\, e^{\imath \boldsymbol{q} \cdot \boldsymbol{R}} = -\boldsymbol{q} \wedge (\boldsymbol{s} \wedge \boldsymbol{q}) e^{\imath \boldsymbol{q} \cdot \boldsymbol{R}} \tag{D.5}$$

Applying these identities to the first part of Eq. (2.40) yields:

$$\begin{aligned}
\nabla \wedge \frac{\boldsymbol{s} \wedge \boldsymbol{R}}{R^3} &= -\nabla \wedge \left(\boldsymbol{s} \wedge \nabla \left(\frac{1}{\boldsymbol{R}} \right) \right) = -\frac{1}{2\pi^2} \int \mathrm{d}\boldsymbol{q} \frac{1}{q^2} (\nabla \wedge (\boldsymbol{s} \wedge \nabla))\, e^{\imath \boldsymbol{q} \cdot \boldsymbol{r}} \\
&= -\frac{\imath}{2\pi^2} \int \mathrm{d}\boldsymbol{q} \frac{1}{q^2} (\nabla \wedge (\boldsymbol{s} \wedge \boldsymbol{q}))\, e^{\imath \boldsymbol{q} \cdot \boldsymbol{r}} \\
&= \frac{1}{2\pi^2} \int \mathrm{d}\boldsymbol{q} \frac{1}{q^2} (\boldsymbol{q} \wedge (\boldsymbol{s} \wedge \boldsymbol{q}))\, e^{\imath \boldsymbol{q} \cdot \boldsymbol{r}}.
\end{aligned} \tag{D.6}$$

For the second part of Eq. (2.40) we obtain:

$$\begin{aligned}
\frac{\boldsymbol{p} \wedge \boldsymbol{R}}{R^3} &= -\boldsymbol{p} \wedge \nabla \left(\frac{1}{\boldsymbol{R}} \right) = -\frac{1}{2\pi^2} \int \mathrm{d}\boldsymbol{q} \frac{1}{q^2} (\boldsymbol{p} \wedge \nabla)\, e^{\imath \boldsymbol{q} \cdot \boldsymbol{r}} \\
&= -\frac{\imath}{2\pi^2} \int \mathrm{d}\boldsymbol{q} \frac{1}{q^2} (\boldsymbol{p} \wedge \boldsymbol{q}) e^{\imath \boldsymbol{q} \cdot \boldsymbol{r}}.
\end{aligned} \tag{D.7}$$

[1]We use the abbreviated notation $R \equiv |\boldsymbol{R}|$ for the modulus of vector quantities.

Using Eqs (D.6) and (D.7) and defining the incoming and outgoing neutrons by plane waves (see Eq. (2.6)) yields the following matrix element:

$$\begin{aligned}
&\langle \boldsymbol{k}'|\nabla \wedge \frac{\boldsymbol{s}\wedge\boldsymbol{R}}{R^3} - \frac{1}{\hbar}\frac{\boldsymbol{p}\wedge\boldsymbol{R}}{R^3}|\boldsymbol{k}\rangle \\
&= \frac{1}{2\pi^2}\int \mathrm{d}\boldsymbol{R}\, e^{-\imath\boldsymbol{k}'\cdot(\boldsymbol{r}+\boldsymbol{R})}\frac{1}{2\pi^2}\int \mathrm{d}\boldsymbol{q}\frac{1}{q^2}\left(\boldsymbol{q}\wedge(\boldsymbol{s}\wedge\boldsymbol{q}) + \frac{\imath}{\hbar}(\boldsymbol{p}\wedge\boldsymbol{q})\right)e^{\imath\boldsymbol{q}\cdot\boldsymbol{R}}e^{\imath\boldsymbol{k}\cdot(\boldsymbol{r}+\boldsymbol{R})} \\
&= \frac{1}{2\pi^2}e^{\imath\boldsymbol{Q}\cdot\boldsymbol{r}}\int \mathrm{d}\boldsymbol{q}\frac{1}{q^2}\int \mathrm{d}\boldsymbol{R}\, e^{\imath(\boldsymbol{Q}+\boldsymbol{q})\cdot\boldsymbol{R}}\left(\boldsymbol{q}\wedge(\boldsymbol{s}\wedge\boldsymbol{q}) + \frac{\imath}{\hbar}(\boldsymbol{p}\wedge\boldsymbol{q})\right) \\
&= e^{\imath\boldsymbol{Q}\cdot\boldsymbol{r}}4\pi\int \mathrm{d}\boldsymbol{q}\frac{1}{q^2}\delta(\boldsymbol{Q}+\boldsymbol{q})\left(\boldsymbol{q}\wedge(\boldsymbol{s}\wedge\boldsymbol{q}) + \frac{\imath}{\hbar}(\boldsymbol{p}\wedge\boldsymbol{q})\right) \\
&= e^{\imath\boldsymbol{Q}\cdot\boldsymbol{r}}4\pi\frac{1}{Q^2}\left(\boldsymbol{Q}\wedge(\boldsymbol{s}\wedge\boldsymbol{Q}) + \frac{\imath}{\hbar}(\boldsymbol{p}\wedge\boldsymbol{Q})\right) \\
&= e^{\imath\boldsymbol{Q}\cdot\boldsymbol{r}}4\pi\frac{1}{Q^2}\left(\boldsymbol{Q}\wedge(\boldsymbol{s}\wedge\boldsymbol{Q}) - \frac{\imath}{\hbar}(\boldsymbol{Q}\wedge\boldsymbol{p})\right) \qquad \text{(D.8)}
\end{aligned}$$

where $\boldsymbol{r}$ is the position of the electron. Actually we have to sum over all the unpaired electrons, thus the final result for the matrix element $\langle \boldsymbol{k}'|\hat{U}|\boldsymbol{k}\rangle$ reads

$$\langle \boldsymbol{k}'|\hat{U}|\boldsymbol{k}\rangle = 8\pi\gamma\mu_k\mu_B\,\hat{\boldsymbol{\sigma}}\cdot\hat{\boldsymbol{W}} \qquad \text{(D.9)}$$

with

$$\hat{\boldsymbol{W}} = \sum_i e^{\imath\boldsymbol{Q}\cdot\boldsymbol{r}_i}\frac{1}{Q^2}\left(\boldsymbol{Q}\wedge(\boldsymbol{s}_i\wedge\boldsymbol{Q}) - \frac{\imath}{\hbar}(\boldsymbol{Q}\wedge\boldsymbol{p}_i)\right). \qquad \text{(D.10)}$$

Inserting Eqs (D.9) and (D.10) into Eq. (2.4) yields

$$\begin{aligned}
\frac{\mathrm{d}^2\sigma}{\mathrm{d}\Omega\mathrm{d}\omega} &= \left(\frac{\gamma e^2}{m_e c^2}\right)^2\frac{k'}{k}\sum_{\lambda,\lambda'}\sum_{\sigma,\sigma'}p_\lambda p_\sigma\langle\boldsymbol{\sigma},\lambda|(\hat{\boldsymbol{\sigma}}\cdot\hat{\boldsymbol{W}})^+|\boldsymbol{\sigma}',\lambda'\rangle\langle\boldsymbol{\sigma}',\lambda'|\hat{\boldsymbol{\sigma}}\cdot\hat{\boldsymbol{W}}|\boldsymbol{\sigma},\lambda\rangle \\
&\quad\times\delta(\hbar\omega + E_\lambda - E_{\lambda'}) \\
&= \left(\frac{\gamma e^2}{m_e c^2}\right)^2\frac{k'}{k}\sum_{\alpha,\beta}\sum_{\lambda,\lambda'}p_\lambda\langle\lambda|\hat{W}_\alpha|\lambda'\rangle\langle\lambda'|\hat{W}_\beta|\lambda\rangle \\
&\quad\times\sum_{\sigma,\sigma'}p_\sigma\langle\boldsymbol{\sigma}|\hat{\sigma}_\alpha^+|\boldsymbol{\sigma}'\rangle\langle\boldsymbol{\sigma}'|\hat{\sigma}_\beta|\boldsymbol{\sigma}\rangle\,\delta(\hbar\omega + E_\lambda - E_{\lambda'}) \qquad \text{(D.11)}
\end{aligned}$$

where $\alpha,\beta = x,y,z$. For unpolarized neutrons we have $p_\sigma = 1/2$, thus the summation over the spin states $\sigma = \pm\frac{1}{2}$ reduces to

$$\sum_{\sigma,\sigma'}p_\sigma\langle\sigma|\hat{\sigma}_\alpha^+|\sigma'\rangle\langle\sigma'|\hat{\sigma}_\beta|\sigma\rangle = \delta_{\alpha\beta}, \qquad \text{(D.12)}$$

which immediately follows by applying Eq. (C.5). Eq. (D.11) thus simplifies to

$$\frac{d^2\sigma}{d\Omega d\omega} = \left(\frac{\gamma e^2}{m_e c^2}\right)^2 \frac{k'}{k} \sum_{\lambda,\lambda'} p_\lambda \langle\lambda|\hat{\boldsymbol{W}}^+|\lambda'\rangle \bullet \langle\lambda'|\hat{\boldsymbol{W}}|\lambda\rangle \times \delta(\hbar\omega + E_\lambda - E_{\lambda'}). \tag{D.13}$$

In the following treatment we shall consider the case of a magnetic target where the orbital contribution to the magnetic moment can be neglected. In fact, this simplification applies to many magnetic systems, e.g., for the case where the orbital moment is suppressed by the crystal-field potential (this effect is called quenching). As a consequence the relevant quantum numbers describing the target are S and M, so that we can rewrite Eq. (D.13) as (note that the matrix elements in Eq. (D.13) are coupled through a scalar product):

$$\frac{d^2\sigma}{d\Omega d\omega} = \left(\frac{\gamma e^2}{m_e c^2}\right)^2 \frac{k'}{k} \sum_{\alpha,\beta} \left(\delta_{\alpha\beta} - \frac{Q_\alpha Q_\beta}{Q^2}\right) \sum_{S,M,S',M'} p_{SM} \langle SM|\hat{V}_\alpha^+|S'M'\rangle \times \langle S'M'|\hat{V}_\beta|SM\rangle \delta(\hbar\omega + E_{SM} - E_{S'M'}). \tag{D.14}$$

with

$$\hat{\boldsymbol{V}} = \sum_i e^{\imath \boldsymbol{Q}\cdot\boldsymbol{r}_i} \hat{\boldsymbol{s}}_i. \tag{D.15}$$

We now divide the summation in Eq. (D.15) into a sum over all atoms j in the target and a sum over all electrons ν of each atom:

$$\hat{\boldsymbol{V}} = \underbrace{\sum_i e^{\imath \boldsymbol{Q}\cdot\boldsymbol{r}_i} \hat{\boldsymbol{s}}_i}_{\text{all electrons}} = \underbrace{\sum_j e^{\imath \boldsymbol{Q}\cdot\boldsymbol{R}_j}}_{\text{all ions}} \underbrace{\sum_\nu e^{\imath \boldsymbol{Q}\cdot\boldsymbol{r}_\nu} \hat{\boldsymbol{s}}_\nu}_{\text{all electrons per ion}} \tag{D.16}$$

where $\boldsymbol{r}_i = \boldsymbol{R}_j + \boldsymbol{r}_\nu$. From quantum mechanics we use the following relation [Condon and Shortley (1977)]:

$$S(S+1)\langle SM|\hat{\boldsymbol{T}}|S'M'\rangle = \langle SM|\hat{\boldsymbol{S}}|S'M'\rangle\langle SM'|\hat{\boldsymbol{S}}\hat{\boldsymbol{T}}|SM'\rangle, \tag{D.17}$$

which we apply to the operator $\hat{\boldsymbol{T}} = \sum_\nu e^{\imath \boldsymbol{Q}\cdot\boldsymbol{r}_\nu} \hat{\boldsymbol{s}}_\nu$ and get:

$$\langle SM|\sum_\nu e^{\imath \boldsymbol{Q}\cdot\boldsymbol{r}_\nu} \hat{\boldsymbol{s}}_\nu|S'M'\rangle = \langle SM|\hat{\boldsymbol{S}}_j|S'M'\rangle \frac{\langle SM'|\sum_\nu e^{\imath \boldsymbol{Q}\cdot\boldsymbol{r}_\nu} \hat{\boldsymbol{s}}_\nu \cdot \hat{\boldsymbol{S}}_j|SM'\rangle}{S_j(S_j+1)} \tag{D.18}$$

with $\hat{\boldsymbol{S}}_j = \sum_\nu \hat{\boldsymbol{s}}_\nu$. The last term of Eq. (D.18) does not depend on the quantum number M, i.e., it does not depend on the direction of the spin and serves as a characteristic value for the magnetic scattering strength of the atom. This quantity is called the magnetic form factor $F_j(\boldsymbol{Q})$, which is obtained by the Fourier transform of the normalized spin-density at the site j. The differential cross-section for magnetic neutron scattering from an S-state ion is therefore given by:

$$\begin{aligned}\frac{\mathrm{d}^2\sigma}{\mathrm{d}\Omega \mathrm{d}\omega} = (\gamma r_0)^2 \frac{k'}{k} e^{-2W(\boldsymbol{Q})} \sum_{\alpha,\beta} \left(\delta_{\alpha\beta} - \frac{Q_\alpha Q_\beta}{Q^2}\right) \\ \times \sum_{j,j'} F_j^*(\boldsymbol{Q}) F_{j'}(\boldsymbol{Q}) e^{i\boldsymbol{Q}\cdot(\boldsymbol{R}_j - \boldsymbol{R}_{j'})} \\ \times \sum_{S,M,S',M'} p_{SM} \langle SM|\hat{S}_j^\alpha|S'M'\rangle \langle S'M'|\hat{S}_{j'}^\beta|SM\rangle \\ \times \delta(\hbar\omega + E_{SM} - E_{S'M'}). \qquad \text{(D.19)}\end{aligned}$$

with $r_0 = e^2/(m_e c^2) = 2.8179 \cdot 10^{-15}$ m the classical electron radius. The Debye-Waller factor has been introduced to take care of the movement of the magnetic ions around their equilibrium position $\boldsymbol{R}_j$. Eq. (D.19) is known as the *master formula* for magnetic neutron scattering and serves as a convenient starting point to evaluate the neutron cross-section for the particular magnetic phenomenon under consideration (see Chaps 7 - 9).

Appendix E

Crystal Lattice and Reciprocal Lattice

The unit cell of a crystal lattice is defined by three basis vectors $\boldsymbol{a}_1, \boldsymbol{a}_2, \boldsymbol{a}_3$. The volume of the unit cell is $v_0 = \boldsymbol{a}_1 \cdot (\boldsymbol{a}_2 \wedge \boldsymbol{a}_3)$. The lattice vector defining a particular unit cell of the crystal is a linear combination of the three basis vectors:

$$\boldsymbol{l} = l_1\boldsymbol{a}_1 + l_2\boldsymbol{a}_2 + l_3\boldsymbol{a}_3, \tag{E.1}$$

where l_1, l_2, l_3 are integers. A crystal lattice with only one atom per unit cell is called a Bravais lattice. There are 14 different Bravais lattices; some examples of Bravais lattices are listed in Table E.1. If the crystallographic unit cell contains n atoms, the crystal lattice can be considered as a superposition of n Bravais lattices with identical basis but not necessarily with identical atoms. The position of the atoms in the unit cell is then given by

$$\boldsymbol{d} = d_1\boldsymbol{a}_1 + d_2\boldsymbol{a}_2 + d_3\boldsymbol{a}_3 \tag{E.2}$$

with $0 \leq d_i < 1$. This is exemplified for the case of the body-centered cubic lattice which consists of two simple cubic lattices shifted by the vector $a\left(\frac{1}{2}, \frac{1}{2}, \frac{1}{2}\right)$ against one another. The body-centered cubic lattice described by the basis vectors of the simple cubic lattice has therefore two atoms per unit cell with position vectors $\boldsymbol{d}_1 = (0,0,0)$ and $\boldsymbol{d}_2 = a\left(\frac{1}{2}, \frac{1}{2}, \frac{1}{2}\right)$.

For each crystal lattice a reciprocal lattice is defined by the relation

$$e^{\imath \boldsymbol{\tau} \cdot \boldsymbol{l}} = 1 \tag{E.3}$$

where

$$\boldsymbol{\tau} = t_1\boldsymbol{\tau}_1 + t_2\boldsymbol{\tau}_2 + t_3\boldsymbol{\tau}_3 \tag{E.4}$$

is a reciprocal lattice vector with integer values of t_1, t_2, t_3. The basis vectors of the reciprocal lattice are related to the basis vectors of the crystal lattice by

$$\boldsymbol{\tau}_1 = \frac{2\pi}{v_0}\boldsymbol{a}_2 \wedge \boldsymbol{a}_3, \quad \boldsymbol{\tau}_2 = \frac{2\pi}{v_0}\boldsymbol{a}_3 \wedge \boldsymbol{a}_1, \quad \boldsymbol{\tau}_3 = \frac{2\pi}{v_0}\boldsymbol{a}_1 \wedge \boldsymbol{a}_2. \tag{E.5}$$

Table E.1 Examples of Bravais lattices.

symmetry	$\boldsymbol{a}_1$	$\boldsymbol{a}_2$	$\boldsymbol{a}_3$	v_0
simple cubic	$a(1,0,0)$	$a(0,1,0)$	$a(0,0,1)$	a^3
body-centered cubic	$\frac{a}{2}(-1,-1,1)$	$\frac{a}{2}(1,-1,1)$	$\frac{a}{2}(1,1,-1)$	$\frac{1}{2}a^3$
face-centered cubic	$\frac{a}{2}(0,1,1)$	$\frac{a}{2}(1,0,1)$	$\frac{a}{2}(1,1,0)$	$\frac{1}{4}a^3$
hexagonal	$a(1,0,0)$	$a\left(-\frac{1}{2},\frac{\sqrt{3}}{2},0\right)$	$c(0,0,1)$	$\frac{\sqrt{3}}{2}a^2c$

Appendix F

The 3-j and 6-j Symbols

The 3-j symbol

For a complete discussion and tabulation of the 3-j symbols see [Rotenberg *et al.* (1959)]. The 3-j symbol is defined by:

$$\begin{aligned}\begin{pmatrix} j_1 & j_2 & j_3 \\ m_1 & m_2 & m_3 \end{pmatrix} &= (-1)^{j_1-j_2-m_3}\sqrt{\Delta(j_1,j_2,j_3)} \\ &\times \sqrt{(j_1+m_1)!(j_1-m_1)!(j_2+m_2)!(j_2-m_2)!(j_3+m_3)!(j_3-m_3)!} \\ &\times \sum_t \frac{(-1)^t}{f(t)} \end{aligned} \tag{F.1}$$

where

$$\Delta(a,b,c) = \frac{(a+b-c)!(a-b+c)!(-a+b+c)!}{(a+b+c+1)!} \tag{F.2}$$

and

$$\begin{aligned} f(t) = t!(j_3 - j_2 + t + m_1)!(j_3 - j_1 + t - m_2)! \\ \times (j_1 + j_2 - j_3 - t)!(j_1 - t - m_1)!(j_2 - t + m_2)! \end{aligned} \tag{F.3}$$

The sum in Eq. (F.1) is carried out over all integers t for which the factorials in $f(t)$ all have nonnegative arguments.

The 3-j symbols are also called Wigner coefficients. They are identical to the Clebsch-Gordon coefficients given in Ref. [Condon and Shortley (1977)]:

$$\begin{pmatrix} j_1 & j_2 & j_3 \\ m_1 & m_2 & m_3 \end{pmatrix} = \frac{(-1)^{j_1-j_2-m_3}}{\sqrt{2j_3+1}} (j_1 m_1 j_2 m_2 | j_1 j_2 j_3 m_3) . \tag{F.4}$$

The 3-j symbols are used for instance to transform the exchange-coupled spin state $\boldsymbol{S} = \boldsymbol{S}_1 + \boldsymbol{S}_2$ from the *individual spin* basis $|S_1 M_1 S_2 M_2\rangle$ into the *total spin* basis $|S_1S_2SM\rangle$:

$$|S_1S_2SM\rangle = \sum_{M_1M_2} \underbrace{(-1)^{S_2-S_1-M}\sqrt{2S+1}\begin{pmatrix} S_1 & S_2 & S \\ M_1 & M_2 & -M \end{pmatrix}}_{(S_1S_2SM|S_1M_1S_2M_2)} |S_1M_1S_2M_2\rangle \tag{F.5}$$

The 6-j symbol

For a complete discussion and tabulation of the 6-j symbols see Ref. [Rotenberg *et al.* (1959)]. The 6-j symbol is defined by:

$$\begin{Bmatrix} j_1 & j_2 & j_3 \\ m_1 & m_2 & m_3 \end{Bmatrix} = \sqrt{\Delta(j_1,j_2,j_3)\Delta(j_1,m_2,m_3)\Delta(m_1,j_2,m_3)\Delta(m_1,m_2,j_3)} \times \sum_t \frac{(-1)^t(t+1)!}{f(t)} \tag{F.6}$$

with

$$\begin{aligned} f(t) = {} & (t-j_1-j_2-j_3)!(t-j_1-m_2-m_3)!(t-m_1-j_2-m_3)! \\ & \times (t-m_1-m_2-j_3)!(j_1+j_2+m_1+m_2-t)! \\ & \times (j_2+j_3+m_2+m_3-t)!(j_3+j_1+m_3+m_1-t)!, \end{aligned} \tag{F.7}$$

$\Delta(a,b,c)$ from Eq. (F.2) and summed over all integers t for which the factorials in $f(t)$ all have nonnegative arguments.

The 6-j symbols are used for instance as Clebsch-Gordon coefficients for the coupling of three magnetic moments with total spin $\boldsymbol{S} = \boldsymbol{S}_1 + \boldsymbol{S}_2 + \boldsymbol{S}_3$. Since the *individual spin* basis $|S_1M_1S_2M_2S_3M_3\rangle$ cannot be transformed into the *total spin* basis $|S_1S_2S_3SM\rangle$ in a unique way, one has to introduce additional spin quantum numbers corresponding to the partial spins $\boldsymbol{S}_{12} = \boldsymbol{S}_1 + \boldsymbol{S}_2$ or $\boldsymbol{S}_{23} = \boldsymbol{S}_2 + \boldsymbol{S}_3$. The basis transformation between the two different coupling schemes is then:

$$|S_{12}S_3SM\rangle = (-1)^{S_3-S_{12}-m}\sqrt{2S+1}\begin{Bmatrix} S_1 & S_2 & S_{12} \\ S_3 & S & S_{23} \end{Bmatrix}|S_1S_{23}M\rangle \tag{F.8}$$

Appendix G

Impulse Approximation

The impulse approximation refers to neutron scattering at large moduli of the scattering vector $\boldsymbol{Q}$. The basic idea is that at large Q, the momentum (impulse) transferred to a single atom from the neutron is much larger than the momentum transferred from interatomic interactions. At high enough Q the energy transferred to a single atom is greater than the interatomic potential energy, and the atoms respond as if nearly free, so that kinetic properties such as the momentum distribution $n(\boldsymbol{p})$ are observed. Essentially we are left with the problem to calculate the cross section for neutron scattering from a single atom.

Dropping the sum over j, j' in Eqs (2.18) and (2.20) yields

$$S(\boldsymbol{Q},\omega) = \frac{1}{2\pi\hbar}\int_{-\infty}^{+\infty} dt\, e^{-\imath\omega t}\langle e^{-\imath\boldsymbol{Q}\cdot\boldsymbol{R}(0)}e^{\imath\boldsymbol{Q}\cdot\boldsymbol{R}(t)}\rangle. \tag{G.1}$$

In order to derive an explicit expression for the expectation value of Eq. (G.1), we solve the corresponding equation of motion:

$$\imath\hbar\frac{\partial}{\partial t}e^{\imath\boldsymbol{Q}\cdot\boldsymbol{R}(t)} = \left[e^{\imath\boldsymbol{Q}\cdot\boldsymbol{R}(t)},\hat{\mathcal{H}}\right], \tag{G.2}$$

where $\hat{\mathcal{H}}$ is the Hamiltonian describing the kinetic energy of the atom with momentum $\boldsymbol{p}$ and mass M:

$$\hat{\mathcal{H}} = \frac{1}{2M}\boldsymbol{p}^2, \quad \boldsymbol{p} = -\imath\hbar\nabla. \tag{G.3}$$

Combining Eqs (G.2) and (G.3) and using the operator theorem $[f(\boldsymbol{r}),\boldsymbol{p}] = \imath\hbar\nabla f(\boldsymbol{r})$ yields for the right hand side of Eq. (G.2) apart from the prefactor $\frac{1}{2M}$:

$$\begin{aligned}\left[e^{\imath\boldsymbol{Q}\cdot\boldsymbol{R}},\boldsymbol{p}^2\right] &= \left[e^{\imath\boldsymbol{Q}\cdot\boldsymbol{R}},\boldsymbol{p}\right]\boldsymbol{p}+\boldsymbol{p}\left[e^{\imath\boldsymbol{Q}\cdot\boldsymbol{R}},\boldsymbol{p}\right]\\ &= -\hbar\boldsymbol{Q}\,e^{\imath\boldsymbol{Q}\cdot\boldsymbol{R}}\boldsymbol{p}-\boldsymbol{p}\hbar\boldsymbol{Q}\,e^{\imath\boldsymbol{Q}\cdot\boldsymbol{R}}\\ &= -e^{\imath\boldsymbol{Q}\cdot\boldsymbol{R}}\left(2\hbar\boldsymbol{Q}\cdot\boldsymbol{p}+\hbar^2\boldsymbol{Q}^2\right).\end{aligned} \tag{G.4}$$

We insert Eq. (G.4) into the equation of motion Eq. (G.2) which has the following solution:

$$e^{\imath \boldsymbol{Q}\cdot\boldsymbol{R}(t)} = e^{\imath \boldsymbol{Q}\cdot\boldsymbol{R}(0)} e^{\frac{\imath t}{2M}\left(2\boldsymbol{Q}\cdot\boldsymbol{p}+\hbar \boldsymbol{Q}^2\right)}, \tag{G.5}$$

thus the expectation value of Eq. (G.1) reads:

$$\langle e^{-\imath \boldsymbol{Q}\cdot\boldsymbol{R}(0)} e^{\imath \boldsymbol{Q}\cdot\boldsymbol{R}(t)}\rangle = e^{\frac{\imath t\hbar Q^2}{2M}} \langle e^{\frac{\imath t \boldsymbol{Q}\cdot\boldsymbol{p}}{M}}\rangle. \tag{G.6}$$

The expectation value on the right hand side of Eq. (G.6) is given by

$$\langle e^{\frac{\imath t \boldsymbol{Q}\cdot\boldsymbol{p}}{M}}\rangle = \frac{\int \mathrm{d}\boldsymbol{p}\, n(\boldsymbol{p}) e^{\frac{\imath t \boldsymbol{Q}\cdot\boldsymbol{p}}{M}}}{\int \mathrm{d}\boldsymbol{p}\, n(\boldsymbol{p})} = e^{-\frac{t^2 Q^2 k_B T}{2M}}, \tag{G.7}$$

where

$$n(\boldsymbol{p}) = e^{-\frac{p^2}{2k_B T M}} \tag{G.8}$$

is the thermal distribution function. Combining Eqs (G.6) - (G.8) yields:

$$\langle e^{-\imath \boldsymbol{Q}\cdot\boldsymbol{R}(0)} e^{\imath \boldsymbol{Q}\cdot\boldsymbol{R}(t)}\rangle = e^{-\frac{Q^2}{2M}\left(k_B T t^2 - \imath \hbar t\right)}. \tag{G.9}$$

We insert Eq. (G.9) into the scattering law Eq. (G.1) and find:

$$S(\boldsymbol{Q},\omega) = \sqrt{\frac{M}{2\pi\hbar^2 \boldsymbol{Q}^2 k_B T}}\, e^{-\frac{M}{2\hbar \boldsymbol{Q}^2 k_B T}\left(\hbar\omega - \frac{\hbar^2 \boldsymbol{Q}^2}{2M}\right)^2}, \tag{G.10}$$

where $E_r = \frac{\hbar^2 \boldsymbol{Q}^2}{2M}$ corresponds to the recoil energy of the atom. Thus the dynamic structure factor corresponds to a Gaussian centered at the recoil energy E_r.

When the correlation function is evaluated in terms of momentum states, Eq. (G.10) transforms to the following form [Glyde and Svensson (1987)]:

$$S(\boldsymbol{Q},\omega) = \int \mathrm{d}\boldsymbol{p}\, n(\boldsymbol{p})\, \delta\left(\hbar\omega - \frac{\boldsymbol{p}^2}{2M} - \frac{\boldsymbol{Q}\cdot\boldsymbol{p}}{M}\right). \tag{G.11}$$

For isotropic scattering Eq. (G.11) simplifies to [Mayers (1993)]

$$S(\boldsymbol{Q},\omega) = \frac{M}{Q} J(y), \tag{G.12}$$

where

$$y = \frac{M}{Q}\left(\hbar\omega - \frac{Q^2}{2M}\right) \tag{G.13}$$

and

$$J(y) = 2\pi \int_{|y|}^{\infty} \boldsymbol{p}\, n(\boldsymbol{p})\, \mathrm{d}\boldsymbol{p}. \tag{G.14}$$

$J(y)\mathrm{d}y$ is the probability that an atom has a momentum component along $\boldsymbol{Q}$ with magnitude between y and $y + \mathrm{d}y$ and is known as the Compton profile. Eqs (G.12) - (G.14) express the "y scaling" property of the neutron cross-section at sufficiently high values of Q. For light atoms like hydrogen and helium the impulse approximation is valid for $Q > 10$ Å^{-1}.

Further reading

- S. W. Lovesey, in *Neutron scattering from hydrogen in materials*, ed. by A. Furrer (World Scientific, Singapore, 1994), p. 19: *Neutron compton scattering*
- G. I. Watson, J. Phys.: Condens. Matter 8, 5955 (1996): *Neutron compton scattering*

List of Symbols

$\hat{A}$, $\hat{B}$	Hermitian operators
$\boldsymbol{a}_1$, $\boldsymbol{a}_2$, $\boldsymbol{a}_3$	basis vectors in real space
$\hat{a}$, $\hat{a}^+$	annihilation and creation operators (phonons and spin-waves)
a, b, c	lattice parameters
A_n^m, B_n^m	crystal-field parameters
B	magnetic flux
b	scattering length
b_{coh}	coherent scattering length
b_{inc}	incoherent scattering length
c	speed of light
c	velocity of excitation
c_{ij}	elastic constants
c_V	specific heat (at constant volume)
D	diffusion constant
D	magnetic anisotropy parameter
d	distance in real space
$d_{\boldsymbol{\tau}}$, d_{hkl}	distance of reflection planes (h,k,l)
$\boldsymbol{d}_\alpha$	position of atom α within unit cell
E	energy of neutron
E_a	activation energy
E_F	Fermi energy
E_λ, $E_{\lambda'}$	initial and final energy of scattering system
e	charge of electron
$\boldsymbol{e}$	unit vector (in real space)
e_{ij}	elastic strain
$\boldsymbol{e}_s(\boldsymbol{q})$	polarization vector of phonon
$\boldsymbol{F}$	force
F	free energy
$F(\boldsymbol{Q})$	magnetic form factor
G	Gibbs free enthalpy
$G(\boldsymbol{r}, t)$	pair correlation function
$G_d(\boldsymbol{r}, t)$	distinct pair correlation function
$G_s(\boldsymbol{r}, t)$	self correlation function
g	Landé splitting factor
$g(\boldsymbol{r})$	pair-correlation function
$g(\omega)$	phonon density of states
$\boldsymbol{H}$	magnetic field
$\hat{\mathcal{H}}$	Hamiltonian
h	Planck's constant
$\hbar$	reduced Planck's constant
(h, k, l)	Miller indices
$\hat{\boldsymbol{I}}$, I	nuclear spin operator, value of
I	intensity

I	moment of inertia
$I(\boldsymbol{Q}, t)$	intermediate pair correlation function
$\hat{\boldsymbol{J}}$, J	total angular moment operator (of ion), value of
J	magnetic exchange parameter, exchange integral
$J(\boldsymbol{q})$	Fourier transformed exchange function
j_{ex}	exchange integral between electrons and $4f$ ions
K	biquadratic magnetic exchange parameter
$\boldsymbol{k}$, $\boldsymbol{k}'$	initial and final wavevector of neutron
k_B	Boltzmann's constant
k_F	wavevector at Fermi level
$\hat{\boldsymbol{L}}$, L	orbital angular momentum operator (of ion), value of
$L(\theta)$	Lorentz factor
l	length in real space
$\boldsymbol{l}_j$	position of j'th unit cell
M	mass of atom (ion)
M	component of angular momentum along quantization axis (quantum number)
m	mass of neutron
m	m-value of supermirrors
$m_{\rm hkl}$	multiplicity of Bragg reflection (h,k,l)
$\boldsymbol{M}$	magnetic moment, magnetization
N	total number of atoms
N_0	number of unit cells
$N(E_F)$	density of states at Fermi level
n	refractive index
$n(\boldsymbol{r})$	local density
n_0	mean density
$n_s(\boldsymbol{q})$, n_q	Bose-Einstein occupation number
$\hat{O}_n^m$	Stevens operators
P	pressure
$\boldsymbol{P}$	polarization vector
$\boldsymbol{P}$	spiral vector
$\boldsymbol{p}$	momentum of neutron
p_λ	probability of state $\lvert\lambda\rangle$
$\boldsymbol{Q}$	scattering vector
$\boldsymbol{q}$	vector in reciprocal space (wavevector of excitation)
$\boldsymbol{q}_0$	magnetic ordering wavevector
R	spin flipping ratio
$\boldsymbol{R}_j$	real space coordinate of atom j
$\boldsymbol{r}$	vector in real space
r_0	classical electron radius
$\hat{\boldsymbol{S}}$, S	spin operator (of ion or neutron), value of
$\hat{\boldsymbol{s}}$	spin operator (of electron)
$s(\boldsymbol{r})$	spin density
$S_{\boldsymbol{\tau}}$, $F_{\rm hkl}$	nuclear structure factor
$S(\boldsymbol{Q}, \omega)$	dynamical structure factor (scattering law)
T	absolute temperature
$\hat{\boldsymbol{T}}$	tensor operator
T_c	critical temperature, transition temperature
T_C	Curie temperature
T_N	Néel temperature
T^*	pseudogap temperature
t	time variable
t	reduced temperature
U	interaction potential
U	internal energy
$\boldsymbol{u}$	displacement from equilibrium position (real space)
V	external potential
V	total volume of sample
v	velocity (of neutron)
v	sound velocity
v_0	volume of unit cell

v_e	velocity of electron
W	energy of system
$W(\boldsymbol{Q})$	Debye-Waller function
w	ionic displacement
w_i	weighting factor of observation i
$w_s(\boldsymbol{q})$	angular frequency of phonon
Y	yield strength
Z	partition function
$\alpha(x)$	distribution function
$\alpha(\omega)$	electron-phonon coupling constant
Γ	linewidth
$\|\Gamma_n\rangle$	crystal-field state
γ	gyromagnetic ratio
γ_c	critical angle of total reflection
Θ_D	Debye temperature
θ	Bragg angle (half of scattering angle)
κ_T	isothermal compressibility
λ	wavelength of incoming neutron
$\|\lambda\rangle$, $\|\lambda'\rangle$	initial and final state of scattering system
λ_L	London penetration depth
$\boldsymbol{\mu}$	magnetic moment of neutron
$\boldsymbol{\mu}$	magnetic moment of ion
μ	attenuation factor
μ_B	Bohr magneton
μ_e	magnetic moment of electron
μ_i	magnetic moment of ion i
μ_N	nuclear magneton
ξ	amplitude of excitation (phonon)
ξ	correlation length
ρ	density
σ	neutron cross-section
σ_{abs}	absorption cross section
σ_{coh}	coherent scattering cross section
σ_{inc}	incoherent scattering cross section
σ_{tot}	total scattering cross section
σ	variance
$\boldsymbol{\sigma}$	Pauli spin operator of neutron
τ	relaxation, jumping time
$\boldsymbol{\tau}$	reciprocal lattice vector
$\boldsymbol{\tau}_1$, $\boldsymbol{\tau}_2$, $\boldsymbol{\tau}_3$	basis vectors in reciprocal space
Φ	neutron flux
Φ_0	flux quantum
ψ	scattering angle
$\|\psi\rangle$	wavefunction (of neutron)
χ	magnetic susceptibility
ω	angular frequency of neutron or excitation (via $E = \hbar\omega$)

Bibliography

Abersfelder, G., Noack, K., Stierstadt, K., Schelten, J. and Schmatz, W. (1980). *Phil. Mag. B* **41**, p. 519.

Allen, P. (1972). *Phys. Rev. B* **6**, p. 2577.

Allen, P. B., Kostur, V. N., Takesue, N. and Shirane, G. (1997). *Phys. Rev. B* **56**, p. 5552.

Allenspach, P., Furrer, A. and Hulliger, H. (1989). *Phys. Rev. B* **39**, p. 2226.

Asmussen, B. and Press, W. (1994). in A. Furrer (ed.), *Neutron scattering from hydrogen in materials* (World Scientific, Singapore), p. 184.

Axe, J. D. and Shirane, G. (1973). *Phys. Rev. Lett.* **30**, p. 214.

Barbara, B., Boucherle, J. X., Buevoz, J. L., Rossignol, M. F. and Schweizer, J. (1977). *Solid State Commun.* **24**, p. 481.

Bardeen, J., Cooper, L. N. and Schrieffer, J. R. (1957a). *Phys. Rev.* **106**, p. 162.

Bardeen, J., Cooper, L. N. and Schrieffer, J. R. (1957b). *Phys. Rev.* **108**, p. 1175.

Becker, K. W., Fulde, P. and Keller, J. (1979). *Z. Phys. B* **28**, p. 9.

Bedell, K., Pines, D. and Zawadowski, A. (1984). *Phys. Rev. B* **29**, p. 102.

Birgeneau, R. J. (1972). *J. Phys. Chem. Solids* **33**, p. 59.

Blundell, S. J. and Bland, J. A. C. (1992). *Phys. Rev. B* **46**, p. 3391.

Bourges, P. (1998). in J. Bok, G. Deutscher, D. Pavuna and S. A. Wolf (eds.), *The gap symmetry and fluctuations in high-temperature superconductors* (Plenum Press, New York), p. 349.

Breitling, W., Lehmann, W., Weber, R., Lehner, N. and Wagner, V. (1977). *J. Magn. Magn. Mater.* **6**, p. 113.

Brown, P. J. (1999). in A. J. C. Wilson and E. Price (eds.), *International tables for crystallography*, Vol. C (Kluwer, Dordrecht), p. 450.

Brüesch, P. (1982). *Phonons: Theory and experiments I* (Springer, Berlin).

Bührer, W., Bührer, R., Isacson, A., Koch, M. and Thut, R. (1981). *Nucl. Instrum. Meth.* **179**, p. 259.

Caglioti, G., Paoletti, A. and Ricci, F. P. (1958). *Nucl. Instrum. Methods* **3**, p. 223.

Cava, R. J., Hewat, A. W., Hewat, E. A., Batlogg, B., Marezio, M., Rabe, K. M., Krajewski, J. J., Peck Jr., W. F. and Rupp Jr., L. W. (1990). *Physica C* **165**, p. 419.

Chang, J., Mesot, J., Gilardi, R., Kohlbrecher, J., Drew, A. J., Divakar, U., Lister, S. J., Lee, S. L., Brown, S. P., Charalambous, D., Forgan, E. M., Dewhurst, C. D., Cubitt, R., Momono, N. and Oda, M. (2006). *Physica B* **385-386**, p. 35.

Chatterji, T. (2006). in T. Chatterji (ed.), *Neutron scattering from magnetic materials* (Elsevier, Amsterdam), p. 25.

Christen, D. K., Tasset, F., Spooner, S. and Mook, H. A. (1977). *Phys. Rev. B* **15**, p. 4506.

Chung, E. M. L., Lees, M. R., McIntyre, G. J., Wilkinson, C., Balakrishnan, G., Hague, J. P., Visser, D. and Paul, D. M. (2004). *J. Phys.: Condens. Matter* **16**, p. 7837.

Clementyev, E. S., Conder, K., Furrer, A. and Sashin, I. L. (2001). *Eur. Phys. J. B* **21**, p. 465.

Coldea, R., Hayden, S. M., Aeppli, G., Perring, T. G., Frost, C. D., Mason, T. E., Cheong, S. W. and Fisk, Z. (2001). *Phys. Rev. Lett.* **86**, p. 5377.

Condon, E. U. and Shortley, G. H. (1977). *The theory of atomic spectra* (University Press, Cambridge).

Cooper, M. J. and Nathans, R. (1967). *Acta Cryst.* **23**, p. 357.

Donnelly, R. J., Donnelly, J. A. and Hills, R. N. (1981). *J. Low Temp. Phys.* **44**, p. 471.

Dönni, A., Furrer, A., Bauer, E., Kitazawa, H. and Zolliker, M. (1997). *Z. Phys. B* **104**, p. 403.

Dorner, B. (1972). *Acta Cryst. A* **28**, p. 319.

Elsenhans, O., Fischer, P., Furrer, A., Clausen, K., Purwins, H. and Hulliger, F. (1991). *Z. Phys. B - Condensed Matter* **82**, p. 61.

Enderby, J. E., North, D. M. and Egelstaff, P. A. (1966). *Phil. Mag.* **14**, p. 961.

Engelman, D. M., Moore, P. B. and Schoenborn, B. P. (1975). *Proc. Nat. Acad. Sci. USA* **72**, p. 3888.

Falk, U., Furrer, A., Kjems, J. K. and Güdel, H. U. (1984). *Phys. Rev. Lett.* **52**, p. 1336.

Feile, R., Loewenhaupt, M., Kjems, J. K. and Hoenig, H. E. (1981). *Phys. Rev. Lett.* **47**, p. 610.

Fermon, C., Ott, F., Gilles, B., Marty, A., Menelle, A., Samson, Y., Legoff, G. and Francinet, G. (1999). *Physica B* **267-268**, p. 162.

Fischer, P., Schobinger-Papamantellos, P. and Kaldis, E. (1976). *J. Magn. Magn. Mater.* **3**, p. 200.

Fischer, P. and Stoll, E. (1968). AF-SSP-21, Tech. rep., EIR Würenlingen, Würenlingen, Switzerland.

Forgan, E. M. (1998). in A. Furrer (ed.), *Neutron scattering in layered copper-oxide superconductors* (Kluwer Academic Publishers, Dordrecht), p. 375.

Freeman, A. J. and Desclaux, J. P. (1979). *J. Magn. Magn. Mater.* **12**, p. 11.

Furrer, A. and Purwins, H. G. (1976). *J. Phys. C: Solid State Phys.* **9**, p. 1491.

Gebhardt, W. and Krey, G. (1980). *Phasenübergänge und kritische Phänomene* (Vieweg, Braunschweig).

Glyde, H. R. and Svensson, E. C. (1987). in D. L. Price and K. Sköld (eds.), *Methods of experimental physics*, Vol. 23 (Academic Press, San Diego), p. 303.

Güdel, H. U., Hauser, U. and Furrer, A. (1979). *Inorg. Chem.* **18**, p. 2730.

Guillaume, M., Allenspach, P., Henggeler, W., Mesot, J., Roessli, B., Staub, U., Fischer, P., Furrer, A. and Trounov, V. (1994). *J. Phys.: Condens. Matter* **6**, p. 7963.

Guillaume, M., Henggeler, W., Furrer, A., Eccleston, R. J. and Trounov, V. (1995). *Phys. Rev. Lett.* **74**, p. 3423.

Häfliger, P. S., Ochsenbein, S. T., Trusch, B., Güdel, H. U. and Furrer, A. (2009). *J. Phys. Condens. Matter* **21**, p. 026019.

Häfliger, P. S., Podlesnyak, A., Conder, K., Pomjakushina, E. and Furrer, A. (2006). *Phys. Rev. B* **74**, p. 184520.

Hälg, B. (1981). AF-SSP-116, Tech. rep., ETH Zurich, Zurich, Switzerland.

Hälg, B., Furrer, A., Hälg, W. and Vogt, O. (1982). *J. Magn. Magn. Mater.* **29**, p. 151.

Häusler, W. (1992). *J. Phys.: Condens. Matter* **4**, p. 2577.

Hayden, S. M., Aeppli, G., Osborn, R., Taylor, A. D., Perring, T. G., Cheong, S. W. and Fisk, Z. (1991). *Phys. Rev. Lett.* **67**, p. 3622.

Hayden, S. M., Mook, H. A., Dai, P., Perring, T. G. and Doğan, F. (2004). *Nature* **429**, p. 531.

Hayter, J. B., Pynn, R. and Suck, J. B. (1983). *J. Phys. F: Metal Phys.* **13**, p. L1.

Heger, G. (2000). in T. Brückel, G. Heger and D. Richter (eds.), *Neutron scattering, Matter and materials*, Vol. 5 (Forschungszentrum Jülich, Jülich), pp. 7–1.

Heidemann, A., Friedrich, H., Günther, E. and Häusler, W. (1989a). *Z. Phys. B* **76**, p. 335.

Heidemann, A., Prager, M. and Monkenbusch, M. (1989b). *Z. Phys. B* **76**, p. 77.

Holstein, T. and Primakoff, H. (1940). *Phys. Rev.* **58**, p. 1098.

Hoppler, J. (2005). LNS-219, Tech. rep., Laboratory for Neutron Scattering, ETH Zurich and PSI Villigen, Villigen, Switzerland.

Huang, Q., Santoro, A., Grigereit, T. E., Lynn, J. W., Cava, R. J., Krajewski, J. J. and Peck Jr., W. F. (1995). *Phys. Rev. B* **51**, p. 3701.

Hüller, A. (1977). *Phys. Rev. B* **16**, p. 1844.

Hutchings, M. T. (1964). in F. Seitz and D. Turnbull (eds.), *Solid state physics*, Vol. 16 (Academic Press, New York), p. 227.

Hutchings, M. T., Withers, P. J., Holden, T. M. and Lorentzen, T. (2005). *Introduction to the characterization of residual stress by neutron diffraction* (Taylor and Francis, Boca Raton).

Jensen, J. and Macintosh, A. R. (1991). *Rare-earth magnetism - structures and excitations* (Clarendon Press, Oxford).

Johnston, D. F. (1966). *Proc. Phys. Soc.* **88**, p. 37.

Kasuya, T. (1956). *Prog. Theor. Phys.* **16**, pp. 45, 58.

Kawano, H., Yoshizawa, H., Takeya, H. and Kadowaki, K. (1996). *Phys. Rev. Lett.* **77**, p. 4628.

Kepa, H., Kutner-Pielaszek, J., Twardowski, A., Majkrzak, C. F., Story, T., Sadowski, J. and Giebultowicz, T. M. (2002). *Appl. Phys. A* **74**, p. S1526.

Kittel, C. (1960). *Phys. Rev.* **120**, p. 335.

Kjems, J. K. and Steiner, M. (1978). *Phys. Rev. Lett.* **41**, p. 1137.

Klotz, S., Strässle, T., Nelmes, R. J., Loveday, J. S., Hamel, G., Rousse, G., Canny, B., Chervin, J. C. and Saitta, A. M. (2005a). *Phys. Rev. Lett.* **94**, p. 025506.

Klotz, S., Strässle, T., Rousse, G., Hamel, G. and Pomjakushin, V. (2005b). *Appl. Phys. Lett.* **86**, p. 031917.

Koehler, W. C., Cable, J. W., Wilkinson, M. K. and Wollan, E. O. (1966). *Phys. Rev* **151**, p. 414.

Korringa, J. (1950). *Physica (Utrecht)* **16**, p. 601.

Larson, A. C. and Von Dreele, R. B. (2000). Report LAUR 86-748, Tech. rep., Los Alamos National Laboratory, Los Alamos, USA.

Lauter, H. J., Lauter-Pasyuk, V., Toperverg, B. P., Romashev, L., Ustinov, V., Kravtsov, E., Vorobiev, A., Nikonov, O. and Major, J. (2002). *Appl. Phys. A* **74**, p. S1557.

Lea, K. R., Leask, M. J. M. and Wolf, W. P. (1962). *J. Phys. Chem. Solids* **23**, p. 1381.

Lippmann, G. and Schelten, J. (1974). *J. Appl. Cryst.* **7**, p. 236.

Loewenhaupt, M. and Steglich, F. (1977). in A. Furrer (ed.), *Crystal field effects in metals and alloys* (Plenum Press, New York), p. 198.

Lovesey, S. W. (1984). *Theory of neutron scattering from condensed matter*, Vol. 2 (Clarendon Press, Oxford).

Lovesey, S. W. and Collins, S. P. (1996). *X-ray scattering and absorption by magnetic materials* (Clarendon Press, Oxford).

Lynn, J. W. (1975). *Phys. Rev. B* **11**, p. 2624.

Mason, T. E., Aeppli, G. and Mook, H. A. (1992). *Phys. Rev. Lett.* **68**, p. 1414.

Mayers, J. (1993). *Phys. Rev. Lett.* **71**, p. 1553.

McIntyre, G. J., Lemée-Cailleau, M. H. and Wilkinson, C. (2006). *Physica B* **385-386**, p. 1055.

McMillan, W. L. (1968). *Phys. Rev.* **167**, p. 331.

Meier, B. H. (1984). *Ph.D. Thesis No. 7620*, Ph.D. thesis, ETH, Zurich, Switzerland.

Meier, B. H., Meyer, R., Ernst, R. R., Stöckli, A., Furrer, A., Hälg, W. and Anderson, I. (1984). *Chem. Phys. Lett.* **108**, p. 522.

Mikeska, H. J. (1978). *J. Phys. C* **11**, p. L29.

Millhouse, A. H. and Furrer, A. (1976). *AIP Conf. Proc.* **29**, p. 257.

Mirebeau, I., Hennion, M., Casalta, H., Andres, H., Güdel, H. U., Irodova, A. V. and Caneschi, A. (1999). *Phys. Rev. Lett.* **83**, p. 628.

Mook, H. A. (1966). *Phys. Rev.* **148**, p. 495.

Murani, A. P., Knorr, K., Buschow, K. H. J., Benoit, A. and Flouquet, J. (1980). *Solid State Comm.* **36**, p. 523.

Nellin, G. and Nilsson, G. (1972). *Phys. Rev. B* **5**, p. 3151.

Nicklow, R. M. (1983). in J. N. Mundy, S. J. Rothman, M. J. Fluss and L. C. Smedskjaer (eds.), *Methods of experimental physics*, Vol. 21 (Academic Press, Orlando), p. 172.

Norman, M. R. and Pépin, C. (2003). *Rep. Prog. Phys.* **66**, p. 1547.

Ochsenbein, S. T., Chaboussant, G., Sieber, A., Güdel, H. U., Janssen, S., Furrer,

A. and Attfield, J. P. (2003). *Phys. Rev. B* **68**, p. 092410.
Orbach, R. (1961). *Proc. Roy. Soc. A* **264**, p. 458.
Ott, F. (2007). *C. R. Physique* **8**, p. 763.
Owen, R. A., Preston, R. V., Withers, P. J., Shercliff, H. R. and Webster, P. J. (2003). *Mater. Sci. Eng. A* **346**, p. 159.
Penfold, J., Weber, J. R. P. and Bucknall, D. G. (1994). in A. Furrer (ed.), *Neutron scattering from hydrogen in materials* (World Scientific, Singapore), p. 65.
Pietsch, U., Holy, V. and Baumbach, T. (2004). *High-resolution x-ray scattering from thin films to nanostructures* (Springer, New York).
Podlesnyak, A., Pomjakushin, V., Pomjakushina, E., Conder, K. and Furrer, A. (2007). *Phys. Rev. B* **76**, p. 064420.
Podlesnyak, A., Russina, M., Furrer, A., Alfonsov, A., Vavilova, E., Kataev, V., Büchner, B., Strässle, T., Pomjakushina, E., Conder, K. and Khomskii, D. I. (2008). *Phys. Rev. Lett.* **101**, p. 247603.
Poirier, J. P. (2000). *Introduction to the physics of the earth's interior* (Cambridge University Press, Cambridge).
Rietveld, H. M. (1969). *J. Appl. Cryst.* **2**, p. 65.
Riste, T. (1970). *Nucl. Instrum. Meth.* **86**, p. 1.
Robinson, R. A., Brown, P. J., Argyriou, D. N., Hendrickson, D. N. and Aubin, S. M. J. (2000). *J. Phys.: Condens. Matter* **12**, p. 2805.
Rodriguez-Carvajal, J. (1993). *Physica B* **55**, p. 192.
Rotenberg, M., Bivins, R., Metropolis, N. and Wooten Jr., J. K. (1959). *The 3-j and 6-j symbols* (MIT Press, Cambridge Massachusetts).
Ruderman, M. A. and Kittel, C. (1954). *Phys. Rev.* **96**, p. 99.
Rüegg, C., Cavadini, N., Furrer, A., Güdel, H. U., Krämer, K. W., Mutka, H., Wildes, A., Habicht, K. and Vorderwisch, P. (2003). *Nature* **423**, p. 62.
Rüegg, C., Furrer, A., Sheptyakov, D., Strässle, T., Krämer, K. W., Güdel, H. and Mélési, L. (2004). *Phys. Rev. Lett.* **93**, p. 257201.
Salzmann, C. G., Radelli, P. G., Hallbrucker, A., Mayer, E. and Finney, J. L. (2006). *Science* **311**, p. 1758.
Santoro, A., Miraglia, S., Beech, F., Sunshine, S. A., Murphy, D. W., Schneemeyer, L. F. and Waszcak, J. V. (1987). *Mat. Res. Bull.* **22**, p. 1007.
Scherm, R., Dolling, G., Ritter, R., Schedler, E., Teuchert, W. and Wagner, V. (1977). *Nucl. Instrum. Meth.* **143**, p. 77.
Schmid, B., Hälg, B., Furrer, A., Urland, W. and Kremer, R. (1987). *J. Appl. Phys.* **61**, p. 3426.
Schönfeld, B., Reinhard, L., Kostorz, G. and Bührer, W. (1997). *Acta mater.* **45**, p. 5187.
Schwahn, D. (2000). in T. Brückel, G. Heger and D. Richter (eds.), *Neutron scattering, Matter and materials*, Vol. 5 (Forschungszentrum Jülich, Jülich), pp. 8–1.
Sears, V. F., Svensson, E. C., Martel, P. and Woods, A. D. B. (1982). *Phys. Rev. Lett.* **49**, p. 279.
Shirane, G. (1974). *Rev. Mod. Phys.* **46**, p. 437.

Shirane, G., Nathans, R., Steinsvoll, O., Alperin, H. A. and Pichart, S. (1965). *Phys. Rev. Lett.* **15**, p. 146.

Shull, C. G. (1968). *Phys. Rev. Lett.* **21**, p. 1585.

Shull, C. G., Wollan, E. O., Morton, G. A. and Davidson, W. L. (1948). *Phys. Rev.* **73**, p. 842.

Sköld, K. and Nelin, G. (1967). *J. Phys. Chem. Solids* **28**, p. 2369.

Sköld, K., Pelizzari, C. A., Kleb, R. and Ostrowski, G. E. (1976). *Phys. Rev. Lett.* **37**, p. 842.

Sköld, K., Rowe, J. M., Ostrowski, G. and Randolph, P. D. (1972). *Phys. Rev. A* **6**, p. 1107.

Squires, G. L. (1996). *Introduction to the theory of thermal neutron scattering* (Dover Publications, New York).

Steele, D. and Fender, B. E. F. (1974). *J. Phys. C* **7**, p. 1.

Steiner, M. (1980). in T. Riste (ed.), *Ordering in strongly fluctuating condensed matter systems* (Plenum Press, New York), p. 107.

Stevens, K. W. H. (1967). *Rep. Progr. Phys.* **30**, p. 189.

Strässle, T., Juranyi, F., Schneider, M., Janssen, S., Furrer, A., Krämer, K. W. and Güdel, H. U. (2004a). *Phys. Rev. Lett.* **92**, p. 257202.

Strässle, T., Klotz, S., Hamel, G., Koza, M. M. and Schober, H. (2007). *Phys. Rev. Lett.* **99**, p. 175501.

Strässle, T., Saitta, A. M., Klotz, S. and Braden, M. (2004b). *Phys. Rev. Lett.* **93**, p. 225901.

Suzuki, K. (1987). in D. L. Price and K. Sköld (eds.), *Methods of experimental physics*, Vol. 23 (Academic Press, New York), p. 243.

Svensson, E. C., Sears, V. F., Woods, A. D. B. and Martel, P. (1980). *Phys. Rev. B* **21**, p. 3638.

Taub, H., Carneiro, K., Kjems, J. K., Passell, L. and McTague, J. P. (1977). *Phys. Rev. B* **16**, p. 4551.

Teixeira, J., Bellissent-Funel, M. C., Chen, S. H. and Dianoux, A. J. (1985). *Phys. Rev. A* **31**, p. 1913.

Tranquada, J. M., Woo, H., Perring, T. G., Goka, H., Gu, G. D., Xu, G., Fujita, M. and Yamada, K. (2004). *Nature* **429**, p. 534.

van Hove, L. (1954). *Phys. Rev* **95**, p. 249.

Vijayaraghavan, P. R., Nicklow, R. M., Smith, H. G. and Wilkinson, M. K. (1970). *Phys. Rev. B* **1**, p. 4819.

Voll, G. and Hüller, A. (1988). *Can. J. Chem.* **66**, p. 925.

Wagner, D. (1972). *Introduction to the theory of magnetism* (Pergamon Press, Oxford).

Warren, B. E. (1941). *Phys. Rev.* **59**, p. 693.

Weber, K. (1967). *Acta Cryst.* **23**, p. 720.

Wybourne, B. G. (1965). *Spectroscopic properties of rare earths* (John Wiley, New York).

Yosida, K. (1957). *Phys. Rev.* **106**, p. 893.

Zhang, S. C. (1997). *Science* **275**, p. 1089.

Zolliker, P., Furrer, A., Meier, B., Graf, F. and Ernst, R. R. (1983). AF-SSP-124, Tech. rep., ETH Zurich, Zurich, Switzerland.

Index

www.ingramcontent.com/pod-product-compliance
Lightning Source LLC
Chambersburg PA
CBHW061236220326
41599CB00028B/5446